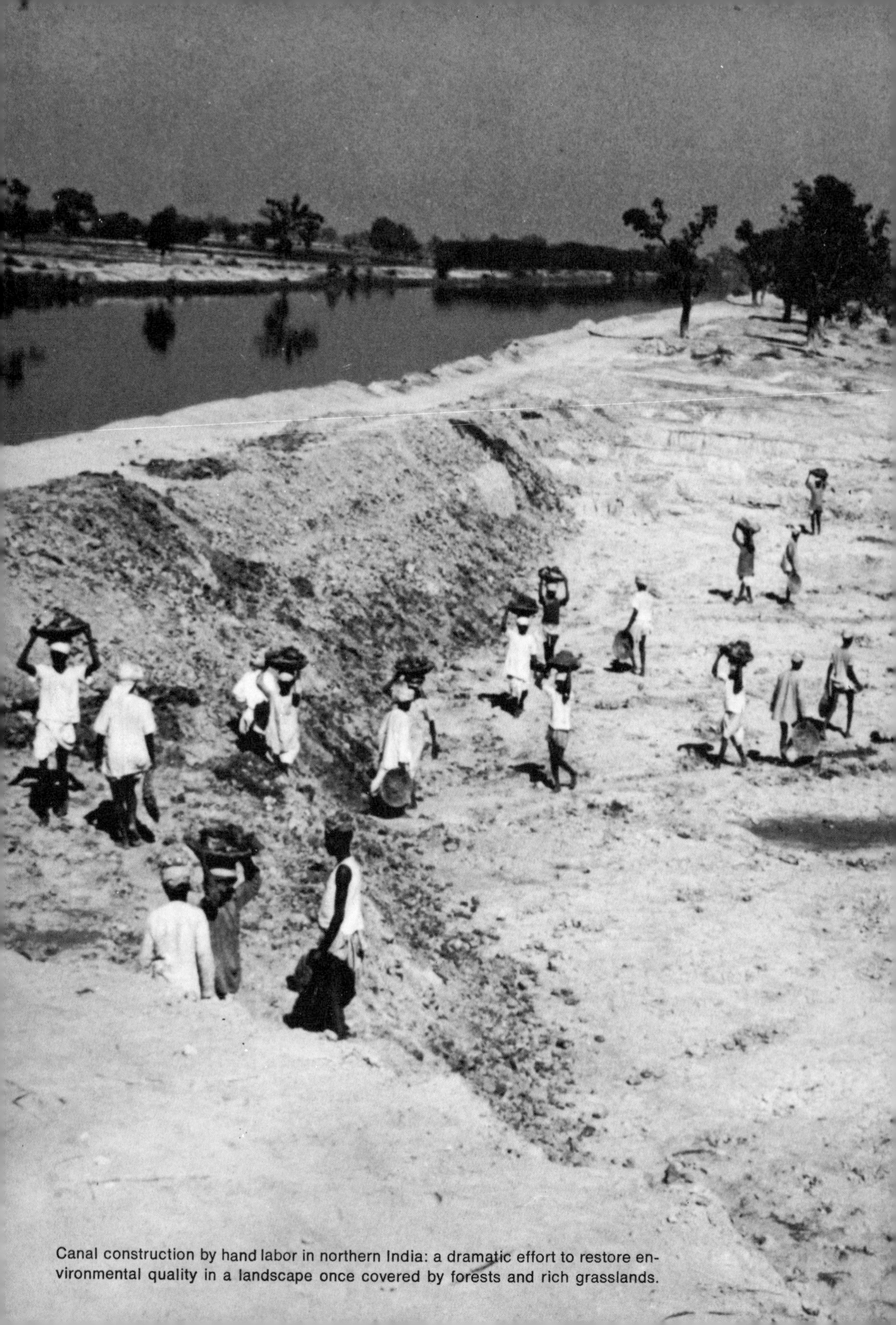

Canal construction by hand labor in northern India: a dramatic effort to restore environmental quality in a landscape once covered by forests and rich grasslands.

ECOLOGY
and the Quality of our
ENVIRONMENT

Charles H. Southwick

The Johns Hopkins University

VNR

VAN NOSTRAND REINHOLD COMPANY

New York Cincinnati Toronto London Melbourne

For My Mother and Father

Van Nostrand Reinhold Company Regional Offices:
New York Cincinnati Chicago Millbrae Dallas

Van Nostrand Reinhold Company International Offices:
London Toronto Melbourne

Library of Congress Catalog Card Number: 71-181443

Manufactured in the United States of America

Published by Van Nostrand Reinhold Company
450 West 33rd Street, New York, N. Y. 10001

Published simultaneously in Canada by Van Nostrand Reinhold Ltd.

Cover and Text design by Saul Schnurman

15 14 13 12 11 10 9 8 7 6 5 4 3 2

Preface

Ecology is a vital and exciting subject for students in many fields. By tradition it has been primarily a graduate course for biology majors. Now the obvious need is to present it to other students. Those in business, law, education, liberal arts, engineering, agriculture, medicine and health, social service, and other related fields can all benefit from the insights and perspectives of ecology. This book is written to meet the needs of these diverse interests. It has evolved from a one-semester course begun in 1967 at The Johns Hopkins University. The primary purpose of that course and of this book is to integrate the principles of ecology with the social and environmental problems of man.

While this book was being written, my work involved several extended trips throughout the world to countries I had visited in the 1950s and 1960s. This provided the opportunity to see the world of the 1970s in terms of changes wrought by ten or twenty years of rapid development. These travels took me throughout eastern and southern Asia, the Middle East, Africa, Europe, and the Western Hemisphere from Alaska to Latin America. Portions of the book were written in Kuala Lumpur, Calcutta, New Delhi, Nairobi, and London. In this age of ecological panic, I entertained the hope that I would find the world in pretty good shape after all. But reassurances were few and far between. Although I did not find full support for the gloom-and-doom hypothesis that the world will end in ten or fifteen years, I did find ample evidence that man still fails to understand his true position on this planet—evidence that the lessons of ecology have not yet taken hold to produce the kind of wisdom we need for our own survival. I found the ecological errors of history still being made with all the speed and conceit of modern technology. I found the need for ecological education to be greater than ever.

Because this book is written primarily for students who are not majoring in ecology, the use of technical terminology has been kept to a minimum. The presentation of ecology and environmental science is in elementary terms, and the concentration is on significant concepts. An attempt has been made to summarize the recent research and relate it to the basic issues of how the world functions or fails to function. Reference and documentation have been provided to enable readers to pursue these critical questions in more detail. Thus, it is hoped that the book will lead to further study, and will arouse, rather than satisfy, a continuing interest in ecology and environmental quality.

In my own ecological education, I am indebted to many teachers, colleagues and students. At the University of Wisconsin, Professors John Emlen, Joseph Hickey, Robert McCabe and Robert Ellarson gave me a solid start in the field of ecology and helped shape my research interests in animal population ecology. At Oxford University in England, Charles Elton, Dennis Chitty, H. N. Southern, Niko Tinbergen, David Jenkins and Peter Crowcroft greatly extended this interest. Other colleagues abroad who have added to my ecological education have been Drs. Francois Bourliere and Mireille Bertrand of Paris; Drs. F. L. Cadigan and I. Muul of Kuala Lumpur; Drs. M. B. Mirza, M. R. Siddiqi, M. F. Siddiqi of Aligarh, India; and Drs. R. K. Lahiri, J. L. Bhaduri, K. K. Tiwari, R. P. Mukherjee, and B. C. Pal of Calcutta. I would also like to express my appreciation to Drs. N. E. Collias, W. F. Crissey, H. A. Hochbaum, and A. W. Stokes, who gave me many benefits during my graduate training from their extensive field experience in ecology.

I am particularly indebted to students and colleagues of The Johns Hopkins University for stimulating discussions and collaboration in ecological teaching and research. These include: Drs. W. J. L. Sladen, Carleton Ray, Edwin Gould, John Oppenheimer, Dwain Parrack, F. B. Bang, K. Shah, Brenda Sladen, Neville and Rada Dyson-Hudson, C. W. Kruse, Loren Jensen, Lloyd Rozeboom, Syd Radinovsky, Byron Tepper, George Schaller, F. S. L. Williamson, Juan Spillett, and many others. All of these people have contributed to the ideas and critical review of the data and opinions developed within this book. Final responsibility for the accuracy and validity of the text lies entirely with me, of course, and none of my colleagues or students can be blamed for the interpretations that are presented herein, or for any inadequacies of the text. Errors which may still occur are entirely my own.

I am indebted to Barbara Hamilton and Jane Entwisle for careful typing and proofreading, to Maureen Myers for assistance with copyright permissions and to Peter Whitehouse for aid with references. Stephen Kraham, Denise Rathbun, and Roberta Bauer at Van Nostrand Reinhold have provided very helpful guidance and editorial counselling.

I am also indebted to the members of my family for their enthusiastic responses to a lifetime of ecological interests: to my parents, Mr. and Mrs. Arthur F. Southwick, my wife, Heather, and children Steven and Karen. They have all shared with me the pleasures and discouragements of being an ecologist.

Charles H. Southwick

BALTIMORE, MARYLAND

Contents

OVERVIEW

RARELY has an academic subject exploded so rapidly in the public mind as ecology in 1969 and 1970. In two or three years, ecology moved from a rather quiet and obscure branch of biology to a major issue of national and international attention. Ecology's influence was felt in every quarter of society from young to old, poor to rich, and farmer to urbanite. Education, business, politics, law, agriculture, engineering, medicine, public health, and even international affairs were all affected by this sudden rise of ecology and environmental concern.

Probably more books and articles on ecology and human affairs were written from 1968 to 1971 than in all the preceding 100 years of ecology's formal existence. The references in Appendix I provide a sample of recent books on ecology and the environment. A complete bibliography of books and articles written on this subject in the last three years would fill several volumes. In early 1970, for example, virtually every magazine and newspaper of general use had some mention of ecology. Popular periodicals such as *Time, Newsweek, Life, Reader's Digest, Forbes, Fortune, National Geographic, Saturday Review, Business Week,* and many others had feature articles or special issues to highlight ecology. Network television and radio joined in to increase this coverage. Ecological organizations proliferated across the nation and around the world. Students in secondary schools, colleges, and universities picked up the banner of ecology with as much or even more fervor than they did the issues of civil rights or the Vietnam war. Rallies, demonstrations, and "Earth Days" on ecologic topics were held throughout the United States and Europe.

By its most active proponents, ecology was hailed as the only hope for the future and the only route to human survival. These proponents felt that it was now absolutely essential for mankind to put ecologic considerations first and foremost in the management of business, industry and agriculture, before pollution enveloped the earth and exterminated all life. It was imperative, they claimed, for man to control his population growth more successfully before human crowding became intolerable. They felt that nothing short of a major reorientation of life style would suffice to stem the rising tide of environmental destruction.

But this sudden expansion of ecological concern also created a backlash. Many critics derided it as an immature, overemotional response to some long-standing problems. They accused the ecologic point of view of grossly exaggerating the problems of pollution and population growth. They felt that the ecology movement had little real substance and all the characteristics of a fad that would quickly disappear and leave little of permanent value. Some detractors openly charged that the ecology movement was a politically inspired subterfuge to direct attention away from the threat of international communism. Others felt it was unpatriotic and detrimental to the American dream of free enterprise. Some black leaders suggested that the ecology movement was a white plot to sidetrack the problems of civil rights. Thus, as in many popular issues, polarities developed, laden with tension and emotion, in which rational discussion became difficult.

The eminent ecologist, Frank Fraser Darling, said in 1970 that the trouble with ecology is that everyone will become sick and tired of the word before they find out what it means. This had the ring of truth. Despite the prominence of ecology as a public issue and newsworthy subject, the fact remained that its prominence was built upon a remarkably weak educational foundation. Although ecology concerns the central relationships of life, it has not been taught as an academic subject at many of our secondary schools, colleges and universities. In short, ecology suddenly burst into public attention, forcing difficult decisions, without the existence of adequate educational background in the public at large necessary to understand, appreciate and evaluate its complex problems and conflicting claims.

Hence this book. It is directed to the questions: What is ecology? Why has it become so important? What relevance does it have to human affairs and the modern dilemmas of man? What should the educated person know about ecology as a science and a philosophy?

THE MEANING OF ECOLOGY

Ecology is easy to define, and its links with human affairs are relatively easy to recognize, at least superficially, but its full scope and applications are difficult to comprehend. *Ecology is the scientific study of the relationships of living organisms with each other and with their environments.* It is the science of biological interactions between individuals, populations and communities. Ecology is also the science of ecosystems—the interrelations of biotic communities with their nonliving environments.[1] These definitions seem simple enough, but their full ramifications for human health and welfare are infinitely complex and by no means fully understood. Too often, environment is still defined in physical terms alone. The biological and social components of the environment of men

[1] The term, *ecosystem*, and other basic ecological terms are defined in the Glossary.

and other organisms are often forgotten or relegated to some other academic category. Too often, ecology and sociology are treated as separate subjects, as are, in fact, the environmental sciences and the behavioral sciences.

The message of ecology is one of synthesis: to take the broader view and to put the pieces back together again. This broad outlook, in an age of specialists, has often been responsible for the scientific disrepute of ecology in the past, and its popular backlash in the present. Ecology, overlapping with other disciplines from chemistry to mathematics and anthropology to zoology, violates traditional academic and scientific boundaries. It seeks to integrate knowledge about man and his environment from the viewpoints of history, current events and future prospects; it attempts to join biology and sociology, and to reunite the behavioral and environmental sciences.

ENVIRONMENTAL QUALITY

This book is also about the quality of our environment. Environmental quality is an elusive term and an ill-defined issue. It means many things to different people. To some, an environment of the highest quality would be an unspoiled wilderness, a habitat of natural order not significantly altered by man. To others, it would be a beautiful city, rich in culture and the works of man. Still others would picture an agrarian or pastoral scene, filled with productive orchards, fields, vineyards and pastures. Many people would symbolize environmental quality in terms of more electricity, more air conditioning, and more labor-saving machines, whereas others would focus attention on clean air, clean water, and natural vegetation.

Despite this variety of choice and interest, there is surprising agreement on those factors which reduce environmental quality. Virtually all agree that trash, dirt, disease, pollution, noise, strife, poverty and degradation all reduce the quality of the environment. All agree that excessive crowding, tensions, fears, anxieties, violence and ill health impair the quality of life. In fact, there might even be agreement on a general operational concept of what constitutes a quality environment. For the time being, we can consider a high quality environment as one offering the most favorable living circumstances for people of diverse interests; an environment conducive to good health and the well-being of all its inhabitants; an environment in which all human needs are most adequately met—the need for solitude as well as sociality, the needs for food, shelter, education, recreation and aesthetic stimulation. To this the ecologist would add that a quality environment is one in which all basic biological variables are intact and healthy—one in which a diverse and stable biotic community is maintained. A quality environment, most ecologists would insist, must involve more than simply human interests. It must consider the living fabric of the entire world, for it is becoming more apparent that what is best for the world as a whole is ultimately best for man.

Perhaps our concern now, and the main force of ecology's popular rise, is our awareness that modern civilization isn't providing us with a quality environment. We get glimpses of the glorious future of modern technology, but we are haunted by the discordant sights, sounds, tensions and fears of the present. We're worried now, not just by war and fanatic violence, but by the air we breathe, the water we drink, the safety of an evening walk, the irritations and ulcerations of urban life, and the stability of our most valued social institutions.

In 1965, the President's Science Advisory Committee issued a report on this theme entitled, "Restoring the Quality of our Environment." It began with an introductory statement by President Lyndon B. Johnson, which read:

> Ours is a nation of affluence. But the technology that has permitted our affluence spews out vast quantities of wastes and spent products that pollute our air, poison our waters, and even impair our ability to feed ourselves. At the same time, we have crowded together into dense metropolitan areas where concentration intensifies the problem.
>
> Pollution now is one of the most pervasive problems of our society. With our numbers increasing, and with our increasing urbanization and industrialization, the flow of pollutants to our air, soils and waters is increasing. This increase is so rapid that our present efforts in managing pollution are barely enough to stay even, surely not enough to make the improvements that are needed.

This report proceeds with an excellent documentation on problems of deteriorating environmental quality in America. It concentrates on pollution, however, and says little about the quality of our social environment. As critically important as pollution is, it should not obscure the combined interaction of social, biotic and physical aspects of the environment. Ghetto life means not only exposure to a low quality physical environment, and a distorted biotic environment of flies, cockroaches, mosquitoes and rats, but it also means exposure to a tense and stressful social environment. Often it is a social environment in which the strongest educational force is not the schoolroom or the stable family, but the street gang, the drug pusher, and the laws of survival in a violent society. It remained for the Kerner Commission of 1968 and the Eisenhower Commission of 1969 (Graham and Gurr, 1969) to highlight these aspects of urban ecology.

Ecology, biology, psychology and sociology are intimately interwoven, and it is best to draw no boundaries between them. This is a fact often forgotten because most students still take separate courses in one or the other. The educational challenge of the present is to put these subjects together in a way they naturally fit, and to eliminate the artificial walls separating them. Throughout the following chapters there are many examples of how closely they interact. We'll see that environmental quality is intimately related to public attitudes. How much do we want to pay for clean air and clean water? What premium do we place on a mature forest, a natural marshland, or a tract of original prairie? Do we prefer an environment of concrete and asphalt, blaring lights and sounds, roadside amusements and constant artificial stimulation, or do we prefer open recreational areas in forests and fields? Do we at least recognize the value of the latter? Are not these items of public opinion, and hence in the realm of psychology and sociology? But they obviously affect the shape and quality of the environment on a massive scale. Do we prefer to build suburban housing as cheaply as possible for maximum economic gain in this fiscal year, or do we prefer to invest more now to insure a better quality environment fifty years from now? If one tries to imagine what some modern suburban developments will look like in fifty years, the penetrating relevance of this question is quite apparent.

PLAN OF THIS BOOK

Ecology and environmental quality are explored in the following chapters according to a logical sequence. Part I examines some areas of ecology which are relevant to human problems: pollution, erosion, pesticide contamination, radioisotope accumulation, rural and urban blight, population growth, urbanization, crowding, group conflict, the ecology of animal and plant invasions, and infectious diseases related to animal populations. These are all areas where ecologic principles are vital to understanding human problems.

Part II presents a history of ecology, highlighting some of its practical, scientific and academic origins. A central theme here is a brief historical review of our attitudes towards our environment: how man has perceived himself in relation to his world. We will consider philosophies which have molded history for many centuries and examine major changes in these philosophies in recent years.

Part III concerns the basic principles of ecology per se: principles which provide the scientific foundations for ecology and environmental biology. We will first consider such topics as the nature of the ecosystem, principles of energy flow in ecologic systems, food chains, ecologic pyramids, trophic levels, ecologic efficiences, ecologic indicators, and ecosystem balance or environmental homeostasis.

In Parts IV and V we will consider population and community ecology: population structure, life table analysis, growth and natural control of populations, social behavior and population limitation, interspecific population dynamics, the organization of biotic communities, and ecologic succession. These principles can then be integrated into the analysis of habitats and major communities. This involves the classification of communities and a consideration of the outstanding characteristics of selected types: tundra, grasslands, coniferous and deciduous forests, deserts, estuaries, and tropical forests. This does not exhaust the full range of natural communities, but it does illustrate the principles of community analysis.

Finally, our subject material will return to its starting point and consider again some of the applications of ecology in human affairs: realistic attitudes toward technology, the conservation of diversity, and the future of man.

Throughout this book, my aim will be to present ecology as an introductory liberal arts subject. I will not consider ecology as a panacea for the world's ills, nor a gospel which can lead to global salvation, but as a developing body of knowledge and set of attitudes which deserve the thoughtful interest and study of all educated people. There is little doubt that it has been neglected as such. Biological education has placed more emphasis on the anatomy of the frog or pig, than on the structure of the ecosystem.

It is abundantly clear that our order of educational and academic priorities has been out of balance, and we are just now becoming aware of our failures and shortcomings. One failure of ecology in our educational programs has been its virtual confinement to professional biologists and those training to become biologists. These are the people who least need formal education in ecology. They are often already aware of its guiding principles and general position in the biological sciences. Those most in need of some ecological background are students who will become the businessmen, engineers, lawyers, political leaders, clergy and teachers of the world; those who have the greatest influence in shaping the present and the future.

At the beginning of the ecological eruption in the late 1960's, an editorial in *Time* (Atlantic Edition, May 10, 1968), addressed this subject:

> In the search for solutions, there is no point in attempting to take nature back to its pristine purity. The approach must look forward. There is no question that just as technology has polluted the country, it can also depollute. The real question is whether enough citizens want action. The biggest need is for ordinary people to learn something about ecology, a humbling as well as fascinating way of viewing reality that ought to get more attention in schools and colleges. The trouble with modern man is that he tends to yawn at the news that pesticides are threatening remote penguins or pelicans; perhaps he could do with some of the humility towards animals that St. Francis tried to graft onto Christianity. The false assumption that nature exists only to serve man is at the root of an ecological crisis that ranges from the lowly litterbug to the lunacy of nuclear proliferation. At this hour, man's only choice is to live in harmony with nature, not conquer it.

Aldo Leopold expressed a similar philosophy almost forty years ago when he said (1933):

> In short, twenty centuries of "progress" have brought the average citizen a vote, a national anthem, a Ford, a bank account, and a high opinion of himself, but not the capacity to live in high density without befouling and denuding his environment, nor a conviction that such capacity rather than such density, is the true test of whether he is civilized.

ECOLOGY
and the Quality of our
ENVIRONMENT

PART I

The Relevance of Ecology to Human Affairs

I

Environmental Deterioration

THE concerns of ecology, environmental quality and human affairs continually intersect like the threads of a fabric. Some of these intersections are obvious, such as pollution and urban crowding; whereas others may be subtle, such as unsuspected behavioral responses to various features of the environment. But all may be potent as ecologic forces impinging upon man. A prominent midwestern university completed two giant high-rise dormitories several years ago which represented novelty and innovation to the architect, but to an ecologist they represented a drastic experiment in social pressures. They were designed on a radial plan which forced social contact in all activities. Privacy was virtually impossible. There was no way in which a student could study by himself, enter or leave his quarters, or even go to the toilet with any assurance of privacy. Traffic patterns seemed to be established with managerial and architectural considerations in mind; little or no consideration was given to the behavioral and social needs of people. As a result, that which the architect and engineer viewed as a triumph of modern technology, the ecologist viewed as a sad commentary on our own lack of self-understanding.

The purpose of this chapter is to highlight some of these important areas where ecology has relevance to human affairs. Thus we shall briefly consider water, air and noise pollution, erosion, pesticide contamination, radioisotope accumulation, and urban and rural blight. In the following chapter, population growth, urbanization, group conflict, the ecology of animal and plant invasions, and the ecology of infectious diseases shall be discussed.

All of these are important aspects of human affairs for which ecology can provide substantial insight. They are not listed in any particular priority or with intentional emphasis, for if I had started with the most important topic, I would certainly have chosen the interactional stage between population dynamics and social behavior. This is such a complex and ill-defined area of human affairs, however, that it would have been unreasonably confusing to begin here. Thus, I have chosen to begin with somewhat more obvious subjects—those of pollution and environmental contamination.

WATER POLLUTION

Water pollution is a broad and generic term with a variety of meanings. It can, in fact, mean almost any type of aquatic contamination between two extremes: (1) a highly enriched overproductive biotic community, as a lake or river enriched with nitrates and phosphates from domestic sewage; or (2) a biotic community with sufficient concentration of toxic substances to eliminate many forms of living organisms or even exclude all forms of life.

Types of water pollution may also be identified by the medium in which they occur (surface water, ground water, soil, etc.), the habitat in which they occur (marine, estuarine, river, etc.), or the source or type of contamination (nutrient, domestic, pesticide, thermal, industrial, etc.). One of the best definitions of pollution is given in the President's Science Advisory Committee (PSAC) 1965 report on "Restoring the Quality of our Environment." It read as follows:

> Environmental pollution is the unfavorable alteration of our surroundings, wholly or largely as a by-product of man's actions, through direct or indirect effects of changes in energy patterns, radiation levels, chemical and physical constitution and the abundance of organisms.

Pollutants of streams, lakes and estuaries come from many sources. Excessive nutrients commonly originate in domestic sewage and run-off from agricultural fertilizer. Certainly the former is the major source of excessive nutrients in most streams and lakes. Toxic chemicals originate in industrial operations, acid waters from mine seepage or surface erosion, and washings of herbicides and insecticides.

Bacterial and Viral Contamination

Water pollution becomes not only an esthetic problem for man, but an economic and medical one as well. Bacterial and viral contamination is a threat for the spread of water-borne diseases such as typhoid, shigellosis or bacillary dysentery, amoebic dysentery, cholera and hepatitis. In many watersheds raw sewage is a serious problem. The Hudson River above New York City receives over 200 million gallons per day of raw sewage (PSAC Report, 1965). Around many estuarine systems and freshwater lakes and even ocean beaches, public swimming areas have been closed in recent years because of high bacterial counts resulting from domestic pollution. During the summer of 1971, the tourist trade at beach resorts in Belgium, France, Spain and Italy was adversely affected by coastal pollution (Novick and Hopper, 1971). Approximately one-half the public bathing beaches of Chesapeake Bay within a twenty mile radius of Baltimore have been closed by public health authorities because of high coliform bacterial counts. These bacteria are taken as indicators of pollution, often, but not exclusively, emanating from domestic sewage. In Florida, oyster beds in Tampa Bay, Pensacola Bay, Indian River and the St. John's River are closed to harvesting because of domestic pollution. Many parts of San Francisco Bay are closed to water contact sports, as are many beaches of Lake Erie near the larger towns and cities. This does not mean that all these communities live on the brink of typhoid or dysentery epidemics, though the potential does exist in times of flooding, but it does mean that human activities and recreation must be curtailed because of environmental deterioration.

There is still much to be learned about the bacterial communities of polluted aquatic systems. Normally we do not think of serious enteric pathogens being in our water systems in any great abundance. Coliform bacteria, the common indicators of domestic pollution, are not pathogenic in their common forms. A recent study of fishes in Chesapeake Bay by Janssen and Meyers (1967), however, showed that the fishes of western shore estuaries of Chesapeake Bay had significant levels of antibodies to a number of enteric pathogens, including *Shigella, Salmonella,* and pathogenic *E. coli,* whereas the fishes of eastern shore estuaries of the Bay did not. The western shore estuaries are subject to much more domestic pollution than are those on the eastern shore. Data of this type are hard to evaluate, but they suggest that enteric pathogens or closely related forms are indeed present in some polluted rivers and estuaries.

The U.S. Public Health Service estimated that a minimum of 40,000 cases of waterborne illness occur every year in the United States (USPHS Environmental Health Service Report, 1970). As specific examples, in Riverside and Medera, California, in 1965, waterborne disease caused 20,000 illnesses and several

Figure 1–1 Water pollution in Lake Michigan. (*From* The Animal and the Environment, *by F. John Vernberg and Winona B. Vernberg. Copyright © 1970 by Holt, Rinehart and Winston, Inc. Reproduced by permission of Holt, Rinehart and Winston, Inc.*)

deaths. Approximately 8 million Americans drink water with a bacteriological content exceeding the recommended standards of the USPHS. Presently about one-half of our nation's 20,000 community water supply systems contain serious defects that make them potentially hazardous.

Effects of Excessive Nutrients

Even with modern sewage treatment plants, water pollution problems are not entirely avoided. Modern plants remove or inactivate bacteria from the effluent water, but such water is still rich in basic nutrients, such as ammonia nitrogen, nitrates, nitrites and phosphates. The best and most efficient sewage plants do not remove these sources of pollution. Such nutrients stimulate plant growth, often in the form of phytoplankton or algae. The enriched waters are thus prone to plankton blooms which may have several undesirable consequences. Some plankton blooms, particularly those of the blue-green algae, produce undesirable odors and tastes in water. Others, such as the dinoflagellate blooms or redtide of the southern coastal regions, produce toxic metabolic products which can result in major fish kills.

Plankton blooms of green algae do not necessarily produce undesirable odors or toxic products, but they can still create problems of oxygen supply in the water. While these blooms exist under abundant sunlight, they contribute oxygen to the water through photosynthesis, but under conditions of continued cloudiness, they consume more oxygen than they produce and lead to oxygen depletion in the waters. Thus, dissolved oxygen may decline rapidly from favorable levels of 10 to 12 ppm[1] to unfavorable levels of 2 to 3 ppm in which fish experience distress and asphyxiation. Highly enriched streams just below sewage outfalls may show a severe reduction in fish populations, as was documented for the Patuxent River of Maryland. (Tsai, 1968). The Potomac River below Washington, D.C. is highly polluted with domestic sewage and has dissolved oxygen levels often less than 1 ppm. This portion of the river displays annual fish kills every May, when fish reach these oxygen-depleted waters during their spring migrations.

Excessive nutrient levels in aquatic systems can cause two other kinds of ecologic consequences. They may lead to extensive growth of aquatic weeds such as Eurasian milfoil, water hyacinth, water chestnut, and many others which have become a worldwide problem. These growths may become so great as to impair fishing, bathing, fish spawning, shellfish production, and even navigation (Sculthorpe, 1967). Hence, excessive plant growths in enriched waters often represent a major economic problem as well as a complete disruption of aquatic ecology.

It has recently been demonstrated that excessive nutrients in water supplies, in the order of 8 or 9 ppm of nitrate nitrogen can cause human disease, as, for example, methemoglobinemia in infants. This is an illness caused by a modified form of normal oxyhemoglobin in the blood, resulting in inadequate oxygen transport by red cells and labored breathing.

Industrial Pollutants

The effects of industrial wastes on aquatic systems could be the subject of an entire book, and we can do no more here than provide a few examples. In Bellingham Bay and the Straits of Juan de Fuca north of Seattle, Washington, sulfite wastes from pulp mills produce abnormal growth of oyster larvae (Bartsch et al., 1967). Figure 1-2 shows a map of the Bay, the figures for a pollution index (Pearl-Benson Index to sulfite wastes), and the percentage of abnormal larvae. Deformed oyster larvae are found many miles from the pulp plants.

Sometimes the effects of pollution are much more dramatic than the modification of larval growth. In 1959, the U. S. Public Health Service started a national survey of pollution-caused fish kills. In 1961, 45 states reported major

[1] 1 ppm = 1 part per million. This is equivalent to 1 milligram per liter or 1 microgram per milliliter.

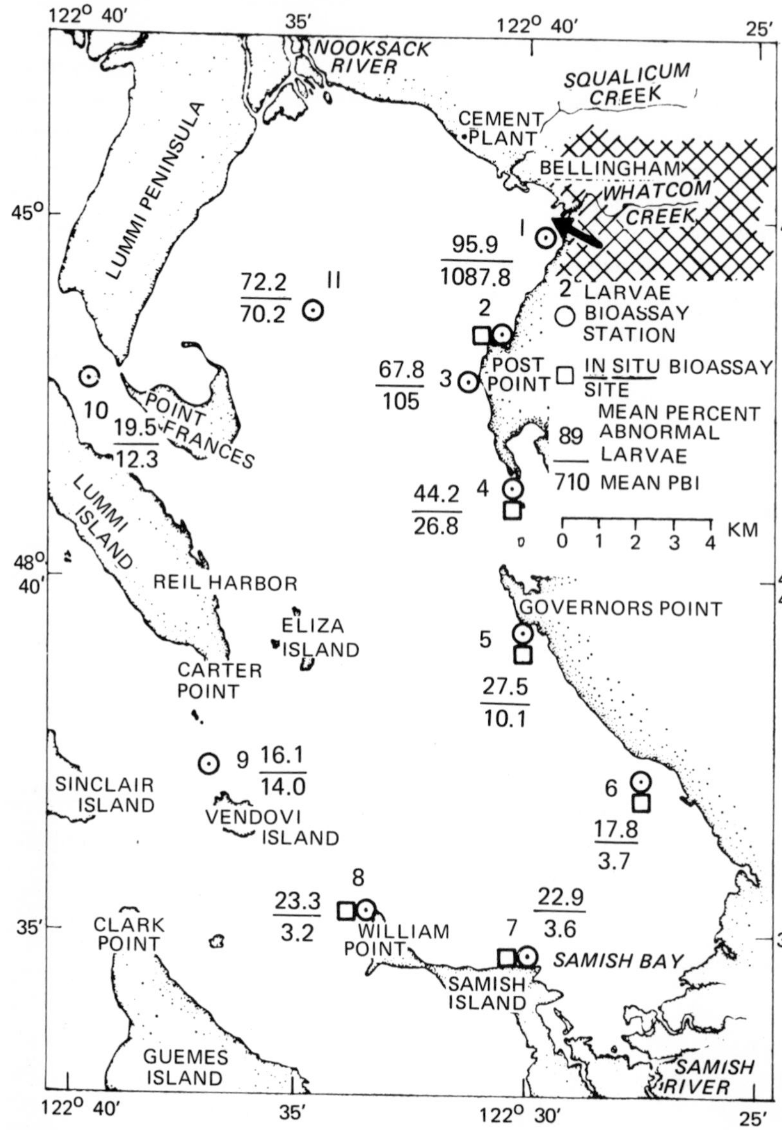

Figure 1–2 Water pollution and abnormal oyster larval growth in Bellingham Bay, Washington. (*From Bartsch et al., 1967. Copyright 1967 by the American Association for the Advancement of Science.*)

fish kills, and in 1962, 38 states reported pollution-caused fish kills totaling 381 separate incidents. Of these 381 fish kills, 43 percent were attributed to industrial wastes, 13 percent to agricultural poisons, 8 percent to domestic sewage, 5 percent to mining operations and 31 percent were of unknown origin.

Some of the various conditions or accidents which caused these kills were as follows: (1) chemicals were used to flush the lines of a power plant (sodium nitrite, hydrazine, ammonium bifluoride, and soda ash); (2) bags of endrin used for insect control were accidentally dumped into a stream; (3) the unauthorized application of copper sulfate was used to control a plankton bloom; (4) outfall from a paper mill contained lignite and sulfite waste liquor; (5) acid mine drainage discharged into a stream; (6) concentrated sulfuric acid released from a cotton seed delinting plant; (7) seepage of the pesticides chlordane and heptachlor from termite treatment project; (8) chromium solution from a plating tank leaked into a sewer system; (9) cyanide released from drains of blast furnaces at a steel company; (10) hot water from a steam generating plant released into a stream. These are merely representative examples of the types of toxic effects which industrial and agricultural operations may have unless rigorous control measures are maintained. Increasing use of powerful chemicals in industrial and agricultural operations increases the risk of environmental damage through accidents.

Occasionally fish kills involve massive populations of aquatic life. In 1962 in San Diego harbor, an estimated 37,800,000 fish were killed by pollution, producing one raft of dead fish 1000 feet long, 10 feet wide and 3 feet deep. In 1963, an extensive fish kill in Chesapeake Bay killed millions of fish from Baltimore to Norfolk, and in 1967 many acres of dead alewife fish washed up along the Lake Michigan shores of the Chicago waterfront. The Chesapeake Bay and Lake Michigan fish kills may have been the result of epizootics of in-

Figure 1–3 Dead alewife fish on a Chicago beach. (*Wide World Photo.*)

fectious disease, however, and were not necessarily the direct result of pollution. Certainly pollution played a role in these outbreaks, however, and may have been a triggering factor.

Another major problem in water pollution is the addition of various ions and chemicals in the water which have toxic effects on plant, animal and human life. Chlorine added to water to control bacteria and algae may persist in streams to cause mortality of plankton and fish. This occurred in Maryland in chlorine-treated waters used to cool an electric power plant on the Patuxent River (Mihursky, 1969). Mercury, as a by-product of industrial operations involved in the production of vinyl chloride, has cropped up as a toxic agent of serious proportions. Mercury is used in many chemical industries, and it also emerges as a by-product of some incinerators, power plants, laboratories, and even hospitals, to the extent of 23 million pounds per year throughout the world (Aaronson, 1971). In Japan, human illness and death occurred in the 1950's among fishermen who ingested fish, crabs and shellfish contaminated with a simple alkyl mercury compound from Japanese coastal industries. This mercury poisoning produced a crippling and often fatal disease known as Minamata disease (Halstead, 1967). The name was derived from Minamata Bay on the southwest coast of Kyushu. Minamata disease was characterized by substantial pathology of the central nervous system and voluntary musculature. Initial symptoms included numbness of the limbs, lips and tongue, impairment of motor control, deafness, and blurring of vision. Cellular degeneration occurred in the cerebellum, midbrain, and cerebral cortex, and this led to spasticity, rigidity, stupor, and coma. The first cases of Minamata disease appeared in Japan in 1953, and of 52 original patients in the villages around Minamata Bay, 17 died, 23 were permanently disabled, and only 3 were able to return to work within 6 months.

Minamata disease has not occurred in the United States as a result of eating coastal fish, but in 1969 it was discovered that fish in Lake St. Clair and Lake Erie contained mercury above permissible levels. The Canadian government immediately banned the commercial sale of fish from these waters until the industrial sources of mercury were removed or corrected, and the mercury levels of fish fell below critical limits. In 1971, the U.S. Food and Drug Administration banned the consumption of tinned swordfish because 80 percent of the samples tested had excessive levels of mercury. Some samples of tuna fish have also shown high levels of mercury, and in 1971 a New York woman who had been dieting by eating tuna every day developed mercury poisoning.

Mercury is just one of many toxic chemicals which may occur in water polluted by industrial or agricultural effluents. Lead, cadmium, and nickel carbonyl are other pollutants which are toxic or pathogenic for man and animals. In fact, more than 12,000 toxic industrial chemicals are in use today. With over one-half of our total available fresh water now being used, one can glimpse the immense ecologic problem confronting modern society.

A basic concern in the overall picture of water pollution is the finite nature of our present surface and ground water supplies. Water use for domestic and industrial purposes is increasing so dramatically, that the time when all available fresh water will be used is now in sight for many communities. Hydrologists estimate the total fresh water supplies of the United States at 515 billion gallons per day, of which we are already using 360 billion gallons per day (Leinwald, 1969). The metropolitan areas of Philadelphia and Wilmington are now using nearly 80 percent of their total available water. This shall reach 100 percent within 10 years at present consumption trends. The great hope, of course, lies in finding new sources and in desalination, but these solutions contain economic and technological problems that remain to be solved. Water pollution and water supply are intimately related because of the necessity for and economics of reuse.

Estuarine and Oceanic Pollution

A final facet of water pollution is the pollution of estuarine and oceanic waters. We have tended to look upon the oceans as so vast that they are virtually unlimited in their ability to accommodate the waste products of terrestrial man. Now there is substantial evidence of global pollution of coastal waters and open oceans. The famous explorers, Jacques Ives-Cousteau and Thor Heyerdahl, both reported in 1970 and 1971 their observations that oceanic waters far from the population centers of the world are seriously polluted. Heyerdahl found waste oil and floating debris in the middle of the central Atlantic Ocean, thousands of miles from land. Cousteau reported a 40 percent decline in marine life in the 50 years of his experience on the world's oceans. He observed that coral reefs are shrinking over the entire world, and more than 1000 species of sea life have become extinct in the last 50 years. It is clear from these studies that modern man has contaminated the far reaches of the earth, and that the world's largest and most basic ecosystems are now reeling under the impact of pollution.

AIR POLLUTION

As serious as water pollution is to the health and welfare of man, in many parts of the world air pollution represents an even more serious threat to human existence. Most of the major cities of the world and many rural areas as well now have serious air quality problems. The relatively small (500,000 people) and modern capital of Malaysia, Kuala Lumpur, a beautiful city surrounded by luxuriant green hills, now has an emerging air pollution problem as its economy begins to prosper and expand. It arises from rapidly increasing automobile, truck and bus traffic, and burgeoning industrial development.

Even the remote capital of Nepal, Kathmandu, which is nestled in the Himalayan mountains north of India, began to show symptoms of air pollution in 1971. Famous for its clear mountain air and startling views of the high Himalaya, the valley of Kathmandu is often hazy with exhaust smoke from rapidly increasing automobile, truck, bus, and airplane traffic. The valley and city of Kathmandu are vulnerable to such pollution because of a natural air inversion —a layer of cold mountain air resting over an enclosed and densely populated valley. Unless control measures are taken now, the magnificent scenery which brings tourists to Nepal, will be viewed through a polluted haze in future years. Air pollution is becoming a problem in cities around the globe.

Ten years ago many engineers and environmental scientists insisted that there were few if any demonstrable ill effects on human health from the levels of air pollution then existing in most cities. They admitted exceptions to this, of course, but these were primarily confined to critical situations such as the famous London smog of 1952 in which 4,000 to 5,000 people died from respiratory distress in a persistent smog, or the crisis in Donora, Pennsylvania in 1948, in which hundreds of people suffered similar fatalities during and after a severe smog condition which hung over the city for several weeks (Leinwand, 1969). Other than these rare circumstances, there was little evidence that chronic levels of urban air pollution represented a public health problem.

Now there is abundant evidence that the levels of air pollution in many cities do represent a major medical problem. This evidence comes from both experimental research with animals and clinical studies in human health. In fact, the health hazards of air pollution in the United States, Japan and parts of Europe are now more clearly documented than are those of water pollution. This is not true, of course, throughout much of Latin America, Africa, and Asia, where impure water still represents a major health problem.

Some of the first clues to the health hazards of air pollution came from clinical observations of increasing rates of emphysema, chronic bronchitis, and respiratory distress in city dwellers, and from experimental studies on laboratory animals exposed to air pollutants. In the latter category of experimental studies, it was noted in Los Angeles several years ago that laboratory mice exposed to ambient air pollution (i.e., normally existing air pollution) developed significant pathology compared to control mice in clean air (Wayne et al., 1968). Aging inbred mice showed increased frequency of pulmonary adenoma, and one strain of mice showed increased mortality of young adult males. Severe smog episodes in Los Angeles caused basic changes in the cellular structure of lung tissue in these animals. Guinea pigs and rabbits developed altered hormone excretion patterns and differential enzyme levels in blood serum in contrast to clean air controls (ibid.).

Pathologic effects of air pollution have also been clearly demonstrated in people. In one study in New York City, children under 8 years of age showed

Figure 1–4 Air pollution in New York City. (*Aero Service Corporation Division of Litton Industries.*)

a prevalence of respiratory symptoms directly related to levels of particulate matter and carbon monoxide (Mountain et al., 1968). In many cities along the Eastern seabord, increasing evidence of dyspnea, bronchitis, cough, sputum production, wheezes, eye irritations and general malaise was elicited as air pollution levels increased (Becker et al., 1968). In Los Angeles, a correlation was shown between carbon monoxide pollution levels and case fatality rates in patients with heart trouble (Goldsmith and Landaw, 1968).

Sulfur dioxide is one of the common gaseous air pollutants which is most injurious to human health. It irritates respiratory epithelium and impairs normal breathing. Most cities have SO_2 levels less than 0.5 ppm, and human effects are not prominent until 0.8 or 1.0 ppm are attained (Battan, 1966). Frequently, ambient levels in cities exceed 0.8 ppm when stagnant air remains for several days and gaseous pollutants accumulate.

Figure 1-5 shows the sudden onset of respiratory distress in people of lower Manhattan following a five-fold increase in levels of sulfur dioxide in the air (McCarroll et al., 1966). On the 16th and 17th of December, 1962, SO_2 levels increased from less than 0.20 ppm to 0.80 and 0.90 ppm. On the 17th, the prevalence of rhinitis in the study population increased from 6 percent to

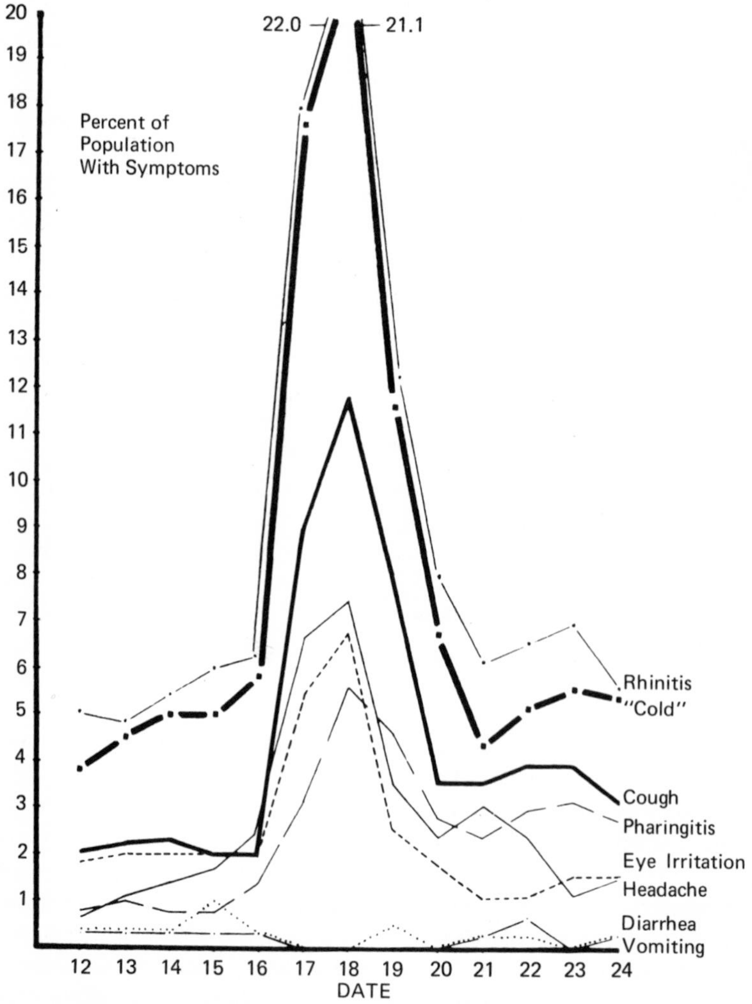

Figure 1–5 Onset of respiratory disease in New York City in December 1962, associated with air pollution episode. (*McCarroll, 1966.*)

22 percent and coughs from 2 percent to 12 percent. Dr. McCarroll's group ruled out an infectious etiology for this episode, and concluded that it was a result of air pollution. Events of this type have become much more prevalent in recent years.

Other investigators have shown correlations between photochemical air pollutants and respiratory distress, emphysema and susceptibility to respiratory infection (Jaffe, 1968). Chronic as well as acute effects have been documented. In the industrial complex of Bayonne and Elizabeth, New Jersey, the death rate from respiratory cancer in males was 35 percent higher in an area of high air pollution, compared to a similar population living in a lower air pollution environment only a few miles away (*Medical World News*, Feb. 3, 1967). On Staten Island, the rate of lung cancer in women in polluted areas was shown to be twice that of women in clear areas (ibid.). Other studies have shown that urban air pollution increases the rate of lung cancer in men three times above that of rural men (Buell, Dunn and Breslow, 1967). This type of evidence could be cited in considerable detail to leave little doubt that air pollution is detrimental to human health. In fact, respiratory ailments related to air pollution—emphysema, chronic bronchitis, lung cancer, and severe asthma—are among the most rapidly increasing health problems in industrialized nations. Various disease problems associated with air pollution were reviewed by Lave and Seskin (1970), who also evaluated the economic costs of respiratory disease attributable to air pollution. They estimated that a 50 percent reduction in U.S. urban air pollution would save the nation $2,080 million dollars per year in health and medical costs.

Air pollution is as complex in origin and type as water pollution. It has been estimated that 164 million metric tons of pollutants enter the United States air every year (Newell, 1971). This pollutant load is composed of a wide range of particulate matter, suspended particles, and gases. In New York City alone, the daily emission of air pollutants includes 3,200 tons of SO_2, 4,200 tons of CO, and 280 tons of particulate dirt. In Philadelphia in 1959, daily releases of air pollutants amounted to 830 tons of SO_2, 300 tons of NO_2, 1,350 tons of hydrocarbons, and 470 tons of particulates.

Particulate matter fall-out affords a dramatic example of dirty air. In many of the world's cities, the average daily air pollution fall-out is in the order 0.5 to 3.0 tons per square mile per day. Some cities have over 4 tons per square mile per day. In Pittsburgh, fall-out was reduced from over 5.5 tons per square mile per day to less than 0.9 tons by a vigorous clean air and smoke abatement program. Up to one percent of urban dust may be lead, which is toxic to humans in fairly low concentrations (Craig and Berlin, 1971). Yet particulate fall-out, while it may be one of the most dramatic forms of air pollution, is certainly not necessarily the most harmful. Most particles which fall out are over 100 micra in diameter, and these seldom if ever reach the alveolar tissue of the lung where irritation occurs. Fine suspended particles and gaseous

Figure 1–6 Industrial air and water pollution. (*Grant Heilman photo from Shepard and McKinley, 1969.*)

substances are the primary agents of respiratory distress. Some of the specific components of air pollution known to have pathologic effects in humans are listed in Table 1-1.

One of the most persistent forms of air pollution which has not responded well to control measures is automobile exhaust. In the late 1960's, over 90 million motor vehicles in the United States produced 66 million tons of carbon monoxide, and 20 million tons of other air pollutants per year (Ehrlich and Ehrlich, 1970). Although various devices have been developed to reduce exhaust emission, these devices are not consistently well maintained by the public. Los Angeles, which has mounted an outstanding campaign of industrial air pollution control, still faces a major smog problem from automobile exhaust. The city has 4,000,000 automobiles for 3,000,000 people. Two-thirds of the area of the central city has been taken over by the automobile in the form of parking space, streets and freeways. One study in Los Angeles showed a direct correlation between air pollution and the frequency of motor vehicle accidents. In the air on crowded freeways, carbon monoxide levels may reach 400 parts per million. Automobile drivers thought to be responsible for acci-

TABLE 1-1 Common Air Pollutants

Pollutants	*Where They Come From*	*What They Do*
Aldehydes	Thermal decomposition of fats, oil, or glycerol.	Irritate nasal and respiratory tracts.
Ammonias	Chemical processes—dye-making; explosives; lacquer; fertilizer.	Inflame upper respiratory passages.
Arsines	Processes involving metals or acids containing arsenic; soldering.	Break down red cells in blood; damage kidneys; cause jaundice.
Carbon monoxides	Gasoline motor exhausts.	Reduce oxygen-carrying capacity of blood.
Chlorines	Bleaching cotton and flour; many other chemical processes.	Attack entire respiratory tract and mucous membranes of eyes; cause pulmonary edema.
Hydrogen cyanides	Fumigation; blast furnaces; chemical manufacturing; metal plating.	Interfere with nerve cells; produce dry throat, indistinct vision, headache.
Hydrogen fluorides	Petroleum refining; glass etching; aluminum and fertilizer production.	Irritate and corrode all body passages.
Hydrogen sulfides	Refineries and chemical industries; bituminous fuels.	Smell like rotten eggs; cause nausea; irritate eyes and throat.
Nitrogen oxides	Motor vehicle exhausts; soft coal.	Inhibit cilia action so that soot and dust penetrate far into the lungs.
Phosgenes (carbonyl chloride)	Chemical and dye manufacturing.	Induce coughing, irritation and sometimes fatal pulmonary edema.
Sulfur dioxides	Coal and oil combustion.	Cause chest constriction, headache, vomiting, and death from respiratory ailments.
Suspended particles (ash, soot, smoke)	Incinerators; almost any manufacturing.	Cause emphysema, eye irritations, and possibly cancer.

dents showed elevated carbon monoxide levels in their blood (Goldsmith and Landow, 1968). Many experts predict that the internal combustion engine must be either outlawed or significantly altered within the next few years if we are to avert air pollution tragedies in our cities. Legislation restricting the use of gasoline and diesel engines by 1975 was introduced into the U.S. Congress in 1970.

There is no doubt that air pollution is a very significant and increasing factor in environmental deterioration. Airplane pilots who have flown for twenty or thirty years report a great increase in "ground haze" and air pollution domes encapsulating cities. Whereas cities were often seen from aerial distances of 30 to 40 miles some years ago, they are now usually enshrouded by air pollution and not visible from more than 5 or 10 miles. The Civil Aeronautics Board has listed "obstructions to vision" by air pollution as a major contributor to aircraft accidents.

In many areas, air pollution has caused dramatic injury to plants, both agri-

Figure 1–7 Automobile traffic—a major source of air pollution. (*U.S. Dept. of Housing and Urban Development.*)

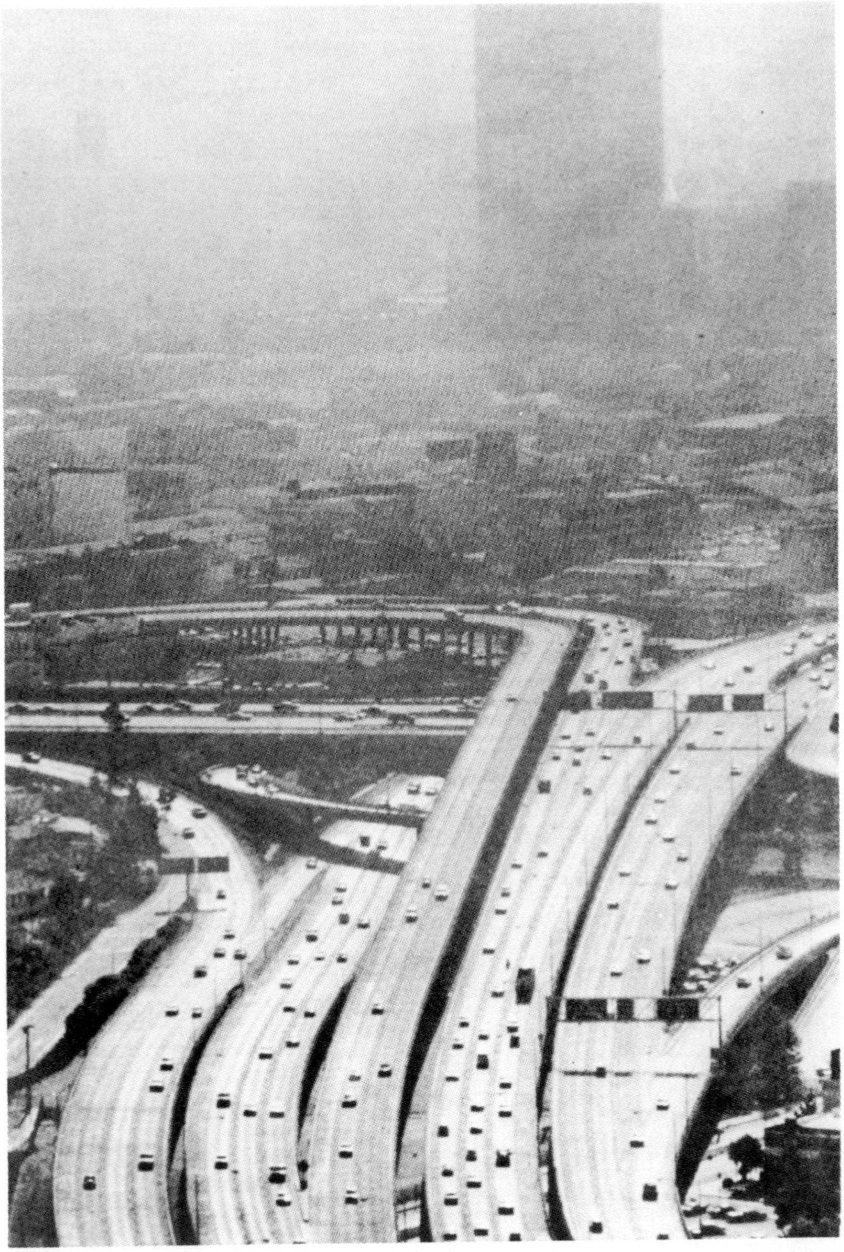

Figure 1–8 Two-thirds of the land area of central Los Angeles has been taken over for automobile use. (*Julian Wasser Photo, from Young and Blair, 1970, National Geographic Society.*)

cultural crops and natural plant communities (Jacobson and Hill, 1970; Waggoner, 1971). Citrus groves, truck garden crops of lettuce, tomatoes, onions and celery, field crops of alfalfa, sweet corn, and tobacco, and even forests of pine, spruce, and deciduous trees have all fallen victims to air pollution in various parts of the world.

Air pollution also takes its toll of buildings and other man-made objects. When moisture accumulates in polluted air, the oxides of sulfur, carbon and nitrogen form weak sulfuric acid, carbonic acid and nitrous acid which are corrosive to metals, stone, paint, rubber, textiles and even some plastics (Battan, 1966). One study estimated that air pollution in 1960 cost the average American family $800 per year in property damage (Leinwand, 1969). Current projections of this figure would certainly put these costs well over $1000 per family per year.

Throughout Europe, many famous buildings, monuments and art treasures of former centuries are deteriorating at an alarming rate due to the erosional effects of air pollution (Young and Blair, 1970). In Athens, the President of the Greek Academy of Sciences, estimated in 1971 that the Parthenon on top of the Acropolis has deteriorated more in the last 50 years than in the previous 2000 years. Athens is normally blessed with clean fresh air, but this depends

Figure 1–9 Air pollution dome in Boston in October, 1969. Concentration of sulfur dioxide in Kenmore Square (*circle*) was 0.35 ppm., a dangerous level. (*H. W. Kendall Photo, Union of Concerned Scientists, from* Environment, *March 1971.*)

entirely on sea breezes. On still, windless days, a noxious air pollution haze quickly forms, even shrouding out the Aegean Sea near Athens.

A serious possibility of world-wide air pollution that was first detected in the late 1960's is the occurrence of stratospheric pollution and a global air pollution veil; that is, a ring of air pollution circling the globe in the northern hemisphere around the latitudes of the U.S., Europe and Japan (M.I.T. Report, 1970). Such a global veil has been detected by satellite photos, and confirmed by both Russian and U.S. scientists. The formations have been temporary, and apparently occurred during periods of unusually stable currents of air circulation around the world. In other words, pollutants were being carried intercontinenally, so that air entering North America contained Japanese contaminants, and air entering Europe contained North American pollutants. Normally, the oceanic mixing of air would break up pollution bands. This is not a major problem now, but it is a serious portent of things to come.

Certainly, many types of air pollution can be controlled by modern technology, but the costs must ultimately be borne by the public through higher prices for industrial goods, higher taxes, reduced profit margins in industry, and more careful monitoring of automobile exhaust control systems. In many cases of environmental quality control, the ultimate responsibility rests with the citizenry at large through established routes of political process. The present cost of air pollution in terms of ill health, agricultural damage, and accelerated deterioration of construction materials and personal goods is very great, but it is so diffuse and indirect that we are not aware of the great price we already are paying for it. When we reach the point where the cost of air pollution is greater than the cost of controlling it, the public must demand appropriate governmental action and be willing to support it.

NOISE POLLUTION

In recent years more attention has been given to noise and unwanted sound as another form of environmental disturbance. Excessive noise has been a part of the industrial environment for a long time—motors, metal presses, riveters, drills, lathes, and heavy machinery of all types have made many factories a din of noise since the beginning of the industrial revolution. Now, however, the public at large is subjected to increasing noise from traffic, airplanes, construction, urban crowding, and electronic entertainment, and we are now aware of much of this sound as a new irritant and source of environmental annoyance.

Sound energy is usually measured in terms of decibels, one decibel being approximately equal to the threshold of hearing in man. A 10-fold increase in sound adds 10 units to the decibel scale, and a 100-fold increase in sound adds 30 units (Ehrlich and Ehrlich, 1970). In a typical urban environment, background noise in a quiet sound-protected room generally runs 40 decibels, while ordinary street noises average 70 to 80 decibels. Around the home, background noise averages 40 to 50 decibels, conversation produces 60 decibels, a garbage disposal 85, a vacuum cleaner 90, a food blender 95, and a power lawn mower 105 decibels. Heavy city traffic at rush hour usually produces 95 to 110 decibels, and a jet aircraft taking off generates 120 to 150 decibels.

Medical science has recently shown that excessive noise can be a significant nervous stress. It can increase irritability and reduce job efficiency. In some cases, it can cause changes in heart rate, blood pressure, and metabolism similar to other types of emotional anxiety and stress. Prolonged noise, above the level of 95 decibels, can also cause hearing loss and early deafness. Workers in noisy factories, construction trades and transportation jobs which have high noise levels are especially subject to hearing loss. The U.S. Public Health Service estimated in 1969 that 7 million Americans were working under noise conditions that could produce permanent deafness.

Noise has penetrated many previously serene environments. Quiet villages of Africa, jungle roads of Southeast Asia, and peaceful lagoons of Polynesia now throb with the sounds of motor bikes, gasoline-powered saws, diesel generators, and outboard motors. The rural life no longer guarantees freedom from the raucous sounds of industrial man.

A special and controversial aspect of noise pollution is the advent of supersonic aircraft, which are scheduled for commercial use in the 1970's. They will produce "sonic booms," loud and forceful reverberations of air as the planes pass overhead. Critics of supersonic planes feel that this will be an intolerable source of sound—a new stress that will cost more in human irritation than it will be worth in reduced transportation times. The proponents of supersonic transportation feel that this is nonsense—that man will adjust to this new technologic advance as readily as he adjusted to the light bulb and motor car. Such a controversy provides another example of the difficulty of assessing the full impact of rapidly advancing technology on the health of man and the quality of our environment.

EROSION

Erosion, a venerable and time-worn subject in the history of conservation, is still a major ecologic force in the world. Whereas there has been improved control of soil erosion from agricultural lands, there has been increasing erosion from newly graded land and excavations related to highway and power-line construction, suburban development, industrial construction, and surface mining.

There is no doubt that excessive silt is a major form of water pollution. Silt laden streams have poor light penetration and often show disruptions of normal ecology and productivity. There may be direct effects on aquatic vegetation, shellfish and plankton feeders, and indirect effects on all aspects of the aquatic community.

It has been estimated that the sediment yield of the Mississippi River watershed in 1952 amounted to 500 million tons per year (Gottschalk and Jones, 1955). Preliminary surveys of the U.S. Soil Conservation Service in the mid-1950's indicated that the rivers of the United States carried about 1 billion tons of sediment to the oceans every year (ibid.). This amounts to more than 300 tons of erosion per square mile per year in the United States, and this does not include all wind erosion.

Some erosion is unavoidable, of course, and should be considered part of natural geologic process. There is no doubt, however, that man's activities have tremendously accelerated erosion. Table 1-2 shows annual soil loss on experi-

TABLE 1-2 Cropping Systems and Soil Erosion[a]

Cropping System or Cultural Treatment	*Average Annual Loss of Soil per Acre (in Tons)*	*Percentage of Total Rainfall Running off Land*
Bare soil, no crop	41.0	30
Continuous corn	19.7	29
Continuous wheat	10.1	23
Rotation: corn, wheat, clover	2.7	14
Continuous bluegrass	0.3	12

[a] Average of 14 years of measurements of runoff and erosion at Missouri Experiment Station, Columbia. (Soil Type: Shelby loam; length of slope: 90.75 feet; degree of slope: 3.68 percent.)

mental plots in Columbia, Missouri, in relation to soil cover. Bare land lost 41 tons per acre per year, whereas a good bluegrass cover lost only 0.3 tons per acre per year (Dasmann, 1968). In hilly areas, bare soil may lose 100 tons per acre per year, whereas forested areas may lose less than 1 ton per acre per year.

Much of the heavy sediment production in the United States comes from direct earth scarring in surface mining and construction activities. No figure is currently available for the acreage of land surface exposed through construction activities, but it is a very sizable amount of land. In 1967 a U.S. Department of Interior publication indicated that 3.2 million acres in the United States were currently exposed in surface mining operations, so this alone would amount to over 120 million tons of sediment production. Many states have laws requiring strip mined areas to be revegetated, but this is often difficult to ac-

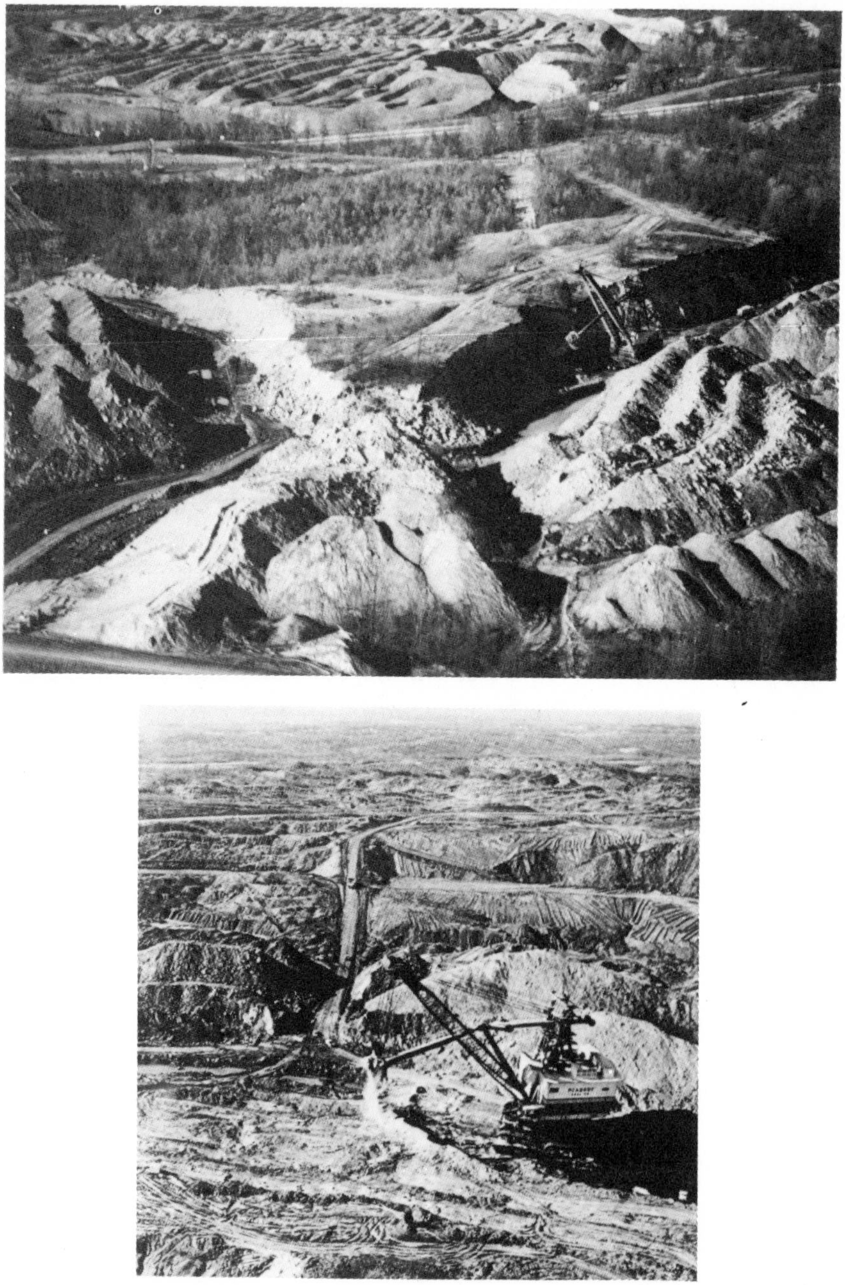

Figure 1–10 Strip Mining operations in Ohio and Kentucky. (*U.S. Dept. of Interior photos.*)

complish and results are highly variable. Strip mining is increasing markedly in many parts of the country, and only token efforts are made to restore these destroyed lands to any semblance of ecologic health.

Although erosion is bad in the United States, it is far worse in many parts of the world. Throughout Africa and Asia the amount of land devastated by erosion is appalling. Vast areas of land in these continents have been denuded of vegetation and soil and now present a configuration of scarred gullies and badlands. Rivers are often heavily silt laden and prone to severe flooding. Coastal waters are muddy several miles from shore due to excessive silt from terrestrial run-off. Even in predominantly forested countries such as Burma and Malaysia, the rivers are muddy brown most of the time and exhibit flash flooding in rain storms. In India, siltation is so severe in some drainage basins that new dams lose 50 percent of their water impoundment capabilities in 15 years. There is no doubt that excessive erosion due to the agricultural, commercial and industrial activities of man is a major factor in environmental deterioration throughout the world.

PESTICIDE CONTAMINATION

The President's Science Advisory Committee estimated in 1967 that "Some 20 tons of DDT residue is contained in the bodies of people of this country; the average amount for each individual is thus about a tenth of a gram." There are also measurable amounts of dieldrin and chlorinated organic pesticides in our bodies. The total amount of DDT in the world's biosphere may total more than one billion pounds (McCaull, 1971). So far there have not been general and well-confirmed effects on human health, but for many of these poisons we do not really know what long term effects might occur. Nor do we have adequate estimates of safe and tolerable levels.

There is abundant evidence that many pesticides widely used in agricultural practice are lethal to fishes and other animals in very small doses. Concentrations of DDT in the range of 0.1 to 0.3 parts per million, and concentrations of endrin in the range of 0.01 ppm can kill some fish. DDT in the concentration of 1 ppm in water for four hours will decrease plankton growth and reproduction by 50 to 90 percent, and 1 part per billion will kill blue crabs in eight days. Heptachlor, lindane or endrin in the concentration of 0.3 parts per billion can kill shrimp in 48 hours.

Certain organisms will concentrate very minute amounts of toxic pesticides in their normal feeding behavior. For example, oysters will retain DDT even when it is present in their water in the trace concentration of 0.01 part per billion, and within 40 days oyster tissue will concentrate DDT 70,000 times. This means that the small amounts of DDT entering the oyster are accumulated, so that DDT concentrations in the oyster increase up to 70,000 times greater than that of the water in which the oyster lives.

This type of ecologic concentration of toxic materials can occasionally produce disastrous effects. This was illustrated in the story of gnat control on Clear Lake, California (Hunt and Bischoff, 1960). In the 1940's and early 1950's, periodic appearances of large numbers of gnats (*Chaeborus astictopus*) were a problem to the residents of Clear Lake, and the resort business was adversely affected. These gnats emerged from April to October, and one survey estimated a total seasonal emergence of 500 gnats per square foot, or a total of 712 billion gnats for the upper part of the lake covering 44 square miles. To control the gnats, DDD was applied at the low concentration of 0.07 ppm. This concentration caused a 99 percent kill of larvae. Chemical treatments were applied in 1949, 1954 and 1957. In the 1954 and 1957 treatments concentrations were increased slightly to 0.05 ppm. Following these applications, western grebes were found dead, and tissue analysis showed concentrations of DDD in them of 1600 ppm, representing a concentration of 80,000 times the application rate. Further study revealed that the DDD had accumulated in insect eating fish, with DDD concentrations in fish running from 40 to 12,000 ppm. The grebes feeding on these fish received lethal doses of the pesticide (Table 1-3).

TABLE 1-3 Pesticide Accumulation in Vertebrate Animals

Location	*Species*	*Pesticide*	*Level of Pesticide*	*Concentration Factor*
Clear Lake, California	Plankton	DDD	12.5 ppm	250
Clear Lake, California	Frogs	DDD	100.0	2,000
Clear Lake, California	Sunfish	DDD	600.0	12,000
Clear Lake, California	Grebes	DDD	1600.0 ppm	80,000
Long Island, New York	Plankton	DDT	0.04 ppm	
Long Island, New York	Shrimp	DDT	0.16	
Long Island, New York	Minnows	DDT	2.00	
Long Island, New York	Ring-billed Gull	DDT	75.00	
Wisconsin	Herring Gull eggs	DDT	200.0	40,000

Similar toxic accumulations of pesticides have been studied in the ring billed gulls of Long Island Sound, herring gulls of Lake Michigan, robins in New Hampshire, and fishes and crabs of Florida (Niering, 1968). Perhaps the most dramatic example of pesticide accumulation occurs in filter-feeding organisms such as oysters, where concentrations of 70,000 ppm have been recorded (Niering, 1968).

Fortunately, human beings are not filter feeders, insectivores, or strictly

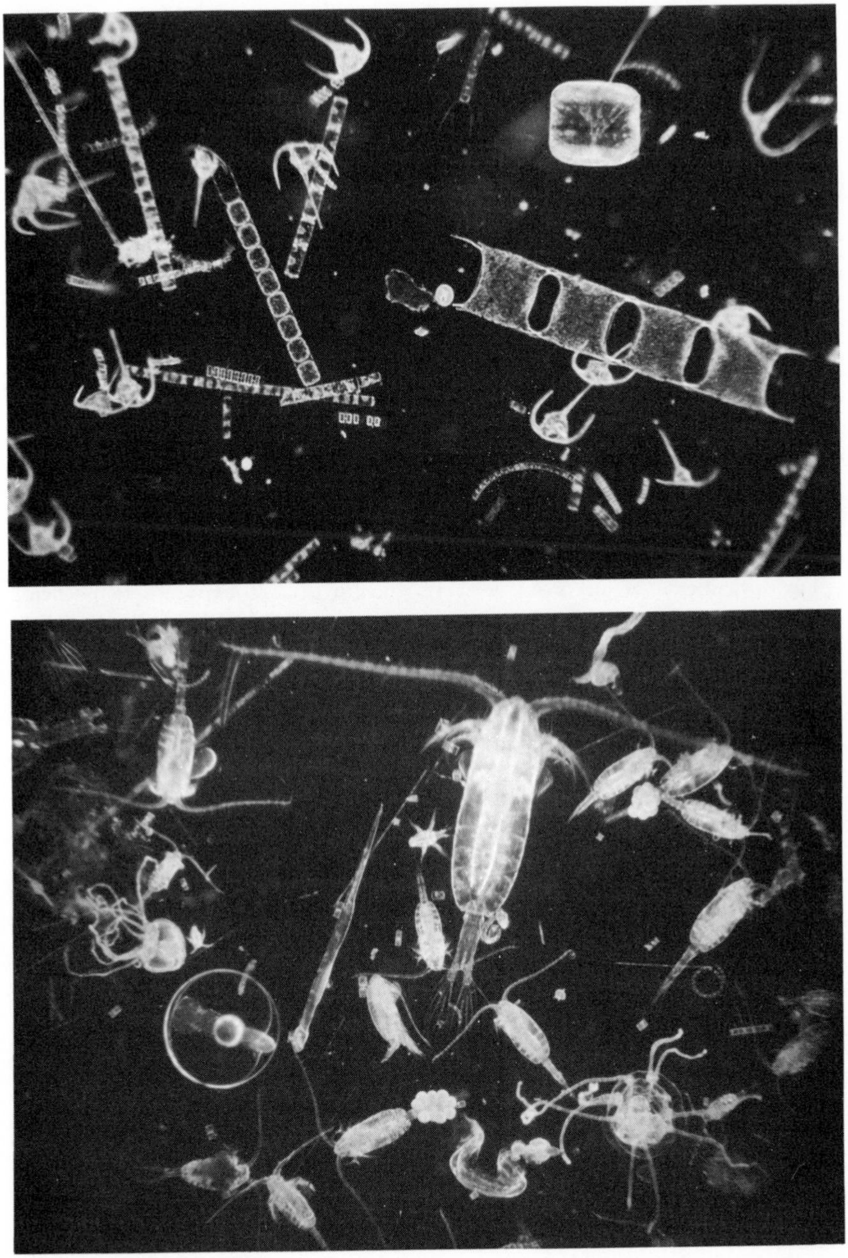

Figure 1–11 Marine plankton populations. Similar organisms are the basis of all productivity in aquatic environments. They may concentrate pesticides, trace elements, and some radioisotopes in the process of growth and reproduction. (*Above*) Plant plankton, primarily algae, magnified 65 times; (*below*) animal plankton, primarily crustacea, magnified 15 times. (*Copyright photographs by Douglas P. Wilson, F.R.P.S., Marine Biological Laboratory, Plymouth, England.*)

piscivores, otherwise similar concentrations would occur in us.[2] We currently ingest pesticides on fruits and vegetables, however, and in a variety of other foods, but probably in such low concentrations that no serious human health problem can be detected at this time. It is known, however, that the average American contains 12 parts per million of DDT in his fatty tissues, though this varies considerably with diet, age, location and other factors (Ehrlich and Ehrlich, 1970). It is also known that people in other countries such as India and Africa where DDT is more frequently used in malaria control may have higher levels of residual DDT in their bodies. Human milk usually contains 0.15 to 0.25 ppm of DDT, which seems like an insignificant amount, but this means that breastfed American babies consume four times the maximum daily intake recommended by the U.N., and five times the DDT content allowed in the interstate shipment of milk.

There is some possibility that these low level concentrations of DDT may be carcinogenic in human tissue over a long period of time. Though the relationship has not been proved, several studies link DDT to cancer. Dr. W. C. Hueper, former Director of the National Cancer Institute, believes that DDT-related cancers could appear within the next 10 to 30 years.

For agricultural or industrial workers who have close exposure to pesticide application, there may be more immediate health effects. Agricultural operators may have 600 to 1000 ppm of DDT in various body tissues. Some fatalities occur every year in the U.S. where workers may get a sudden accidental exposure to high concentrations of dust or sprays. There is also some recent evidence that continued exposure to pesticides produces chronic pathology. A post-mortem study of pesticide residues in agricultural and industrial workers in Hawaii showed, "Subjects with the highest total residues in the tissues were those with evidence of emaciation, a variety of carcinoma, and extensive focal or generalized pathology of the liver" (Casarett et al., 1968). This study measured DDT, DDE, DDD, dieldrin and heptachlor epoxide. Over the long term, certain pesticide residues in the human body may cause genetic effects by damage to DNA molecules (Niering, 1968).

The entire subject of pesticide contamination came to public attention in 1962 with the publication of *The Silent Spring,* by Rachel Carson. This book pointed to dangers of pesticide accumulation in natural communities, and it presented an alarming picture of trends. Many scientists felt that the book overstated the case against pesticides, but possibly only in point of time. Some of the dire predictions made by Rachel Carson have actually occurred. For example, 70 percent of the robin population in Hanover, New Hampshire, died in 1963, after DDT spraying for Dutch Elm disease. There has been a long term decline of ospreys and bald eagles along the east coast of the

[2] Filter feeders are animals which obtain food particles by filtering or straining small particles of food from large quantities of water which are passed through a part of the alimentary system.

United States due to pesticide accumulation which upsets calcium metabolism and makes eggs nonviable. DDT reaches ospreys and eagles by the food chain of aquatic organisms and fish. Food chains refer to series of living organisms, connected by providing a source of food or nutrition for each successive member. For example, bass which feed on minnows, which feed on insects, which feed on algae, constitute a food chain. Food chains are discussed more fully in Chapter 9. The peregrine falcon in Eastern United States is thought to be extinct as a breeding bird, largely through pesticide poisoning (Hickey, 1969).

Although many alarming predictions in Miss Carson's book have fortunately not yet developed, her writings served to alert the public to the dangers of indiscriminate use of persistent poisons. Several states and some European countries have passed legislation banning the use of DDT, but care must be taken to insure that newer generation pesticides are not equally dangerous to the ecosystem. Our present systems of agriculture and disease control (especially mosquito control) require pesticides, but it is now imperative that ecologically safer control systems be developed and utilized. Perhaps one of the main problems was that DDT was too successful in its initial years; it seemed to be the perfect pesticide, and it created the feeling that our problems of insect control were solved for all time. Research on other methods of insect control virtually came to an end. Even when DDT-resistant strains of flies and mosquitos appeared in the 1940's and 1950's, the full magnitude of problems with chemical insecticides was not foreseen, and we continued their use in massive quantities. In other words, we boxed ourselves into a corner on the use of DDT. Our systems of agriculture and mosquito control soon came to require DDT and related pesticides. In these terms, we became "ecologically addicted" to chemical pesticides since we became dependent upon them. Now a sudden and complete withdrawal of their use would certainly be damaging to agricultural production and public health. Yet their continued indiscriminate use can be ecologically damaging to the total environment. We must therefore find new methods of insect and disease control which are less dangerous to the total ecosystem and more selective in their effects. In the meantime, we must regulate the present use of chemical pesticides much more carefully and pursue basic research on pest control more vigorously.

RADIOISOTOPE ACCUMULATION

Radioactive isotopes of elements such as calcium, cesium, strontium, phosphorus, iodine, zinc and many others are formed as direct or indirect products of radiation. Within biotic communities and ecosystems, these radioactive substances may become dispersed or accumulated, depending upon the biologic activity of the element and the period of radioactivity of the isotope.[3] The pathways of radioactive elements in ecosystems have been the subject of intensive research for 25 years, and much of this work has been reviewed in a book edited by Schultz and Klement (1963) and various papers written by Woodwell (1967; 1969).

One radioisotope of greatest biological interest is strontium-90, because it normally occurs in radioactive fall-out. It retains its radioactivity for a long time (with a half-life of 28 years), and it behaves like calcium in biochemical cycles. Thus it is absorbed by plants, ingested by animals, and is deposited in bone tissue close to blood-forming tissue. Strontium-90 can accumulate in natural biological systems as shown in Table 1-4. This table shows that musk-

TABLE 1-4 Strontium-90 Concentrations in the Food-Chain of Perch Lake, Ontario[a]

Component of the Ecosystem	*Counts of Radioactivity per minute per gram of wet weight*	*Concentration Factor*
Water	0.04 cpm	1
Bottom sediments	8.2 ± 0.6	180
Aquatic plants	12.9 ± 0.6	280
Fresh water clams	33.7 ± 1.7	730
Minnows and small fish	43.8 ± 3.1	950
Muskrats	162.5 ± 35.1	3,500

[a] Data from Ophel, 1963.

rats concentrate strontium-90 3500 times above the levels of the water in which they live. Grazing animals concentrate strontium-90 by ingesting it through grass and forage, and it can then be passed on to man through milk. Man may also ingest strontium-90 and other radioisotopes directly through vegetables and fruits.

Radioactive phosphorus is another isotope which can readily accumulate in plants and animals. Ducks and geese in the Columbia River valley in the vicinity of Hanford atomic plant showed concentrations of phosphorus-32 of 7500-fold, adult swallows showed concentrations of 75,000 and young swallows, 500,000 (Odum, 1959). Jack rabbits showed concentrations of iodine-132 to the extent of 500-fold. This degree of concentration occurred through the natural food chains which we shall be discussing later in this book.

Although these isotopes accumulate in human tissues as well as those of plants and animals, it is controversial at the present time as to whether current

[3] All radioactive isotopes release atomic particles in a process known as decay. Some isotopes decay or lose their radioactivity in a matter of seconds or minutes, whereas others may require thousands of years for decay. For example, radium isotope 226 emits radioactivity for over 3,000 years, whereas carbon 11 decays entirely in 40 minutes.

levels represent serious health hazards to man. Some scientists such as medical physicists Drs. John W. Gofman, Arthur Tamplin, and Ernest Sternglass, feel that man's radiation exposure is already sufficient to produce serious disease (leukemia and bone tumors), genetic damage, and infant mortality (Ehrlich and Ehrlich, 1970; Gofman and Tamplin, 1970). Other scientists refute these as irresponsible fears. The fact remains that we know very little about the long-term effects of such isotopes in critical tissues like bone marrow, endocrine glands, and reproductive organs.

It is known that each increment in radioactivity to which living cells are exposed produces a direct increment in mutation rate. There are no threshold values for radiation below which mutational damage may not occur. Thus there is no such thing as a "safe dose" of radiation. Since most mutations are detrimental to an organism, it is important to recognize that any and all additions to the radiation load received by an organism, including man, are potentially harmful.

Many of the newly developing nuclear power plants will release tritium, a heavy radioactive isotope of hydrogen, and other radioactive isotopes such as krypton-85 in their effluents. Tritium is bound in heavy water and it then enters biologic systems. In living organisms it enters the cycles of organic synthesis and can be picked up by DNA molecules in the nucleus, where its low energy beta particle radiation may produce genetic damage. According to the U.S. Atomic Energy Commission and the International Commission on Radiation Protection, tritium and krypton from power plants will not represent a hazard to man or other organisms if kept within prescribed limits. Many ecologists feel, however, that we do not know enough about the pathways of these isotopes in ecologic food chains to fully predict all of their potential consequences. This becomes a far more difficult scientific question than the effects of tobacco smoking or thalidomide use since natural ecosystems cannot be fully duplicated in the laboratory, and it also becomes a more difficult ethical question since the individual will have no freedom of choice about the matter once nuclear plants are built in large numbers. Some of the benefits and problems of nuclear plants have been discussed in a book by Bryerton (1970).

The entire subject of radiation ecology is a topic requiring far more research, and it will become increasingly important as the pressures for greater industrial uses of atomic energy increase. As in many areas of ecology, what we do not know at the present time may be considerably more dangerous to us in the long run than what we do know.

SOLID WASTES

Nowhere is the intemperate nature of our affluent society more evident than in the creation of solid wastes. Every man, woman, and child in the United States now generates about 5.3 pounds of solid wastes per day in the form of garbage, bottles, tin cans, waste paper, plastic containers, used appliances, junked automobiles, etc. This, along with the daily accumulation of agricultural wastes, generates over two billion tons of solid wastes per year in the United States (USPHS, 1970). This is an incredible amount of debris which must be put somewhere. If it is burned it creates intolerable air pollution, if it is dumped into lakes and rivers it creates water pollution and unsightly junk piles. Hence, it must be buried or dumped in the ocean. The latter is obviously unwise, for already the open ocean is cluttered with bobbing plastic bottles and globules of waste oil.

Some of the specific components of our solid wastes in the United States are relatively indestructible, and this includes 55 billion cans, 26 billion bottles, 65 billion metal and plastic bottle caps, and 7 million automobiles thrown away each year. Obviously current patterns and trends in solid waste occumulation cannot continue. We must convert to better technology and a different philosophy of resource use before we become buried in our own debris. Biodegradable packaging materials and more conservative uses of hard goods may offer some solutions, and more constructive, imaginative uses of sanitary landfills may offer other partial solutions. Ultimately, however, different patterns of economy and life style must be developed if we want to maintain any semblance of environmental quality in the face of our mounting junk piles.

RURAL AND URBAN BLIGHT

It is difficult to say anything scientific about landscape blight, although we all know that it is a form of pollution which exists to an alarming extent. Many sections of the country show a roadside landscape of dilapidated billboards, abandoned business enterprises, scarred fields, junkyards, and refuse depots. Since colonial days man's tendency has been to exploit, discard and move on, so that the rusting and rotting shambles of rapid economic turnover litter our landscape.

We have much to learn in this regard from the Europeans. Many European countries have maintained higher population densities with considerably less despoliation of the environment. Surely these countries have some junk yards and disposal pits as we do, but they are fewer, smaller and more tastefully concealed, and there is greater concern for the integrity of the environment. Many parts of the European countryside retain a beauty that has been lost in the American scramble for economic progress.

Throughout the mid-1960's, Mrs. Lyndon B. Johnson made environmental beauty her cause célèbre, and she devoted long hours of personal work and attention to the improvement of our nation's landscape. She emphasized her belief in a major relationship between ecology and behavior when she stated:

> Ugliness is an eroding force on the people of our land. . . . It seems to me that one of the most pressing challenges for the individual is the depression and the tension resulting from existence in a world which is increasingly less pleasant to the eye.

This relationship is difficult to "prove" scientifically, but it clearly represents the empirical experience of many citizens as well as scientists trained in the environmental and behavioral sciences.

At least two factors contribute conspicuously to landscape blight in both urban and rural areas. These are poverty on one hand, and excessive economic exploitation on the other. Both are evident in the American landscape and many of the developing nations of the world as well. Whenever economic exploitation is geared toward maximum financial gain in the shortest period of time, considerable landscape blight can result, for such an economic policy is characterized by cheap construction, blatant advertising, and excessive land despoliation without the added expense of returning used landscape to a more healthy state. Likewise, poverty conditions are characterized by overuse of landscape without the care or the capital necessary to maintain and restore exhausted resources. Thus those who decry all economic development may provide no more of an adequate solution to the ecologic problems of an area than those who propose maximum exploitation. It again becomes a matter of balance, in which planned economic development must be closely linked with wise ecologic management of natural ecosystems.

Although landscape blight is primarily a matter of ethics, esthetics, and economics, science can play its condemning role as well. Our scarred landscapes create unfavorable erosion and run-off, our forgotten roadsides harbor ragweed, poison ivy, tin cans and broken glass, and our burning trash heaps contribute to air pollution. We should now be at the stage where both science and ethics cause us to re-examine this entire matter. The subject was discussed brilliantly by Aldo Leopold (1949) in an essay entitled, "The Land Ethic." This essay becomes more pertinent to the human dilemma each year. It should be required reading in all college or high school biology courses. One brief quotation will indicate its trend of thought:

> There is as yet no ethic dealing with man's relation to land, and to the animals and plants which grow upon it. Land, like Odysseus' slave-girls, is still property. The land relation is still strictly economic, entailing privileges but not obligations.
>
> The extension of ethics to this third element in human environment is, if I read the evidence correctly, an evolutionary possibility and an ecological necessity.

By no means do I wish to belittle the many-faceted conservation movement in the United States, nor did Leopold wish to do so. There are abundant signs of progress through the sincere efforts of many individuals and groups. Conservation has taken effective hold of numerous problems, and has achieved many solutions in the protection and maintenance of landscape and wildlife. But the sheer accelerating force of economic progress has prevailed at many points, and the concerns of the ecologist have often been over-ridden. The game between economic profit and the integrity of the landscape is usually decided in economic terms. This order of priorities must change, and there are fortunately some signs in recent years that such changes can occur.

2

Population Dynamics and Social Behavior

ONE of the major areas of interest and concern for ecology is the scientific study of populations. This includes the analysis of population structure and composition, population growth patterns, the natural control of populations, and the interactions between populations of man, animals, and other living organisms.

Population studies are fundamental to many human problems in agriculture, pest control, wildlife management, and economic planning. The purpose of population research is to understand the interplay of factors influencing populational change, and ultimately to predict and regulate the course of populations. Man obviously wants to limit or reduce populations of undesirable organisms such as mosquitos, rats, parasitic worms, allergenic weeds, and pathogenic organisms, whereas he may wish to maximize within reasonable limits valuable and productive animals for esthetic or economic reasons. The greatest challenge of all, as well as scientifically and philosophically the most difficult area of human endeavor, is understanding and regulating human population change.

WORLD POPULATION GROWTH

There is no doubt about the fact that human populations are drastically out of balance. Figure 2-1 shows the sharply accelerating pattern of human population growth which has prevailed since the seventeenth and eighteenth cen-

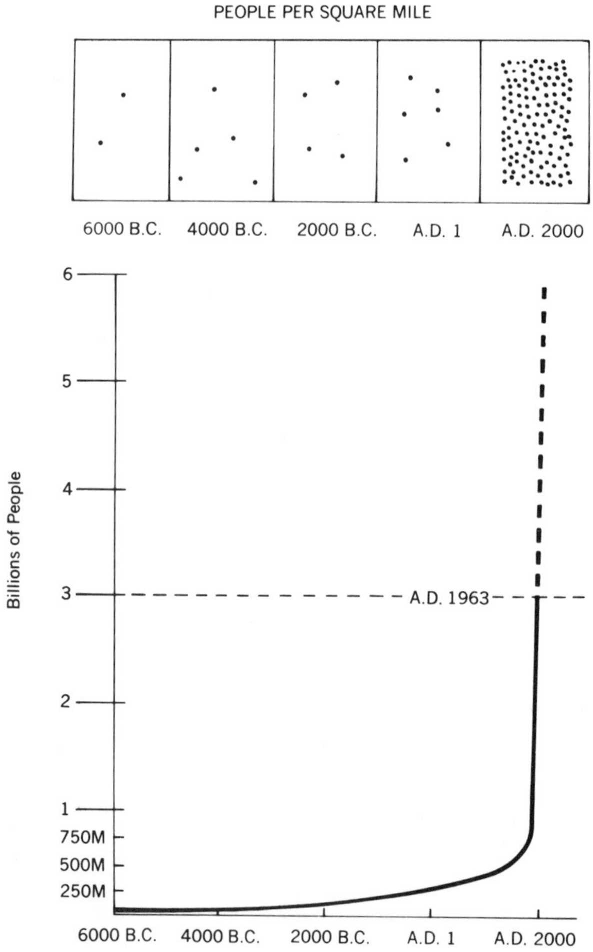

Figure 2–1 World population growth. (*United Nations chart, from* The Animal and the Environment, *by F. John Vernberg and Winona B. Vernberg. Copyright © 1970 by Holt, Rinehart and Winston, Inc. Reproduced by permission of Holt, Rinehart and Winston.)*

turies. For thousands of years prior to the seventeenth century, the population of the world was less than 500 million people. Human populations were limited by high mortality, and even in the great civilizations of Egypt, Greece and Rome, expectation of life probably did not exceed 30 years (Dorn, 1962). The classic checks on human populations were disease, famine and wars. In the fourteenth century, at least 25 percent of the adult population of Europe died in epidemics of bubonic plague, and between the years 1348 and 1379 A.D., England's population was reduced almost 50 percent by plague (Zinnser, 1935; Ehrlich and Ehrlich, 1970). Cholera, typhus, small pox, malaria, yellow fever, and sleeping sickness have all been among the great killers of man throughout history. Several of these diseases, especially malaria and yellow fever, severely limited the abundance and distribution of people in tropical countries until the advent of modern medicine and public health.

Asia has traditionally been the great stage for famine, though starvation deaths also occurred in Europe. Walford (1878) listed 200 famines in Great Britain between 10 and 1846 A.D. The famous Irish famine of 1850 resulted in the death of more than 12 percent of the adult population between 1846 and 1851. In China, in 2000 years preceding the 20th century, over 1800 separate famines were recorded, many of them killing millions of individuals. More recently, 4 million famine deaths occurred in China in 1920–21, more than 5 million in Russia from 1918 to 1934, and over 2 million in India in 1943 (Ehrlich and Ehrlich, 1970). Although modern agriculture and transportation have alleviated famine mortality, malnutrition continues to be a significant demographic force, particularly since it accentuates illness and mortality from infectious disease.

War has also been a major mortality factor throughout history, not only by causing direct battle casualties, but also by stimulating disease and famine. Great epidemics and massive crop failures have often been triggered by the disruptive conditions of warfare. The casualties of war have frequently involved as many civilians as soldiers. Some historians believe that the Thirty Years War in Europe between 1618 and 1648 resulted in the death of 30 percent of the inhabitants of Germany and Bohemia (Ehrlich and Ehrlich, 1970).

Despite the combined actions of disease, famine and warfare, several major developments in the sixteenth, seventeenth and eighteenth centuries stimulated population growth and eased the heavy mortality burdens of man. The opening of the New World provided a vast new area of space and natural resources for economic exploitation. The rise of modern medicine laid the foundations for the control of infectious diseases. The industrial revolution greatly increased the productivity and mobility of man. These factors combined to accelerate human population growth and the world population curve began to rise upward.

Between the birth of Christ and the seventeenth century, the world's population took approximately 1600 years to double. The next doubling was achieved within 200 years, and the next, from 1 to 2 billion people, within 80 years. The most recent doubling, from 2 to 4 billion people will be completed by 1975, a span of 45 years, and at current rates of growth, the world's population will double again within another 35 years.

Currently the world's population is increasing at approximately 2 percent per year—a net increase of 70,000,000 people per year, or more than 1,300,000 per week. Every day of every week, the world's population increases by more than 170,000 people. All of this is primarily a product of man's increased life span and reduced mortality through modern medicine and agriculture. Man's fertility patterns were basically established during periods of much higher mortality rates thousands of years ago, and they have remained high in the intervening years although mortality has declined substantially.

Rates of population increase are highly variable, of course, in different parts of the world. In general, annual rates of increase in western Europe and the United States are in the range of 0.5 percent to 2.0 percent. In many parts of Africa and Asia, annual rates of increase in the range of 2.0 to 3.0 percent are common. In Latin America, many countries are increasing at annual rates between 3.0 and 4.0 percent. The developing countries least able to afford rapid population growth are experiencing the greatest growth.

What does this mean in terms of human welfare and environmental quality? It usually means that many countries have excessive demands for products and services which cannot be met. It is exceedingly difficult, if not impossible, for those countries with the highest rates of population growth to meet the requirements of their people for food, housing, jobs and the physical amenities of life, and it is virtually impossible for them to maintain adequate services in education, medicine, public health and social welfare. Thus we emerge with such problems as: (1) 60 percent of the world's people are still inadequately fed, receiving a per capita calorie intake less than that recommended by the UN (Lowdermilk, 1963; Paddock and Paddock, 1967); (2) 80 percent of the children of India show evidence of malnutrition; (3) there are only 9 acres of land surface available for every person on the earth, and only 16 percent of this is available for agricultural production; thus only 1.4 acres of potential cropland are currently available for each person and this is decreasing each year (UN, Statistics of Hunger Report); (4) many countries, especially in Asia and Africa, still have an illiteracy rate exceeding 70 percent; (5) in many developing nations, there is only one physician for every 5,000 to 10,000 people, a ratio which makes adequate medical care impossible.

This list could be extended considerably to indicate that many parts of the world are currently living under severe population pressure. While the western nations gain strikingly in prosperity, most of the world's peoples face increasing poverty and population pressure.

When the global picture is considered, the outstanding fact exists that we cannot provide a quality life or maintain a quality environment for the majority of the world's peoples at the present time, let alone provide for more than 1,300,000 new people every week.

Ecologically, this represents a highly unstable condition. It cannot persist indefinitely without major corrective or compensating changes. These grim facts coupled with increasing global communication, rapid social and technologic change, and rising expectations throughout all nations add fuel to the fires of social and political instability. As noted by Eric Hoffer in *The Ordeal of Change* (1963):

> . . . a population subject to drastic change is a population of misfits—unbalanced, explosive, and hungry for action. In other words, drastic change, under certain conditions, creates a proclivity for fanatical attitudes, united action, and spectacular manifestations of flouting and defiance; it creates an atmosphere of revolution.

Again we see the intimate relationships between ecology and human affairs. I am not implying that ecology can provide the answers to all of these problems, but surely a failure to recognize their ecologic background invites certain disaster. We will explore many of these topics in greater detail in subsequent sections—the objective here is to illustrate the relevance of ecology to these problems.

URBANIZATION

Not only is there excessive population growth throughout most of the world, but these populations are becoming disastrously concentrated in and around major cities. In the United States in 1800, only 5 percent of the population lived in towns larger than 2,500 people. In 1960, over 65 percent of the population lived in towns of that size, and now 73 percent live in cities of over 100,000. By the year 2000, the population of the United States will total at least 350 million, of which 80 to 90 percent will live in urban areas (Day, 1968). In construction and urban development, we are devouring 4000 acres of U.S. landscape per day.

The primary growth in population will not necessarily be within the city centers, but in a vast, formless and largely unplanned urban sprawl around the cities. Similar changes are occurring throughout most of the world. In 1800, in England and Wales, only 20 percent of the population lived in cities; now more than 80 percent lives in cities of over 100,000.

Throughout the world, many urban centers have experienced phenomenal growth in the twentieth century. Calcutta developed from less than 850,000 in 1901 to 2.1 million by 1941, 6.5 million by 1965 and nearly 8 million by 1972. The metropolitan area of Los Angeles developed from 102,000 in 1900 to over 3 million in 1965—a 30-fold increase in less than 3 generations. In two generations (from 1910 to 1960) Albuquerque, New Mexico, increased 20-fold, and Phoenix, Arizona nearly 40-fold (Dasmann, 1968). Mexico City tripled in size in one generation (1940 to 1965) from 1.4 million to 5.2 million, and Sao Paulo,

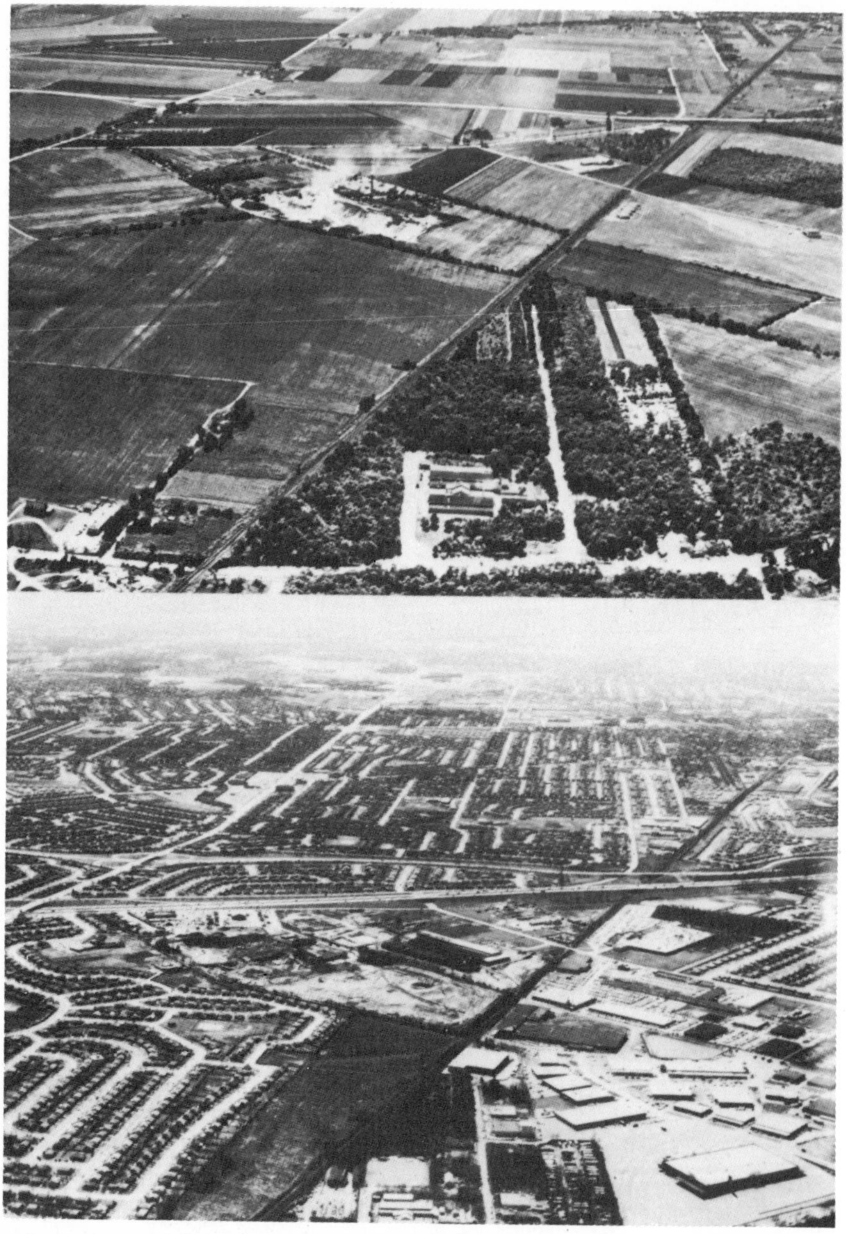

Figure 2–2 Urban and suburban growth and development consumed 4,000 acres per day of U.S. landscape in the late 1960s and early 70s. (*SKYVIEWS*.)

Brazil, showed a similar pattern. These latter cities are still among the fastest growing in the world.

The above census figures refer to the metropolitan areas and not the city limits per se. It should be pointed out that one problem of many cities has been a net loss of people within the city boundary. This has occurred as people have

Figure 2–3 Unplanned urban sprawl in Los Angeles. (*William A. Garnett Photo, from Science, Sep. 1970. Copyright 1970 by the American Association for the Advancement of Science.*)

migrated to the suburbs, leaving behind a deteriorating inner city and declining tax revenues. This local movement should not obscure the broader picture of great urban growth throughout the world by urban sprawl and unplanned aggregation. The fact remains that urban areas as distinct from the political boundaries of cities per se are dramatically increasing in number and size.

Urban sprawl has also been accompanied by changes in the economic and racial composition of city populations. In the United States, affluent whites have fled the central city for the suburbs, and increasing percentages of blacks and other economically disadvantaged people have remained in or entered the inner city. Many factors have been responsible for these migrations. Upper class whites have left the inner city because of dissatisfaction with city living—overcrowding, pollution, noise, crime, high taxes for property owners, deteriorating real estate, poor schools, and so forth. The poor have had to remain in the inner city because of economic inability to move, housing discrimination in better neighborhoods, the high cost of commuting, or a variety of other reasons. Concurrently, people from poverty-stricken rural areas have migrated into cities seeking jobs. Often, jobs are unavailable for those without education and skills, and the cities' welfare rolls are increased. All of these migrations have contributed to the economic and social deterioration of cities—the loss of tax revenues, sky-rocketing welfare costs, increased narcotics addiction and expenditures for law enforcement, fire protection, medical care and other social services.

Obviously these trends in both quantitative and qualitative terms cannot continue. They must taper off, or reach equilibrium at some point. There is already ample evidence that optimal size in human terms has been exceeded in many cities. As stated in a feature article entitled, "The Sick, Sick Cities," in *Newsweek* magazine (March 17, 1969):

> New York and America's other great urban centers are gripped in an agonizing crisis of confidence so profound that it prompts wise men to wring their hands and sends cowards running for cover.

Facing the realities of harsh ghetto life with inadequate housing, poor sanitation, ill health, and a breakdown of social services, giant cities are increasingly becoming a "behavioral sink" of human despair. The *Newsweek* article continued to characterize urban environments as:

> Choking in air so polluted that it filters out a quarter of the sun's light; stifled by traffic jams; plagued by strikes that cripple essential services; victimized by muggers who fill the streets with fear, America's cities daily appear to confirm Thomas Jefferson's dour conviction that they would be "penitential to the morals, the health and the liberties of man."

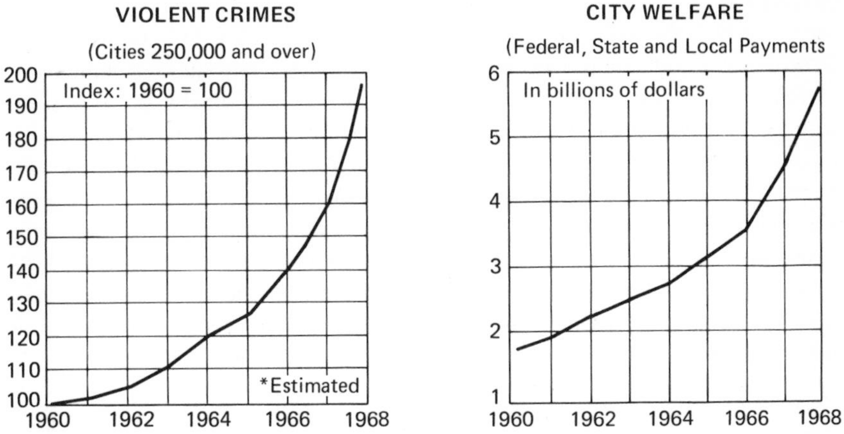

Figure 2–4 Increasing crime rates and welfare costs in American cities in the 1960s. *(Bob Weiss Associates, Inc., New York.)*

As an example of mounting urban problems that presently defy all attempts at solution, Figure 2-4 shows the increase in violent crimes and welfare rolls in American cities within an eight-year period.

Despite these problems of urban areas, people continue to flock toward the cities of the world by the millions as if irresistibly drawn. Obviously urban centers have an attraction and fascination for many people that rural areas can never provide. It is doubtful, however, if the main driving force of urban expansion lies in the cultural and historical excitement of cities; rather it is more likely to be found in the realities of modern economics. As such it may be very close to the struggle for survival, and thus far more of an ecologic phenomenon than we realize. Even in Calcutta, where the resident population density already exceeds 75,000 per square mile, and there are hundreds of thousands of homeless people barely clinging to existence, people continue to pour into the city. They come even in the face of mounting urban violence, averaging 10 murders per day in 1971. They come in hopes of a job, a livelihood, a welfare handout, a better life, or merely the chance to survive. Although they may be victims of mistaken perception and misplaced hopes, they at least feel they can find food and some form of collective security within the city. The disillusionment is massive and the city totters on the edge of social and physical collapse. Unless some urban trends can be reversed, Calcutta may be a glimpse of the future as other cities continue to grow beyond reasonable limits of size and crowding.

It is no wonder that the question, "Can these cities survive?" is being asked more frequently and penetratingly than at any time in man's history. This is fundamentally an ecological question, and the solutions, if they exist, must ultimately involve the broad perspective and systematic analysis of ecology.

CROWDING

Crowding is a conspicuous ecologic force impinging upon modern man. It results from more than a simple increase in population density—it is also a product of activity patterns, psychological needs, and social contact. Crowding implies a force, a pressure, and a psychological reaction. No clear definition of crowding is possible at the present time; it remains an operational concept with many subjective qualities. Crowded populations are, however, easier to recognize than to define, and if they are studied more carefully, better definitions may become possible. For the time being, we might define crowded populations as those in which there seem to be excessive numbers of individuals per unit space in relation to the activity of the individuals and the quality of the environment.

With an operational concept of this type, ecological studies have shown that some rodent populations may be "crowded" in relation to their social behavior and habitat requirements at a density of 5 individuals per acre, whereas others of the same species may not be "crowded" until they have a density of several hundred per acre. Species differ in this regard, and so also do individuals and social groups within the same species. In man, a western cattle rancher may feel "crowded" at a density of 10 homesteads per square mile, whereas a modern suburbanite may feel relatively "uncrowded" at a density of several hundred homes per square mile. The western rancher ranges several square miles per day over what he considers his personal property, and he does not welcome potential competitors in his territory. Conversely, the suburbanite restricts his daily movements, apart from going to work and to shop, to a lawn around his home, and he does not view neighbors as a competitive threat. Hence, in both animal ecology and human sociology, it is important to distinguish between population density per se, and crowding, since the latter depends so much on behavioral factors.

The spatial requirements of man are obviously very variable, and result from both individual differences and social conditioning. It is impossible to state what the optimal spatial and social needs of man are—those of the Japanese city dweller are much different than those of the Kansas farmer. Nevertheless, it is safe to assert that man is now subjecting himself to increased crowding through the combined forces of population growth, urbanization, transportation and mass communication.

Man's perception of crowding has skyrocketed upward in recent years. The growth of urban traffic, and the increased numbers of people in public buildings, schools, hospitals, stores, exhibitions, recreational areas, etc., have all contributed to the feeling of population pressure. Through the mass media of radio, television and newspapers, individuals are vastly more aware of worldwide events. Social pressures at any point of the globe are quickly felt by all nations. Local crime, violence, and warfare now become worldwide worries in

Figure 2–5 Urban crowding as exemplified on this Hong Kong street is a conspicuous trend throughout the world. (*Environment Photo, March, 1970.*)

a matter of hours. The people of one continent feel threatened by those of another continent thousands of miles away. The inevitable result has been a great magnification of social anxiety and population pressure. The world has thus become more limited in its psychological space, and man more crowded in a social and behavioral sense.

How does one evaluate crowding as an ecologic force? It definitely requires knowledge of at least three major parameters: (1) population density, (2) environmental structure, and (3) social behavior in relation to each of the former. Observational studies of crowded populations are possible in both animals and man, but in both, especially man, it is often impossible to isolate crowding as the key variable. Many other factors are interwoven. Crowded human populations are often so characterized by poverty, malnutrition, lack of educational and recreational opportunities, various environmental hazards, and unstable social patterns—that it is virtually impossible to separate crowding itself as the one key variable. The essential question becomes, how does one measure and analyze crowding per se apart from all its related phenomena? It may also be important to distinguish between crowding of various origins. Since man often seeks social contact and stimulation, does not crowding contain constructive and creative social elements, or is it inevitably a factor in increased social problems and environmental deterioration?

From man himself, the answers to these questions are confusing and uncertain. The behavioral sciences of man have just begun to investigate human spatial behavior. More direct clues have come from animal studies where experimental work has been possible. The simpler social systems of many animals have permitted more direct observation of behavior, more ready identification of variables, and more penetrating analyses of cause and effect relationships. In animal studies it has been possible to distinguish between extrinsic and intrinsic crowding; that is, crowding imposed on the animals as an external force, as in captivity, compared to intrinsic crowding in which dense aggregations result from the social behavior of the animals. A similar distinction is sometimes possible in human populations, but frequently it is difficult to discern if people are crowded by choice or by the external pressures of economics and society.

It was demonstrated more than 30 years ago by Professor W. C. Allee and his students at the University of Chicago that most species of animals have optimal levels of crowding, above which and below which deleterious effects occur (Allee, 1949; 1958). These studies showed that certain levels of crowding are necessary to maintain normal biologic function and social activity. Goldfish, for example, grow most rapidly and live longest in social groups of certain densities compared to isolated fish or fish in smaller groups. Fresh water shrimp survive longest and tolerate harmful environmental conditions most successfully at certain optimal states of crowding (Southwick, Sladen and Reading, 1964). Similarly, bobwhite quail survive harsh winters best and reproduce most successfully in the spring in coveys of certain optimal size (Stoddard, 1932). Beyond optimal levels of crowding, however, benefits are quickly lost and various forms of pathology occur. Growth may be stunted, life span shortened, and various diseases or abnormal behaviors appear.

Most studies on crowded animal populations in the last 20 years have concentrated primarily on the harmful effects of overcrowding. Such effects usually involve both physiologic and behavioral abnormalities. The physiologic effects of overcrowding center around endocrine imbalances of the adrenal-pituitary system (Barnett, 1964; Christian and Davis, 1964). These imbalances can lead to a great variety of disease conditions including gastric ulcers, hypertension, nephrosclerosis, arteriosclerosis, and increased susceptibility to infectious diseases (Selye, 1950). Prolonged exposure to crowding and its associated stresses have produced Addison's disease (hypoadrenalism or adrenal exhaustion) myasthenia (muscular weakness), and a variety of serious metabolic disorders leading to shock, coma, and even death. These phenomena are discussed further in Chapter 13 under the topic, "Social Stress and Physiological Response." It should be pointed out that these stress effects are quite variable, however, and do not consistently reflect the same patterns in different individuals and various social groups.

Behavioral pathologies also appear in experimental animals as a result of crowding, and these are sometimes manifested more quickly than are the physiologic changes. Overcrowded animals often show increased aggression and violence, abnormal sexual behavior, disruption of normal nest building and maintenance, breakup of stable social groups, disappearance or alteration of normal social roles, parental desertion of young, and frequently cannibalism of young (Brown, 1953; Calhoun, 1962; Myers, 1964; Southwick, 1955, 1967, 1971; Bates, 1968). These topics will be discussed in more detail in Chapter 14 under the subject of social behavior and population dynamics.

The leading question, of course, is whether or not similar processes occur in man as a result of overcrowding. There are so many variables in man which affect health and behavior, that this becomes a very difficult question. There is no doubt that some striking correlations of disease and behavior are associated with urban concentrations. Many human problems seem to be so accentuated in crowded populations that they virtually constitute an "inner city syndrome"—disease problems such as tuberculosis, venereal disease, and emphysema; maternal problems such as high prenatal loss, parturitional difficulties, and high infant mortality; behavior problems such as alcoholism, drug abuse, mental illness, and criminal assault; and social problems such as high divorce rates and social instability.

As specific examples of these generalities, Faris and Dunham (1939) showed in Chicago that unusual concentrations of schizophrenia occurred in the populations crowded into central Chicago around the loop. More recently Myers and Bean (1968) found a high prevalence of mental illness in the lower socioeconomic classes and dense urban ghettos of several other cities. The three volume report on Manhattan entitled, "Mental Health in the Metropolis," also found unusually high prevalences of mental illness in crowded inner city inhabitants (Srole et al., 1962). McHarg (1969) found high rates of infectious disease, mental illness, drug abuse, and violent crime in the high density areas of inner city populations in Philadelphia.

The problems of urban man seem so strikingly similar to those effects produced in overcrowded animal populations, it is tempting to conclude that similar processes are occurring. But, as stated previously, so many factors are interacting with crowding in urban man—poverty, malnutrition, lack of educational and recreational opportunities, various environmental hazards, and unstable social patterns—that it is virtually impossible to separate cause and effect relationships. There are enough solid exceptions to the principle of "high density produces high pathology" that one must be cautious in carrying this dogma too far. For example, the young nation of Singapore is obviously crowded—over 2,000,000 people on an island of 224 square miles, making it one of the most densely populated nations of the world—yet it enjoys remarkable good health. The death rate is one of the lowest in the world (5.5 per 1000), and the perinatal mortality is lower than that of Britain and the United States. Major factors in this good record may be its youthful population (40 percent of the Singapore population is under 15 years of age), and the well planned interspersion of parks and greenbelts with residential and commercial areas. One does not feel as "crowded" in Singapore as in Hong Kong, Tokyo, New York or Calcutta. Such variables demonstrate the difficulty of generalizing broadly about crowding or population density as a pathologic influence. Thus, the precise role of crowding in the "inner city syndrome" must remain a cloudy issue until more and better research studies are available.

An encouraging development of recent years is the awareness on the part of some architects, city planners, anthropologists and others of the relationships between our environment and mental health. In *Architectural Environment and our Mental Health* (1968), Moller explores the impact of physical space and structure on the psychological well-being of man. He noted that "more buildings will be built in the next ten years than have accumulated since the beginning of civilization." [1] And he wisely emphasized that, "We should recognize that if we continue to permit the structuring of a physical environment which is essentially hostile, to which man must somehow try to adapt, the cost to individuals and to society will be far too great—in terms of mental illness, delinquency, poor motivation, and the fulfillment of our capabilities for creative work and community usefulness."

Anthropologists have also been interested in the spatial behavior of man, and the book by Professor E. T. Hall (1966) entitled, "The Hidden Dimensions," was one of the first to highlight some of the behavior patterns of man relating to social spacing. Hall showed the existence of "personal space" in individuals, and pointed out cultural differences in the extent and importance of personal space. More recently a book edited by Esser (1971) reviewed the interest of psychiatrists, sociologists, biologists and psychologists in the relationships of ecology and environment on human behavior. It emphasized how little we know about ourselves in these vital matters.

Although most of the data on human spatial behavior are preliminary, it is a hopeful sign that several professional and scientific disciplines are now giving attention to this topic. We can be assured that the ecology of crowding will become increasingly vital to man's health and welfare.

[1] A statement I have not been able to confirm from other sources.

GROUP CONFLICT AND VIOLENCE

It may seem presumptuous and unwarranted to include the study of group conflict within the concerns of ecology. Traditionally this has been the domain of history, economics, political science, psychology, sociology and related behavioral sciences. Ecology has had very little to say about group conflict, other than in animal populations. Perhaps this is one problem in itself—a failure by ecologists and non-ecologists alike to concern themselves with ecologic factors in human conflict.

There is, however, increasing awareness of environmental factors in conflict and violence. The anthropologist B. J. Siegel (1969) has pointed out that violence is "one among several strategies of social response to environmental threat." Both the Eisenhower report (Graham and Gurr, 1969) and the Kerner report (1968) emphasized ecologic conditions and social change as underlying factors in violence. The final section of the Eisenhower report is entitled, "Ecological and Anthropological Perspectives," and it considers the relationships between overcrowding, human aggression, and defensive responses to environmental and social threats. The medical psychologist, G. M. Carstairs, concluded in this report (p. 741): "In summary, it seems that overpopulation only aggravates the widespread threat to social stability presented by masses of our population who are basically unsure of their personal future, who have lost confidence in their chance of ever attaining a secure place in their community."

In the general conclusions of the Eisenhower report, the Commission observed (p. 779):

> Probably the most important cause of major increases in group violence is the widespread frustration of socially deprived expectations about the goods and conditions of life men believe theirs by right. Men's rightful expectations have many sources, among them their past experience of gain or loss, ideologies of scarcity or abundance, and the condition of groups with which they identify.
>
> New expectations and new frustrations are more likely to be generated in times of social change . . . nations undergoing the most rapid socio-economic change also are likely to experience the highest levels of collective violence.

The Kerner report (1968) emphasized depressing ecologic and social conditions in inner city ghettos as preludes to violence (p. 204): "The ghettos too often mean men and women without jobs, and schools where children are processed instead of educated, until they return to the street—to crime, to narcotics, to dependency on welfare, and to bitterness against society. . . ."

Bitterness toward established society is certainly a major factor in fanatic violence. In recent years, multiple murders, bombings, hijackings and urban violence have often had the common thread of extreme bitterness against prevailing social conditions. In many cases, the perpetrators of fanatic violence have felt that they had no other recourse—that they were finally boxed in by social pressures so great that only violence offered any possibility of achieving their goals. This suggests that fanatic violence is indeed a symptom of excessive population pressure and social anxiety—an outpouring of ecologic and psychologic forces in man analagous to cannibalism in animal populations. It is unwarranted at this stage to consider the ecologic aspects of violence as proven

Figure 2–6 Bitterness and rage toward established society is frequently a common denominator in human violence. (*UPI Photo by Dennis Connor.*)

and clearly understood relationships, but these are certainly relationships that must be investigated more carefully.

The relevance of ecology to warfare is less generally recognized, though it may be even more direct and intimate. Certainly warfare is an exceedingly complex and highly institutionalized form of group conflict, but its awesome complexity need not obscure the fact that ecologic factors are prominent in the origins of war. Spatial ambitions, resource requirements, intergroup competition, and power struggles between groups have often been causal elements in the history of warfare. In basic terms, these stem from environmental and social interactions between peoples. As the economist Kenneth Boulding observed (1965):

> The philosophy of public health envisages man as a species in an ecological framework. This ecological framework consists not only of other species but also of man's own institutions and artifacts. His houses, sewers, weapons, corporations, nations, and churches must be regarded as species in the grand ecosystem, just as wheat, rabbits, and field mice are. We must think of this whole system as exhibiting a dynamic course through time, the future of which is a function in some degree of its existing state at the present moment. We are certaintly not in anything like an ecological equilibrium, and it is hard to predict at the present time what such an equilibrium would be.
>
> The ecological point of view, to which the public health movement is deeply committed, can make an important contribution to the study of war and peace itself, and to the conquest of war.
>
> In the last resort, the epidemiology of war is not tremendously different from the epidemiology of malaria. Just as malaria is the produce of a biological system that includes man and the Anopheles mosquito, so war is the produce of a social system that includes independent national armed forces . . . and [various] conflict systems.
>
> The study of the ecology of war and peace, therefore, begins with the study of the ecology of conflict itself.

Quincy Wright, in his monumental *Study of War* (1942) pointed out that there were 278 major wars in the world from 1480 to 1940, a span of 460 years. This represented an average of six wars every decade with an average duration of 3.6 years per war. Of these wars, Wright classified 48 percent as "balance of power" wars; that is, conflicts over dominance status and hierarchies between nations; 28 percent were civil wars, that is, intragroup conflicts; sixteen percent were imperialistic wars, that is, wars of territorial expansion; only eight percent of the total were defensive wars, that is, communicative errors where both groups considered themselves attacked and entered the war in a feeling of defense.

The contribution of ecology to our understanding of social institutions as complex and culturally involved as warfare lies in the systems analysis approach which is characteristic of ecology. That is, a systematic consideration of all fundamental relationships of social groups to their environment (the major steps of systems analysis are discussed in Chapter 6). This point of view requires that we consider the physical, biotic and social aspects of the environment as interactional and not distinct. Ecological questions which arise concerning nations in conflict are: What are their basic resource requirements, their population trends, their perceived environmental and social needs, and their physical and social ambitions?

The analysis of Vietnam in the 1950's and early 1960's by ecological thought certainly led to vastly different conclusions than its analysis by traditional political and military considerations, as I recall distinctly on six visits to Southeast Asia from 1959 to 1971. Ecologically, Vietnam could be seen in the early 1960's as an intergroup conflict within a nation; a nation possessing a number of ecologic and social problems around which there were understandable pressures and disputes. Even those portions of North and South Vietnam involved in the conflict, however, possessed a degree of ecologic and cultural unity not shared by other nations. Ecologically, this conflict should have been resolved by the Vietnamese people. Both the environment and the social values within the conflicting groups were foreign to other nations. Hence, the conflict was considerably extended and intensified over many years by the involvement of external powers from both the West and the East. The opportunities for mistaken judgments were magnified by applying simplistic theories of international relations to the conflict. We had no way, for example, of even recognizing the enemy in the normal daily life of the Vietnamese villager, not to mention the difficulty of measuring the social values and cultural ambitions. Both Presidents Eisenhower and Kennedy recognized this and warned staunchly against direct military involvement by the United States. These views were replaced, however, by more traditional political and military thought, in which many basic environmental and social considerations were ignored, and the conflict was perceived in Western terms as a war between good and evil. This oversimplified view consistently led to errors of military judgment and one of the costliest wars in America's history.

The history of East Pakistan in 1971 provided another example of the vital impact of ecologic factors in war. Pakistan was created as a geographically divided nation on August 14, 1947, on the basis of political convenience and not ecological reality. It was formed as a Muslim nation by partition from Hindu India, but East and West Pakistan remained divided by 1000 miles of Indian soil, by distinctly different climactic and ecologic influences, and by different cultural backgrounds as well. West Pakistan was dominated by Punjabis of the desert, and East Pakistan was populated by Bengalis of the Ganges delta. Ecologically and anthropologically, East and West Pakistan had every basis for being two states. Only politics and religion kept these distinct communities together—an illogical union destined for tragedy.

The tragedy broke forth in full measure in 1971. A free election in Pakistan gave an overwhelming mandate for political office to Sheik Mujib Rahman, an East Pakistani who received over 90 percent of the votes in East Pakistan. His party was politically unacceptable to the ruling political leaders of West Pakistan, a minority group in the nation but the one with political and military control. The West forced a delay in the opening of Parliament in March of 1971, and this provided the final spark for violent revolt. The people of East Pakistan declared national independence under the banner of "Bangla Desh," and proclaimed Sheik Mujib Rahman as their leader. A tragic and bloody war of atrocities ensued. Poorly equipped "Freedom Fighters" of Bangladesh began a guerilla war against the modern army of West Pakistan which moved to crush the revolt with massive military action. Villages were razed and thousands of people were slaughtered as jet warplanes, tanks and heavy firepower struck against the 75 million people of Bangladesh. The revolt was initially crushed, producing immeasurable death and destruction. Both sides reacted with unreasonable violence and anger—undoubtedly a product of years of frustration, poverty and population pressure. The conflict led to another mass refugee movement, equalling that of 1947. More than 7 million East Pakistani refugees flooded into overcrowded West Bengal (a part of India) from April to June 1971. The refugee camps became tragic sinks of disease, starvation and despair. Food and shelter facilities were overwhelmed, and cholera, pneumonia and bronchitis broke out in epidemic proportions.

East Pakistan provides another classic case in which ecologic forces overpowered the political foolishness of man. One of the most crowded and impoverished regions on earth, East Pakistan was ripe in 1970 and 1971 for Malthusian forces. It contained 75,000,000 people in only 55,000 square miles —a population density greater than that which would prevail if all the world's people were forced into the United States. In November of 1970, East Pakistan suffered a mortality of a half million people from a severe cyclonic storm which struck the lower Gangetic delta. These were mainly people crowded into submarginal habitats by severe population pressure. Then just four months later, the violent conflict of "Bangla Desh" left a battered nation—a nation with millions of refugees overwhelmed by disease, starvation, and intense bitterness, and a landscape torn apart by the ravages of war. Ecologically Bangladesh is an example of the interaction of population pressure and political frustration. Human tragedy was compounded by a failure to recognize and understand the ecologic forces at work.

ALIEN ORGANISMS: THE ECOLOGY OF INVASIONS

Throughout the last 300 years there has been an accelerating exchange of plants and animals between continents. The carefully balanced biotic communities which evolved over tens of millions of years on each major land mass have been drastically intermixed and often disrupted. Some of this has occurred by accidental introductions, but much of it has resulted from the purposeful hand of man. The starling (*Sturnus vulgaris*) of Europe was purposefully introduced into New York in 1890, and the carp (*Cyprinus carpio*) of Asia was brought to California in 1872. Both have spread through North America and have produced a series of unfortunate changes in natural communities. They have become serious pests and have displaced more desirable native species. Carp, in particular, have contributed to habitat deterioration in lakes and streams by altering the bottom conditions and creating muddy water.

The majority of intercontinental transplants have been equally undesirable and costly. Alien organisms are often freed of their normal checks and balances, and they may find "vacuum niches"; that is, vast stores of food and habitat with none of their natural competitors, diseases, and predators. Generally, the worst agricultural pests of each continent are alien plants and animals. Thus, in North America, the European corn borer (*Pyrausta nubialis*), the Japanese beetle (*Popilla japonica*), the boll weevil (*Authonomus grandis*), the gypsy moth (*Porthetria dispar*), and the Mexican bean beetle (*Epilachna varivestis*) are all accidental aliens, and the economic damage wrought by

Figure 2–7 The desert locust, *Schistocerca gregaria,* in the Somali Republic, Africa, in August, 1960. This species has also invaded Mexico and has produced severe crop damage. Swarms may consist of up to 1000 million individuals, capable of consuming 3000 tons of food per day. (*Photo by A. J. Wood, from Owen, 1966. Courtesy Dr. A. Bellehu, Desert Locust Control Organization.*)

these and related insect pests per year is a multi-billion dollar figure. Similarly, some of the worst agricultural pests of Europe and Asia have come from North America.

Insects are by no means the only intercontinental pests which have caused ecologic and economic damage throughout the world. Examples may be found in almost every class of plants, animals and microorganisms. In Australia, the European rabbit (*Oryctolagus cuniculus*) has been a multi-million dollar pest of the Australian rangeland. In New Zealand, the red deer (*Cervus elaphus*) of Scotland devastates forests and orchards. In Asia and the Pacific world, the giant African land snail (*Achatina fulica*) is a serious pest of agricultural crops.

Disease organisms, as well, have spread rapidly from continent to continent with serious results. In the Middle Ages, plague and typhus spread throughout the world in great pandemic waves. Potato blight which became severe in Ireland in the mid-nineteenth century circled the globe with tragic effects. Influenza exhibited a great worldwide sweep in the First World War and it continues to reverberate between continents, now aided by modern air travel. A new type of cholera (*Vibrio cholerae* type El Tor) moved westward from Indonesia and the Philippines to India and the Middle East in the 1960's and activated this serious disease in areas where it had been well under control. In 1970, this wave of cholera moved into eastern Europe and Africa (Hirschhorn and Greenough, 1971). Many other cases of ecologic invasions have been brilliantly documented in a book by Elton (1958). A popular presentation of this subject has been given by Laycock (1966).

It would be unfair to claim that all introductions have been unfortunate. Certainly many successful and desirable transplants have been made. The ring-necked pheasant (*Phasianus colchicus*) of Asia was brought into the United States about 1875, and has become one of our favorite game birds. The striped bass (*Roccus saxitilis*) of the Atlantic coast has been transplanted in Pacific waters and has been a very desirable addition. Many other examples of successful alien plants and animals could be offered, but in the worldwide balance sheet they are substantially outnumbered by the regrettable errors which have occurred through the intentional or accidental hand of man.

ECOLOGY AND INFECTIOUS DISEASE

One of man's favored dreams of the future has been freedom from infectious disease. With the advent of modern sanitation, antibiotics, vaccinations and insecticides, some of those dreams have recently seemed to be within grasp. Thus, in the United States we have eliminated the active transmission of smallpox, cholera, diphtheria, and malaria—all common diseases less than 100 years ago.

In the 1950's the World Health Organization began the task of worldwide malaria eradication, a job which was considered feasible with modern residual insecticides such as DDT to eliminate the vector mosquitos. Vigorous mosquito campaigns in many tropical countries have achieved local eradication of malaria, but for the world at large the goal of malaria eradication seems more remote now than it did 10 years ago due to certain ecological and evolutionary events. Although intensive mosquito campaigns eliminated 99.99 percent of the mosquitoes in many regions, a small proportion of some mosquito populations survived and led to the development of insecticide-resistant strains. DDT resistance in *Anopheles* mosquitos was first noted in Greece in 1951, but now virtually all continents have resistant strains of mosquitos. More recently, strains of the malaria organism itself, *Plasmodium falciparum,* have developed resistance to modern antimalarial drugs such as chloroquine and primaquine. Thus, malaria is returning to certain parts of the world, especially southeast Asia, and is presenting very serious medical problems in control and clinical treatment. In 1968, even peaceful Ceylon, a nation thought to be an outstanding example of malaria eradication in the late 1950's, experienced more than a million new malaria cases, and in parts of Africa, 100 percent of the children are now infected with malaria by 5 years of age in areas previously free of malaria.

Other infectious diseases are on the increase rather than decrease. Trypanosomiasis, or African sleeping sickness, has recently been increasing in certain African countries because newly independent governments have been unable to support the expensive control programs maintained by colonial powers. Schistosomiasis, also known as snail fever or bilharziasis, is increasing markedly throughout many parts of Africa and the Middle East because it follows new irrigation projects which create favorable habitats for the snails which are intermediate hosts. We don't know enough about snail ecology to effectively control schistosomiasis.

We needn't go to the exotic reaches of Africa or Asia to find examples of infectious diseases which are difficult or impossible to eliminate because of the realities of ecological systems. In the United States we have a number of infectious diseases which are maintained in natural biotic communities and can be transmitted to man by various ecologic routes. These include bubonic plague, encephalitis, rabies, Rocky Mountain spotted fever, histoplasmosis, and tularemia. These belong to a group of diseases known as zoonoses; that is, infectious diseases which naturally infect both animals and man.

Bubonic plague, for example, caused by the bacterium *Pasturella pestis,* is extensively reservoired in wild rodents in western United States, as it is throughout much of Africa and Asia.[2] Normally it remains in ecologic balance with its rodent hosts (ground squirrels of the genus *Citellus* in the United States, gerbils of the genus *Tatera* in India, and gerbils of the genus *Gerbillus* in South Africa), but occasionally it may be fleaborne to domestic rodents, such

[2] The term "reservoir" in disease ecology refers to a host population in which an infectious organism is maintained, frequently without producing serious disease in the reservoir host.

as domestic rats, and then on to man. This chain of transmission is most likely to occur when rodent populations are displaced, as in cases of drastic habitat alteration, or when they suffer extensive mortality, and their fleas seek alternate hosts. If, for example, wild rodent populations which are infected with *Pasturella* increase markedly they may expand their ranges to come into contact with domestic rodents. Then if a sharp decline in the rodent population occurs, through either natural or artificial means, rodent fleas may bite humans and transmit the bacillus.

It is no coincidence that plague outbreaks in human populations most often occur in war-torn nations when massive habitat alterations take place and both human and animal populations are subjected to disruptive movements. Ecologic instability always increases the probability of drastic changes in biotic communities.

Yellow fever is caused by a mosquito-borne virus which is extensively reservoired in wild monkeys in Latin America and Africa. In Latin America the virus is normally transmitted from monkey to monkey by a treetop dwelling mosquito of the species *Haemagogus spegazzini.* This mosquito lives in the high canopy of forest, and normally does not come near man. In forest clearing activities, however, such as road construction through forests, these mosquitos can be swept to the ground when trees are felled, and they may then bite people. Most of the sporadic cases of yellow fever in Central and South America in the last 20 years have been transmissions by *Haemogogus,* and not the traditional vector of human yellow fever, *Aedes aegypti.*

Encephalitis is a severe viral disease reservoired in wild birds, and occasionally transmitted to horses and humans by mosquitos. The probability of transmission at any given time and place is a function of quantitative relationships between bird, mosquito, horse, human and virus populations. We do not know enough fundamental ecology at the present time to effectively predict when and where encephalitis may next be transmitted to man.

Rocky Mountain spotted fever is caused by a rickettsial organism extensively reservoired in small mammals such as squirrels, wood mice, opossums, raccoons and others. It is transmitted by tick bites from animal to animal and to man. The disease is not confined to the Rocky Mountain region, but is known to occur in many other areas of the United States, including the Maryland and Virginia environs of Washington, D.C. Without undue disruption of animal populations, the rickettsial organisms normally remain in balance with their hosts, and human infections are rare.

Does the presence of these infectious diseases in our biotic communities mean we should eliminate all birds and mammals? By no means, for in the first place this would be impossible, and in the second, it would be highly dangerous and might well magnify the probabilities of human disaster. It would create further instability in the ecosystem, and even more violent perturbations in the relationships of hosts and pathogens.

The conquest of infectious disease has been so dramatic in some cases—with smallpox, diphtheria, polio and a few others—that we tend to believe in the unlimited ability of modern medicine and public health to cope with any infectious disease. But the facts of the matter are that we are a long way from conquest of many infectious diseases. Some viruses mutate faster than our ability to develop vaccines. The Hong Kong flu of 1967 literally swept around the globe, with a worldwide attack rate of 20 to 30 percent, and we were unable to stop it. Fortunately, it was a mild virus, with very low fatality, but the next mutation which could arise on any continent might be considerably more virulent. This possibility was highlighted by Dr. Joshua Lederberg in recent Senate hearings (Lederberg, 1969).

It is entirely possible that a combined synergism between air pollution and respiratory air-borne viruses might represent a very serious health hazard to man. We know that many air pollutants, such as SO_2, reduce ciliary motility and increase the chances of respiratory infection. This action, coupled with a newly evolved influenza virus, could represent a medical problem of the first order, against which modern medicine is poorly equipped at the present time. With modern jet air travel, new virus infections can spread from continent to continent before medical scientists can identify the virus.

In later chapters, we will discuss the great significance of maintaining ecologic diversity and stability in the environments of man. We are just beginning to comprehend the necessity and complexity of what has long been called the "balance of nature."

SUMMARY

In the last two chapters we have discussed a number of major areas in which ecology has relevance to human problems. These have been somewhat arbitrarily grouped into two categories: *Environmental Deterioration,* including water, air and noise pollution, erosion, pesticide contamination, radioisotope accumulation, and urban and rural blight; and *Population Dynamics and Social Behavior,* including population growth, urbanization, crowding, group conflict, alien organisms and the ecology of infectious diseases.

Probably the greatest danger of this approach is mental indigestion. "Where," one might well ask, "are the limits of ecology? If these are the concerns of ecology, does not ecology try to claim dominion over everything?"

A partial answer to these questions might be phrased as follows. The limits of ecology are the limits of man and the living world—ecology should include everything within the framework of life. Ecology, in fact, is the great synthesis, the theme around which all disciplines concerned with man and the living world must be oriented. Our various scientific and educational disciplines become far more meaningful when they are related to the theme of man's relations to himself and his environment. Then history, anthropology, sociology, economics, religion, psychology, biology, chemistry and virtually all of the humanities and sciences acquire far more significance. They become oriented around man and his universe; yet they avoid an exclusively anthropocentric point of view. They see man as a member of a global ecosystem, and not as its sole master. They assist more effectively in relating the past, present and future; they provide perspective and insight in seeing the shortcomings and successes of the present. The next three chapters will attempt to examine the history of ecology and the history of man with this perspective.

PART 2

Historical Aspects of Ecology

3

Early Man and the Rise of Agriculture

HUNTER-GATHERERS AS ECOLOGISTS

EARLY man was a successful practicing ecologist. He survived in a rich and competitive biotic community, and his relationship to this community was continually intimate. He was by no means the strongest, the swiftest, nor the hardiest among his congeners, but he possessed several distinct advantages. He had a close effective social organization, he had unusual manipulative ability, and he had, of course, an emerging intelligence. Thus he developed tools and fire in early Paleolithic time (500,000 to 1,000,000 years ago), and he accumulated knowledge at a faster rate than other primates.

Much of this knowledge was ecological. It was knowledge of his environment and the most effective use of it. It was detailed knowledge of food and water resources, some of which can still be seen in the "primitive" peoples of today. The Kalahari bushmen, for example, can find water in a barren desert where other men would surely die of thirst. The Australian aborigine can locate grubs and lizards in the Australian deserts far better than a modern biologist. Elton (1933) pointed out that, "The Arawak of the South American equatorial forest knows where to find every kind of animal and catch it, and also the names of the trees and the uses to which they can be put." Bates (1960) found on the Micronesian atoll of Ifaluk that the native people had a detailed ecologic knowledge of the plants on the islands and the use to which each could be put in terms of food, medicine, construction and ornament. They also possessed detailed knowledge of the reefs and sea around them. In all societies of hunter-gatherers there was strong selective pressure against the nonecologist. This rigorous natural selection prevailed throughout the two million years of Paleolithic man, and before that the 10 to 15 million years of Pliocene and Miocene hominids and protohominids, the forerunners of man.

There is increasing evidence that early man was well adapted to his environment. The popular conception that he barely clung to life through a precarious and difficult struggle is definitely misleading (Lee and DeVore, 1968). Recent studies of both ancient and modern hunter-gatherers have shown that they frequently have had an abundant life, with very ample resources and sustenance. Some present-day primitive peoples are declining, of course, but usually as a result of modern forces—a deteriorating environment or competition from agricultural peoples. Other studies on former hunter-gatherers have shown that malnutrition was rare, starvation infrequent, and chronic diseases, as we know them today, of relatively low incidence—all evidence of sound ecologic balance in their way of life (Dunn, 1968; Neel, 1970). There was, however, high infant mortality, primarily through infectious disease, and high "social mortality" (infanticide, geronticide, warfare, etc.), and these served as primary mechanisms of population regulation.

The ecological ability of early man does not necessarily mean that he was a good conservationist. Guthrie (1971) has pointed out that the concept of early man living in blissful harmony with his environment is a serious distortion of facts. Early man knew enough practical ecology to survive and even prosper, but he exploited his environment at every opportunity. He was a persistent forager and relentless hunter whose primary goal was survival. He was often nomadic "in part because prolonged habitation in any one area depleted game and firewood and accumulated wastes to the extent that the region was no longer habitable " (Guthrie, 1971). Some scholars have felt that the overzealous hunting of early man contributed to the widespread extinction of animals in the Pleistocene Epoch[1] (Martin and Wright, 1967). Although this is a controversial hypothesis, there is some evidence that early man was

[1] The Pleistocene Epoch is also known as the Ice Ages, that period from 10,000 to 600,000 years ago, in which great populations of large mammals, including wooly mammoths, mastodons, giant pigs, royal bison, camels, horses, giant armadillos and great ground sloths roamed the temperate regions of North America and Europe.

not always conservative in hunting or environmental protection. For example, plains Indians of North America were known to have killed many more bison than they could utilize by driving them over cliffs. Apart from some of these extravagances, early man did feel a close kinship with nature, and considered himself a small part of a vast scheme.

As pointed out by Lee and DeVore (1968): "Cultural man has been on earth for some 2,000,000 years; for over 99 percent of this period he has lived as a hunter-gatherer. Only in the last 10,000 years has man begun to domesticate plants and animals, to use metals, and to harness energy other than the human body." The civilizations of agricultural and industrial man have a long way to go before they can match the longevity of primitive man.

PASTORAL MAN AND THE DOMESTICATION OF ANIMALS

With the rise of civilization and its elaborate divisions of labor, more niches for the nonecologist became available. The weaver, potter and tool-maker did not require the same broad ecological knowledge as the hunter and gatherer. The most significant ecological achievement of civilized man was, of course, the domestication of plants and animals for greater productivity and control over the means of subsistence. Thus, the development of pastoral and agricultural life, in the Neolithic period of 10,000 B.C. to 6,000 B.C., altered the entire pattern of human existence. Permanent villages became established, intergroup cooperation and trade routes developed, and a demand arose for a new type of ecologic knowledge, that associated with the husbandry of plants and animals. Economics displaced ecology as the vital key to survival success.

The beginning of recorded history, about 3,000 B.C., showed civilizations in Egypt and Mesopotamia with cities and a high degree of vocational specialization. The Bronze Age, shortly after 3,000 B.C., and the Iron Age, starting about 1,000 B.C., accelerated agriculture, exploration and conquest, but did little for man's general ecological knowledge. Environmental exploitation expanded without a comparable increase in environmental harmony. Major land changes occurred. Forests were cut, fields cleared, pastures grazed and plowed, and the landscape was carved to fit the new economic demands of man. A general increase in aridity occurred throughout much of the world then occupied by civilized man, and some of the great land barrens of our modern world probably developed during this time (Marsh, 1864; Sauer, 1938; Sears, 1935; Thomas, 1955–1956).

There have been, in general, two theories to explain the development of these great land barrens, especially those in the Middle East and North Africa. The theories differ in their views on the role of man. One has viewed man as the victim of climatic change, and the other has viewed man as the perpetrator of climatic change. Thus, in one view, man had to adapt to desert conditions imposed upon him by great continental forces of climate and geography, and in the other view, man himself was the major desert-making force.

The "man-as-victim" view has probably prevailed in most academic circles until recent years, not so much because of positive evidence to support it, but mainly because it seemed more logical and definitely kinder to the man's ego. There is now increasing evidence from archaeology and ecology that man's role as a desert-making force has been underestimated. Rather than being a passive victim of climatic change, man's influence in inducing this change has been significant and perhaps critical. This view does not assert that man's activities have been the sole forces in desert formation, but it does assert that they were influential in accelerating and intensifying desert formation.

THE RISE OF AGRICULTURE

Agricultural man set into motion several major forces which had significant geographic and meteorologic consequences: deforestation, overgrazing, intensive burning, and land scarring. Geographic consequences came about through increased erosion, soil loss and declining water tables; meteorologic consequences resulted from reductions in atmospheric humidity and cloud cover, increased heat reflectivity and a lowering of rainfall. Ecological studies have shown that forests help to maintain the level of rainfall necessary for their own existence, and that deforestation results not only in an immediate lowering of ground water levels, but also in longterm lowering of rainfall. Forests recycle moisture back into their immediate atmosphere by transpiration where it again falls as rain. Transpiration return from an acre of forest may reach 2,500 gallons per day (McCormick, 1959), and thus create a natural system of water reuse. If the forest is removed, this natural reuse cycle is broken, and water is lost through rapid run-off. Grasslands perform the same function to a lesser degree, and operate with a smaller amount of available moisture. It is interesting to examine Middle Eastern history in light of this principle.

CIVILIZATIONS OF THE MIDDLE EAST, NORTH AFRICA AND THE MEDITERRANEAN

The span of history from 5,000 B.C. to 200 A.D., which we know primarily as the period of great civilizations—Sumeria, Babylonia, Assyria, Phoenicia, Egypt, Greece and Rome—was also a period of unprecedented environmental disturbance. We tend to concentrate our attention on the superb achievements of these civilizations in literature, art, government and science, while we virtually forget their incompetence in land management. Perhaps this is not surprising since we seem to ignore our own shortcomings in the same area. Nevertheless, the fact remains that these golden civilizations prospered at the expense of their environments. They left a landscape which has never recovered, and a legacy to future civilizations which ushered forth a period of dark ages for more than a thousand years.

As specific examples of the destructive nature of these civilizations, we can cite critical landscape changes in many areas of the Middle East, North Africa and Mediterranean lands. As late as 7,000 B.C., the headwaters of the Tigris and Euphrates Rivers were covered with forests and grasslands (Saggs, 1962). In fact, most of the area now occupied by Iran and Iraq was productive and well-watered (Sauer, 1938). Domestic cattle appeared in the seventh millenium B.C., probably around 6,300 B.C. (Perkins, 1969). Thus herds of domestic cattle, sheep and goats found very favorable pasture in these virgin grasslands. Their great success provided a major stimulus for the developing civilizations of Mesopotamia and Sumeria, but it also provided the first significant onslaught on these grasslands and their adjacent forests. The herdsmen prospered, utilizing the stored capital of thousands of years, and the upsurge of prosperity was followed by more people and larger herds. Forests were cut to provide additional pasture and more land was exposed to the increasing livestock populations. Residential communities developed around 4,000 B.C., and further deforestation occurred to provide timbers for developing cities. Elaborate agricultural practices ensued with irrigation canals extending the reach of the rivers. At first the erosion and silt load of the canals was manageable, and the alluvial soil along the rivers was remarkably fertile. With passing generations, however, siltation increased and it became necessary to occupy great numbers of slaves and laborers with the job of maintaining irrigation channels free of silt (Dasmann, 1968). After 3,000 B.C., the increased silt of the Tigris and Euphrates filled in the Persian Gulf 180 miles from its origin. All of these changes—loss of vegetation and soil, lowered water tables, declining agricultural productivity, diminishing rainfall, and the added economic burden of siltation in the irrigation canals—were significant factors in the fall of the great Babylonian empire (Saggs, 1962). Successive waves of invaders—Kassites, Elamites, Assyrians and eventually Persians—conquered this tragic land and increased its devastation.

At the same time of the Sumerian and early Babylonian empires in the Middle East, there was a thriving civilization in the Indus valley of present-day Pakistan. A prosperous and advanced culture existed at the site of Mohenjo-Daro in the Province of Sind (Wallbank, 1958). There is evidence of an urban civilization, with well-planned streets, dwellings, municipal halls, palaces and a well-engineered drainage system. An advanced governmental structure apparently existed. There was skillful use of bronze, copper, silver and lead, and there was also beautifully glazed pottery, delicate jewelry and carefully woven cotton textiles. This prosperous civilization came to an abrupt end about 2,500

B.C. No one knows the reason for its total collapse, but there was again a loss in the means of subsistence. The Sind today is mostly desert, except in those portions bordering the Indus River and its related irrigation canals.

Man, over a period of 3,000 years in the sixth to the third millenia B.C. had become a major force in environmental change. Although his powerful civilizations and glorious cities had flourished, they were intimately dependent upon the land which gave them birth. When the pastoral and agricultural riches of the land were destroyed, the civilizations could no longer be supported, decline began, and they became more vulnerable to invading armies.

In Africa, similar patterns have been found. Even the Sahara within written history has not been as extensive as it is now. In Egypt and the Sudan, Davidson (1959) found abundant evidence of former productivity within historical times:

> Even as late as the third millenium large numbers of cattle are known to have found grazing in lower Nubia (formerly Ethiopia, now part of Sudan) where, as Arkell says, "desert conditions are so severe today that the owner of an ox-driven water-wheel has difficulty in keeping one or two beasts alive throughout the year." And anyone who has traveled in these dusty latitudes will have noticed how the wilderness of sand and rock that lies to the west of the Nile, far out upon the empty plains, is scored with ancient wadi beds which must once have carried a steady seasonal flow of water, but are now as dry as the desert air.

Throughout Saharan Africa, there are "lost cities" of former grandeur in areas now too arid to support life (Davidson, 1959). It would be extravagant and incorrect to claim that man's activities produced the Sahara desert, for it was largely the result of great climatic shifts over the course of geologic history. But the evidence is clear that man extended Saharan conditions and forced once productive and well-watered areas to become barren wastelands.

The Sahara is still actively extending its arid conditions—in some areas at the rate of one mile per year (Rienow and Rienow, 1967). In recent centuries, over 250 million acres (390,000 square miles) in Africa have been converted from agricultural and pastoral production to nonarable desert, primarily through the agent of man and his livestock. Similarly, the great Thar desert of western India has increased its size by 60,000 square miles within the last 100 years (Ehrlich and Ehrlich, 1970).

According to Sauer (1938), the second great period of human destruction upon the landscape occurred during the latter days of Rome. Phoenicia, Greece and Rome all prospered on the riches of the Mediterranean lands. One of the best accounts of this Roman exploitation is provided by Marsh (1864), who pointed out:

> The Roman Empire, at the period of its greatest expansion, comprised the regions of the earth most distinguished by a happy combination of physical advantages.

Figure 3–1. The remnants of a once-forested hillside in North Africa. Pollen deposits indicate that this range of hills was covered by coniferous forest before the time of Roman expansion. Now only the inorganic skeleton of a former ecosystem remains—the vegetation and soil have been lost.

> The provinces bordering on the principal and secondary basins of the Mediterranean enjoyed a healthfulness and equability of climate, a fertility of soil, a variety of vegetable and mineral products, and natural facilities for the transportation and distribution of exchangeable commodities, which have not been possessed in an equal degree by any territory of like extent in the Old World or the New. The abundance of the land and of the waters adequately supplied every material want . . . the luxurious harvests of cereals that waved on every field from the shores of the Rhine to the banks of the Nile, the vines that festooned the hillsides of Syria, of Italy, and of Greece. . . .

Even much of North Africa was rich and productive, clothed in cedar forests which provided further wealth for Rome. But Marsh hastened to add a more recent description of these lands, as he observed them in the mid-nineteenth century during his tenure as United States Ambassador to Italy.

> If we compare the present physical conditions of the countries of which I am speaking, with the descriptions that ancient historians and geographers have given of their fertility and general capability of ministering to human uses, we shall find that more than one-half of their whole extent—including the provinces most cele-

brated for the profusion and variety of their spontaneous and their cultivated products, and for the wealth and social advancement of their inhabitants—is either deserted by civilized man and surrendered to hopeless desolation, or at least greatly reduced in both productiveness and population. Vast forests have disappeared from mountain spurs and ridges; the vegetable earth accumulated beneath the trees by the decay of leaves and fallen trunks, the soil of the alpine pastures which skirted and indented the woods, and the mould of the upland fields, are washed away; meadows, once fertilized by irrigation, are waste and unproductive. . . .

Besides the direct testimony of history to the ancient fertility of the regions to which I refer—Northern Africa, the greater Arabian peninsula, Syria, Mesopotamia, Armenia, and many other provinces of Asia Minor, Greece, Sicily, and parts of even Italy and Spain—the multitude and extent of yet remaining architectural ruins, and of decayed works of internal improvement, show that at former epochs a dense population inhabited those now lonely districts.

It appears, then, that the fairest and fruitfulest provinces of the Roman Empire, precisely that portion of terrestrial surface, in short, which, about the commencement of the Christian era, was endowed with the greatest superiority of soil, climate and position, which had been carried to the highest pitch of physical improvement, and which thus combined the natural and artificial conditions best fitting it for the habitation and enjoyment of a dense and highly refined and cultivated population, is now completely exhausted of its fertility, or so diminished in productiveness, as, with the exception of a few favored oases which have escaped the general ruin, to be no longer capable of affording sustenance to civilized man.

The decay of these once flourishing countries is partly due, no doubt, to that class of geological causes, whose action we can neither resist nor guide, and partly also to the direct violence of hostile human force; but it is, in far greater proportion, the result of man's ignorant disregard of the laws of nature. . . .

WORLD-WIDE ENVIRONMENTAL DESTRUCTION

The same phenomenon has been discussed at length by Paul Sears (1935), Fairfield Osborn (1948), and Carl Sauer (1967). They have shown that the environmental destructiveness of man has not been limited to the Middle East, Africa and the Mediterranean, but has, in fact, occurred throughout China, India, sub-Saharan Africa and the New World. In India, for example, the original vegetation of Rajasthan and the southern borders of the Gangetic basin—areas which are now arid and semidesert—was deciduous forest and far richer in natural soil moisture than at the present time (Champion, 1937).

In northern India, the magnificent city of Fatehpur Sikri, just 23 miles west of Agra, was built in the sixteenth century, only to be abandoned within two decades due to a lack of water. It was overcome by the arid landscape its people helped to create. Looking at Fatehpur Sikri today in the midst of an oven-like desert, it is hard to envision the productive and well watered environment which gave it birth 400 years ago.

In North America, the present desolate shifting sand areas bordering the Colorado river were rich grassland pastures as late as the seventeenth century (Sauer, 1967). Mexico was extensively forested over vast areas now occupied by dry and rocky plains (Osborn, 1948). Of all these destructive changes wrought by man, Sears (1935) writes:

> Wherever we turn, to Asia, Europe, or Africa, we shall find the same story repeated with an almost mechanical regularity. The net productiveness of the land has been decreased. Fertility has been consumed and soil destroyed at a rate far in excess of the capacity of either man or nature to replace.

Fairfield Osborn, in *Our Plundered Planet* supports these views and points out that destructive land use practices have not been confined to ancient peoples and former centuries, but have in fact been very recent phenomena as well. This can be seen in Australia, where forest destruction, overgrazing and unlimited burning, have greatly accelerated wind erosion, and have thus extended the Australian deserts within the last 100 years. The present-day expansions of the Sahara and Thar deserts are largely due to excessive grazing and land misuse. The American dustbowl of the 1920's and 1930's is another well-documented story of land misuse, changing a once fertile grassland into a dusty pit of ecologic tragedy. Throughout much of present-day Latin America and in many tropical or subtropical regions of the world, the dangerous game of "slash and burn" agriculture is actively practiced. This denudes forested tracts rapidly, permits a few years of productive agriculture, and leaves a wake of destruction. The forest that may have taken 1,000 years to develop can be destroyed in two or three years. In the tropics, where natural soils are thin and very susceptible to rapid leaching and erosion, the entire organic ecosystem may be irreparably damaged in less than a decade. We have either ignored the lessons of ecologic history or are too blinded by the economic pressures of life to appreciate their importance to us.

Considerable progress in a positive direction has been made in recent years by the well-publicized "Green Revolution." Many countries throughout Asia, Africa and Latin America have made dramatic advances in agricultural production through the development of irrigation, new varieties of seeds and increased fertilizer production (Horsfall, 1970). India, for example, became self-sufficient in food grain production in 1971 for the first time in more than 20 years. The Philippines became self-sufficient in rice production in 1970 (Athwal, 1971). It remains to be seen, however, if this progress of the Green Revolution can be maintained or if it will simply provide additional spurts to population growth. More will be said of this in Chapter 17, but for the present time, the Green Revolution does represent a hopeful indicator of advancement in restoring the quality and productivity of the agricultural environment in a number of nations.

Figure 3–2 Devastated farmlands of the American dust bowl, converted from productive grassland to desert through human land misuse. (*Photo by U.S. Soil Conservation Service, from Dasmann, 1968.*)

SUMMARY

There is little doubt that many environments throughout the world have been rendered barren and inhospitable by excessive pressures from the axes, plows and hoofed animals of man. The pioneer civilizations altered their own biotic and physical environments and displayed man's ability to trigger ecologic changes leading to his own downfall.

In the decade of the 1970's there is very little evidence that man recognizes this aspect of his own history or realizes its applicability to his present predicaments. In country after country around the world I have seen the conservation movement losing ground, either through neglect or ridicule. In Malaysia and Indonesia, deforestation is occurring at increasing rates, and the people believe that their forests are unlimited. In India, overgrazing and hydroelectric power development schemes are invading the few National Parks and Forest Reserves which still exist. In the Himalaya mountains, the landscape is being carved within an inch of its life to support a burgeoning and none-too-healthy human population. In Africa and Latin America, slash and burn agriculture is encroaching upon natural areas which have existed in ecologic balance for thousands of years.

Much of this environmental destruction is done in the name of economic development. The pressures of increasing populations place a greater burden than ever on the finite blanket of life that covers this earth. That blanket is getting torn and shredded in thousands of places. Like a wound or burn on the skin, its ability to heal depends upon the extent of the lesion and the health of the surrounding tissue. A point can be reached where the healing capacity is lost.

The Green Revolution offers some hope that the trends of environmental destruction which prevailed for many centuries can be reversed, at least in agricultural environments. This requires knowledge and skill in the development of new strains of crops, in the management of soil and water, in the control of pests, and in other areas of agricultural science. Whether the progress of the Green Revolution can be maintained is still an open question—its great promises are faced with many problems. At best, the Green Revolution concerns only the agricultural environment and does not involve natural environments, forests, urban areas or many other habitats vital to man. Exploitation and deterioration of these areas are still occurring throughout much of the world.

4

Man's Attitudes Toward His Environment

EARLY MAN

WE can imagine that the world was a baffling place to early man. His attitudes towards his environment were probably shaded by reverence, awe and fear—fear of storms, disease, wild animals, earthquakes, volcanos, celestial phenomena and countless other events which were not understood. This primitive sense of fear and loneliness has been poetically described by the eminent Bengali writer Tagore:

> When in the depth of the night
> in the phantasmal light of the sick-bed
> appears your wakeful presence,
> it seems to me
> that the countless suns and stars
> have guaranteed my little life:
> then I know that you will leave me
> and the fear spreads from sky to sky,
> the fear of the terrible indifference
> of the All.[1]

[1] "Fear" from the anthology, *Boundless Sky* by Rabindranath Tagore. Calcutta: Visva-Bharati, 1964.

Amid the accoutrements of modern civilization it is impossible to project ourselves fully into the mind of early man, but perhaps we can visualize the awesome nature of the African veldt or the Malaysian forest throughout the cycles of storm and drought, heat and cold, night and day, disease and health. It is no wonder that mysticism and religion arose in virtually all peoples to help explain the unexplainable. Man felt himself to be a close integral part of his world and subject to its whims. He recognized his dependence upon nature and held it in awe. A unity of life existed and man considered himself one small part of a vast and complex scheme of nature. Again Tagore conveys this feeling in the song offerings of Gitanjali, one of which reads:

> The same stream of life that runs through my
> veins night and day runs through the world and
> dances in rhythmic measures.
>
> It is the same life that shoots in joy
> through the dust of the earth in numberless blades
> of grass and breaks into tumultuous waves of leaves
> and flowers.
>
> It is the same life that is rocked in the ocean-
> cradle of birth and death, in ebb and in flow.
>
> I feel my limbs are made glorious by the
> touch of this world of life. And my pride is from
> the life-throb of ages dancing in my blood this moment.[2]

These are modern writings, of course, since Tagore was born in Calcutta in 1841 and died in 1941, having received the Nobel prize for literature in 1913 and British Knighthood in 1915. Many of his writings, however, capture the ancient philosophies of India, and they may, in fact, mirror the sensitivities of early man toward his environment.

THE JUDAIC-CHRISTIAN TRADITION

Gradually, throughout the rise of agricultural civilizations in the last 10,000 years, man has increased his confidence and ability to control his means of subsistence. Religion became more a means of establishing a favored relationship with the unknown, and less a means of survival per se. Landscape became property and the concept of proprietary ownership developed. The Judaic-Christian philosophy in particular portrayed land as the God-given property of man (White, 1967). Nature existed for man, to be used by him and exploited for his own good. Man no longer considered himself merely one part of the biotic community, interacting with it in a humble and mutually beneficial

[2] From *The Song Offerings of Gitanjali* by Rabindranath Tagore. New York and London: MacMillan and Co., 1964.

way. He was now its owner and master, capable of treating it like any other property. According to White (1967), Chrisitianity arose as "the most anthropocentric religion the world has seen." The egocentricity of man could be seen, however, long before the time of Christ. For example, the famous passage from Genesis 1:28 exemplifies the favored status which religion bestowed on man:

> And God blessed them, and God said to them, "Be fruitful and multiply, and fill the earth and subdue it; and have dominion over the fish of the sea and over the birds of the air and over every living thing that moves upon the earth."

The writer of Deuteronomy (8:7–9) continued this theme:

> For the Lord your God is bringing you into a good land, a land of brooks, of water, of fountains and springs, flowing forth in valleys and hills, a land of wheat and barley, of vines and fig trees and pomegranates, a land of olive trees and honey, a land in which you will eat bread without scarcity, in which you will lack nothing, a land whose stones are iron, and out of whose hills you can dig copper.

This was the Promised Land and man was now the master, for he represented God on earth: (Genesis 1:26)

> And God said, "Let us make man in our image after our likeness: and let them have dominion over the fish of the sea, and over the fowl of the air, and over the cattle, and over all the earth, and over every creeping thing that creepeth upon the earth."

These aspects of the Judaic-Christian tradition are considered by White (1967) to be the philosophic basis for the complete exploitation of land for the benefit of man. Nature, in fact, existed for no other reason than to serve man. Hence, White calls this, *The Historical Roots of Our Ecologic Crisis,* and he feels that our modern problems of land destruction and pollution originated in the Judaic-Christian philosophies.

This specific thesis might be contested on several grounds: (1) much ecologic damage had occurred long before the first formulation of the book of Genesis in the Old Testament (the Yahwist text of Genesis was written between 850 B.C. and 750 B.C., according to Anzou, 1963), and (2) misuse, exploitation and destruction of the landscape have occurred with equal severity in non-Christian parts of the world.

We have seen in the previous chapter how the destructive influence of man upon his environment was by no means confined to the areas of Judaism and Christianity. It occurred throughout broad reaches of Asia and Africa, certainly in China, India, and the Sudan, in areas subject to highly varied religious influences, and even in the Middle-East long before the time of Judaic-Christian influence.

Hence, it is doubtful if the ecologic ills of the world can be blamed on any single religious or philosophic foundation. Nonetheless, the persistence and growth of the exploitation of land as property, existing solely for the use of man, was certainly fostered by the Judaic-Christian concepts of land ownership.

The geographer Carl Sauer believed that there were three great periods of habitat destruction in the history of man: First, a period at least three to four thousand years ago, when the great herds of pastoral man and the early successes of agriculture caused extensive land scarring, erosion, and irreversible aridity in Africa, the Middle East and Asia. Second, the latter days of Rome and the disorderly period immediately following, when many of the Mediterranean lands were despoiled. Third, the transatlantic expansion of European commerce and peoples into the New World, when rapid and disastrous land exploitation occurred throughout the Americas. Sauer documented these views in an important book first published in 1938 and reprinted in a recent anthology (Leighly, 1967).

Does this mean that each major advance of man carries with it the possibility of some destructive force? Keeping in mind the domestication of hoofed animals, the expansion of the Roman empire, the invasion of the New World, the development of gunpowder, the rise of modern medicine, the internal combustion engine, and now the advent of the atomic age—all major technologic achievements which also contained destructive capabilities, one is tempted to believe the above statement may have sober elements of truth.

THE INDUSTRIAL REVOLUTION AND THE BACONIAN CREED

The industrial revolution of the eighteenth and nineteenth centuries, and certainly the technologic revolution of the twentieth century, continued to foster the ideas of proprietary ownership and economic exploitation of the land. Man has generally viewed land as existing for his own benefit and profit potential. The widespread practice of shifting cultivation and the agricultural techniques of slash, burn, cultivate, and then move on, have characterized much New World agriculture until recent times, and tragically these practices still prevail in tropical regions.

The technologic revolution has gone several steps further in molding the attitudes of man. The Baconian creed, that scientific and technologic knowledge means control and power over nature, arose from the writings of Francis Bacon in the early seventeenth century, but it has found its fullest expression in the last one hundred years. The developments of electrical energy and the internal combustion engine gave man unprecedented power for transportation, production, and environmental modification. These new tools, as products of scientific technology, furthered the belief that science held sovereignty over nature.

THE RISE AND FALL OF TWENTIETH CENTURY OPTIMISM

During the twentieth century the Baconian creed has given man the optimism that he can achieve all things. He can control and eliminate disease, produce unlimited food, develop comfortable housing and clothing for all, and enter a golden age of technology and economic prosperity in which his only remaining problem would be the use of leisure time.

This twentieth century optimism was temporarily deflated by two world wars and the great economic depression of the early 1930's, but it came surging back in the late 1940's. Franklin Delano Roosevelt said in 1944 that "we shall expand indefinitely," and he was referring to unlimited expansion in terms of prosperity and well-being for all.

A contrary philosophy was expressed by Aldo Leopold (1949), William Vogt (1948), and Fairfield Osborn (1948), which suggested that man was out of balance with his environment and would soon pay the consequences. These men insisted that only strict attention to the principles of ecology and the view that man was part of a sensitive and intricate ecosystem could save us from a program of self-defeat. Fairfield Osborn stated this philosophy well when he wrote (1948):

> The question remains. Are we to continue on the same dusty perilous road once traveled to its dead end by other mighty and splendid nations, or, in our wisdom, are we going to choose the only route that does not lead to the disaster that has already befallen so many other peoples of the earth? (page 193)

He concluded the book by stating:

> There is only one solution: Man must recognize the necessity of cooperating with nature. He must temper his demands and use and conserve the natural living resources of this earth in a manner that alone can provide for the continuation of civilization. The final answer is to be found only through comprehension of the enduring processes of nature. The time for defiance is at an end.

The fears and warnings of Osborn, Vogt and Leopold seem more pertinent today than when they were first written more than twenty years ago. Nonetheless, these views were not widely accepted by educated people throughout the world. They were held primarily by ecologists and conservationists who have continued in a state of depression over the ecologic condition of the world, and who spend long hours commiserating with each other, basically convinced that the real message of ecology will never get accepted before irreversible harm has been done. The problem is, however, that it is so hard to demonstrate that irreversible harm has been done. In many cases the necessary research has not yet been accomplished, or if it has, the message lies buried in technical reports. In other cases, the problem requires a sense of historical perspective and ecologic outlook that is not often acquired in current educational programs. It takes more than the written word of popular publications to convince the pragmatic businessman and politician that things are really as bad as ecologists say they are. Man still has great dreams of making the desert bloom and controlling the unruly river with hydro schemes. Each success is quoted far more often than numerous failures.

So the tide of optimism about the power of science and technology to erase man's problems continued to expand. In 1955 the editors of *Fortune* compiled a volume entitled, *The Fabulous Future,* in which they invited prominent business leaders to express their views about the future of man. Most of the essays were clear extensions of the Baconian Creed, and most expressed in glowing terms the prospects of the future. David Sarnoff, then Chairman of the Board of RCA, stated this creed as follows:

> There is no longer any margin for doubt that whatever the mind of man visualizes, the genius of science can turn into functioning fact.

Under a section entitled, "Things to Come," Mr. Sarnoff envisioned:

> Other sources of energy—the sun, the tides, and the winds—are certain to be developed beyond present expectations. New materials by the score—metals, fabrics, woods, glass—will be added to the hundreds of synthetics and plastics already available through our capacity to rearrange the structure of matter.
>
> Fresh water, purified from the briny seas, will enable us to make deserts flourish and to open to human habitation immense surfaces of the globe now sterile or inaccessible.

Mr. Sarnoff emphasized, however, that the big question is man himself. Can he realize his potential? Can he meet the social ills of modern nations? Sarnoff felt that we could attain a technologically utopian world if social tolerance and mutual understanding were first achieved.

Crawford Greenawalt, President of duPont Company, indicated a similar optimism based on the caution of a social contingency. "If we are successful in recapturing that atmosphere of freedom, incentive, and self-respect, the future will be boundless indeed." Mr. Greenawalt emphasized personal freedom and the capitalistic system as the great provider. He felt that America's progress was primarily due to our political and economic system, and not to our abundant and relatively virgin natural resources. His conclusion was one of optimism:

> Fortunately we do not have to visualize the precise shape of things to come to be able to predict with confidence another 25 years or more of growth and prosperity.

The ecologist would criticize this as being too provincial a view, limited to the United States, and much too limited in its time dimension. Twenty-five years in the history of man is a mere blip on the scanning screen of history.

George Humphrey, Secretary of the Treasury in 1955, limited his predictions to the "free-world":

> The present and future of the free-world people look good. An America of confidence, prudence and imagination will mean . . . a future finer than our minds can even dream.

Henry Luce, Editor-in-Chief of Time-Life, Inc., also foresaw a splendid future for man; a life of unlimited prosperity and super-abundance:

> Today we can glimpse the end of poverty. . . . Abundance has become visibly the norm of life in America. Having been achieved in America as a human reality, this economy of abundance is likely to become the global condition of man's life on earth.

Although these preceding statements were from leaders of business and government in the mid-1950's, scientists were also expressing similar views of the future. At times they were somewhat more guarded by "ifs, ands, and buts," but on the whole optimism shone through. An example would be the writings of the eminent scientist, Harrison Brown (1954):

> It is clear that the nations of the West possess sufficient resources and productive capacity to catalyze a successful world development program at the present time. Our physical ability to bring about successful transition is not one of the unknowns. We have the ability to do it; whether we have the vision and the will is another matter." (p. 248)

> I can imagine a world within which machines function solely for man's benefit, turning out those goods which are necessary for his well-being, relieving him of the necessity for heavy physical labor and dull, routine, meaningless activity. The world I imagine is one in which people are well fed, well clothed, and well housed. Man, in this world, lives in balance with his environment, nourished by nature in harmony with the myriads of other life forms which are beneficial to him. He treats his land wisely, halts erosion and overcropping, and returns all organic waste matter to the soil from which it sprung. He lives efficiently, yet minimizes artificiality. It is not an overcrowded world; people can, if they wish, isolate themselves in the silences of a mountaintop, or they can walk through primeval forests or across wooded plains. In the world of my imagination, there is organization, but it is as decentralized as possible, compatible with the requirements for survival. There is a world government, but it exists solely for the purpose of preventing war and stabilizing population, and its powers are irrevocably restricted. The government exists for man rather than man for the government.

> In the world of my imagination the various regions are self-sufficient, and the people are free to govern themselves as they choose and to establish their own cultural patterns. All people have a voice in the government, and individuals can move about when and where they please. It is a world where man's creativity is blended with the creativity of nature, and where a moderate degree of organization is blended with a moderate degree of anarchy.
>
> Is such a world impossible of realization? Perhaps it is, but who among us can really say? At least if we try to create such a world there is a chance we will succeed. But if we let the present trend continue it is all too clear that we will lose forever those qualities of mind and spirit which distinguish the human being from the automaton. (p. 258)

Dr. Brown considered unbridled population growth as the most devastating trend standing in the way of achieving a balanced and prosperous world. Although he believed that science and technology could meet most of our physical needs, he shared with Sir Charles Galton Darwin (1953) the fears that human populations might not be stabilized in an orderly way, and will again be controlled someday by high death rates. This ambivalent faith and fear was clearly expressed in the following paragraph from his final chapter:

> In our survey of the situation in which man now finds himself, we see that, although our high-grade resources are disappearing, we can live comfortably on low grade resources. We see that, although a large fraction of the world's population is starving, all of humanity can, in principle, be nourished adequately. We see that, although world populations are increasing rapidly, those populations can, in principle, be stabilized. Indeed, it is amply clear that, if man wills it, a world community can be created in which human beings can live comfortably and in peace with each other. But it is equally clear that the achievement of this condition will require the application of intelligence, imagination, courage, unselfish help, planning, and prodigious effort. And it is equally clear that the time for decision is the present. With the consumption of each additional barrel of oil and ton of coal, with the addition of each new mouth to be fed, with the loss of each additional inch of topsoil, the situation becomes more inflexible and difficult to resolve. Man is rapidly creating a situation from which he will have difficulty extricating himself. (p. 265)

This viewpoint of guarded optimism represented sound scientific thinking and reasonable philosophy as well. It placed the burden squarely on man and his institutions. It saw in world affairs the potentials of both a tragic and a glorious future for man.

THE ONSET OF DESPAIR

In general, throughout the 1950's and early 1960's, optimism prevailed, and the balance tipped decidedly toward the vision of a marvelous future for man.

Economic prosperity was skyrocketing in many countries of the world. America was on the verge of its "New Frontier." The conquest of space was catapulting man's dreams toward distant goals. The world was not yet enmeshed in any major wars. Medicine was making great strides with the anticipated conquest of malaria, cholera and polio. The future looked bright. The thoughts of Henry Luce in 1955 seemed to be on the verge of realization.

Then the bubbles of bright prospects began to break. The United States gradually became bogged down in the distressing war in Vietnam. The Middle East erupted again into serious military action. America lost three of her outstanding leaders by assassination. Violence erupted in her cities. The War on Poverty became a hollow phrase. India experienced recurring famines. Africa became the scene of tragic wars in some of her most advanced countries. Tension increased throughout the world. A deep and pervading pessimism began to set in. The bright and glorious future of science and technology was tarnished even among the most hopeful and entirely scuttled among many of the world's peoples.

This pessimism was clearly expressed in many writings of 1966 and 1967. The editors of the *Johns Hopkins Magazine* produced an issue in 1967 built around the theme, "The War Against Suffering." Its preface contained these stark facts:

> Two-thirds of the human race live in the underdeveloped areas of the world.
>
> Most of them drink unsafe water, prepare food dangerously, dispose of wastes recklessly, and live in unfit dwellings.
>
> In some parts of the developing world, every living individual is afflicted with intestinal parasites.
>
> In some parts of the developing world, half of the children born never reach their fifth birthday.
>
> Within the developing world at least 400 million people suffer from trachoma, a curable disease that causes progressive loss of sight.
>
> Within the developing world, despite a massive campaign by the World Health Organization to eradicate malaria, 400 million people live in malarious areas where no eradication program is under way. In some of these areas, most people have the disease.
>
> Schistosomiasis, a debilitating parasitic disease, afflicts 200 million people—more than the population of the United States. In some areas, the chances of escaping the disease are only one in five.
>
> Leprosy afflicts more than 10 million people; fewer than one in five are receiving any kind of treatment. In the countries where leprosy is prevalent, 740 million people are exposed to the risk of infection.

Cholera, a disease on the decline in the 1950's, is spreading. The number of reported cases tripled between 1960 and 1964.

Venereal diseases are spreading into the underdeveloped areas of the world, to an extent not yet measured.

This is, in part, the health situation amid two-thirds of the world's people. By the year 2000, if present demographic trends continue, it could be the story of four-fifths of the world's people.

An equally grim picture of the world's condition was drawn in an article on poverty in *Scientific American* in the late 1960's (Simpson, 1968). This article pointed out that two-thirds of the world's people live in countries whose annual per capita income is less than $300. The world's largest democracy, India, has a per capita annual income of less than $100. This is less than one-seventh the definition of poverty in the United States. One might say that the cost of living is less in India than in America, but this is a common misconception. Many staples of life, such as rice, flour, salt and sugar are actually more expensive in India than in the United States. In Calcutta, one of India's most crowded cities, 77 percent of the population has less than 40 square feet of living space per person. In Pakistan, a survey conducted by the NIH Office of International Research, indicated that 46 percent of the households surveyed had an inadequate caloric intake. Simpson concluded that the rich nations and people of the world are getting richer, while the poor get poorer.

A similarly pessimistic view of global trends was composed in December, 1966 by a Committee of biologists working under the auspices of the International Biological Program, sponsored by the National Research Council. The introduction to their program statement read as follows:

The gravest problem facing man in a peaceful world is the establishment of a reasonable harmony between a stabilized world population and the environmental resources upon which that population depends.

Man, like other organisms, lives in and depends upon environment. That environment, it is increasingly clear, is both limited and vulnerable. Millions of human beings at present exist with inadequate food. Vast quantities of wastes reduce the quality of our environment.

Mankind is engaged in a program of self-defeat. Human population increases at an unprecedented rate, and it has proved difficult to stabilize population in those countries where stability is most needed. Technological power is altering environment at an increasing rate, often in ways which are to man's disadvantage—erosion, occupation of prime farm land by housing, atmospheric pollution, and the widespread use of poisons intended for pests. The product of increase in population and increase in the technological power of that population is an increase in rate of detrimental change in the environment on which man depends.

There is no present reason for confidence that man will succeed in bringing about

> a harmony between his population and environment. It does not seem that technology alone can produce a long-term solution. Technological progress can increase food production and improve environmental conditions, but such benefits are only palliatives if population meanwhile increases. Technological expedients may also produce unforeseen environmental detriment, as in the case of currently increasing pollution problems. Solution of the problems of man's relation to his environment is possible only through difficult, long-range decisions on limitation of both population and the technological modification of environment.

These views are a far cry from those expressed by those business leaders writing on "the fabulous future" in 1955. Who is right? Are we engaged in a "program of self defeat," or are we entering a golden age where "abundance is likely to become the global condition of man's life on earth"?

THE ECO-ACTIVISTS

In the last few years there has been a remarkable ground swell of public interest in the issues of ecology and environmental quality. Popular opinion has become acutely aware of pollution, crowding and landscape deterioration. Among young people particularly, there is a renewed interest in trying once again to live in harmony with nature, and to slow down or stop the despoliation which has accompanied America's progress so far. On August 22, 1969, *Time* noted that a new type of student activist, which the magazine called the "eco-activists" were taking aim at the environment as well as at the establishment. Students throughout the country began campaigning for clean air and water, demonstrating against blatant polluters, supporting local conservation campaigns, and arousing public interest in environmental and social issues. *Time* noted, "One reason for youthful concern with environmental damage is simple: the young will have to live with it." Many students felt that nothing short of completely altered philosophies and life styles could stem the tide of environmental destruction and social disintegration in America. Their views found expression in the best-selling book by Charles Reich, *The Greening of America,* which addressed itself to the rising sensitivities and consciousness of young people to the social and environmental problems of modern technologic societies. The general public, however, has been less concerned. The popular American folk singer, Pete Seeger, chartered the 96 foot slope *Clearwater* in August of 1969 to sail up the Hudson River, making frequent stops to dramatize the sad, polluted condition of this mighty river. He met substantial citizen apathy, however, and the crowds he drew seemed more interested in the television regalia than in the message of pollution.

Segments of the business community, however, have expressed a new interest in environmental quality. Advertisements emphasized pollution abatement, and frequent press releases publicized the money being spent on pollution control. Some of this undoubtedly reflected genuine concern for the American landscape and the American public, but much of it had the hollow ring of advertising propaganda. Corporate enterprises were obviously aware of increasing legal liabilities which could arise from environmental damage; they were acutely aware of their images, and how much the current market value of any corporation depends upon image; and they were well aware of profit potentials in the pollution control field. Despite this plethora of motivations, the rising interest in the business community in environmental quality was a most encouraging trend. The hope is that it will be sincerely based on ecological facts, that it will recognize the urgency of environmental harmony, that it will not place unreasonably blind faith in pure technology to solve pollution problems, and that it will provide the courage to tell stockholders that profits are down this year because more money was spent on improving the environment.

As Ian McHarg pointed out in *Design with Nature* (1969), economic progress is not inevitably in conflict with good conservation if we observe certain ecologic principles. If we insist on denuding watersheds and then build suburbs on flood plains, we can obviously expect floods and ecologic catastrophies of this sort. If we insist on seashore development by building on the primary sand dunes facing the ocean, we can predict with a high degree of accuracy that we will sooner or later face critical damage from ocean storms and hurricanes. If we so overpopulate a rural area with bad septic tanks that we create ground water pollution, we can certainly anticipate a water quality crisis. As McHarg (1969) observed, we can avoid many of these problems by ecologic planning, by learning to design our man-made environments to be compatible with nature rather than in conflict with her.

How persistent the new interest in ecology will be remains to be seen. If based on sheer emotionality it may wither and die like any fad. If based on sound ecological principles, it should prosper and increase its influence. The public at large, through its attitudes and actions, will determine the outcome of this critical issue. The costs and sacrifices may well be great, as, for example, when the requirements of pollution control cause the termination of an industry and produce what Ralph Nader has called the "environmentally unemployed." Critical evaluations will have to be made to achieve a new balance of economic and ecologic considerations. Life styles and philosophies will have to change, and the world will certainly require the best education, knowledge and wisdom man can muster to achieve the balanced and productive life for which all people strive.

5

Academic Origins of Ecology Prior to the Twentieth Century

The first author of the term "ecology" is uncertain, though several writers used it in the mid-nineteenth century. Many biologists grant credit to the Germany biologist Ernst Haeckel, who used the term in 1866 to refer to the interrelationships of living oranisms and their environment. A few years earlier, in 1857, Henry David Thoreau also spoke of ecology among other fields of biology and natural history, but he apparently did not provide a definition of the term (Kormondy, 1969). In any case, the origins of ecologic knowledge are much older than either Haeckel or Thoreau, and date to the origins of biology itself even though the word was not used until the nineteenth century.

THE ORIGINS OF BIOLOGY

The beginnings of ecologic knowledge for modern man lay in the development of biology and medicine as scientific disciplines. The fourth century Greek, Hippocrates (460–377? B.C.), often called the "Father of Medicine," emphasized certain ecologic factors in the genesis of health and disease. His work, "On Airs, Waters and Places," was environmental in emphasis:

> Whoever wishes to investigate medicine properly, should proceed thus: in the first place to consider the seasons of the year, and what effects each of them produces (for they are not at all alike, but differ much from themselves in regard to their changes). Then the winds, the hot and the cold, especially such as are common to all countries, and then such as are peculiar to each locality. We must also consider the qualities of the waters, for as they differ from one another in taste and weight, so also do they differ much in their qualities. In the same manner, when one comes into a city to which he is a stranger, he ought to consider its situation, how it lies as to the winds and the rising of sun; for its influence is not the same whether it lies to the north or the south, to the rising or to the setting sun. These things one ought to consider most attentively, and concerning the waters which the inhabitants use, whether they may be marshy and soft, or hard, and running from elevated and rocky situations, and then if saltish and unfit for cooking; and the ground, whether it be naked and deficient in water, or wooded and well watered, and whether it lies in a hollow, confined situation, or is elevated and cold; and the mode in which the inhabitants live, and what are their pursuits, whether they are fond of drinking and eating to excess, and given to indolence, or are fond of exercise and labour, and not given to excess in eating and drinking.

Aristotle (384–322 B.C.), sometimes referred to as the "Father of Biology," classified animals according to their habits and habitats; e.g., gregarious vs. solitary, carnivorous vs. herbivorous, resident vs. migratory, etc. Although he was a good naturalist he could not yet be considered an ecologist for he did not really consider interrelations of plants, animals and their environments other than in the most superficial terms. Theophrastus (372–287? B.C.), a student of Aristotle, was regarded by some scholars as the first ecologist, though it should be emphasized that the word ecology was not used until much later. Nonetheless, Theophrastus discussed plants and their environments in a moderately systematic way. He studied plant types and form in relation to altitude, moisture and light exposure.

About the time of Christ, the Hebrews had some ecological notions, but of a very general and rather obvious sort. An example is the Parable of the Sower (Luke, 8:4–8) which illustrates the dependence of seed germination on the soil.

In Rome, Pliny the Elder (23–79 A.D.) was a naturalist with a system of classification somewhat like that of Aristotle; that is, he classified plants and animals by their habits and habitats. He was a great compiler and cataloger of facts, but most of this was basic natural history and could not be considered truly ecological in orientation.

Beginning slowly in the twelfth century and accelerating rapidly in the sixteenth century, natural history flourished as a great scholarly endeavor. The description and cataloging of animals became a favorite scientific activity. Albertus Magnus (1193–1280) was perhaps the leading naturalist of his time who wrote on plants and their environments. In the sixteenth century, Konrad Gesner (1516–1565), Aldrovandi (1522–1605), and Cordus (1515–1544) all began to relate biology and natural history to geography and chemistry even though the latter had barely begun as a science.

It remained for Robert Boyle (1627–1691), sometimes called the first modern chemist, to relate the origins of chemistry to vital processes in plants and animals. He did experiments on the effects of low air pressure on a variety of animals from frogs to chickens and mice. In one series of experiments he showed the ability of ducks to withstand a shortage of air longer than a chicken. He wrote:

> We put a full-grown duck (being not then able to procure a fitter) into a Receiver, whereof she fill'd, by our guess, a third part or somewhat more but was not able to stand in any easy posture in it; then pumping out the Air, though she seemed at first (which yet I am not too confident of upon a single tryal), to have continued somewhat longer than a Hen in her condition would have done; yet within the short space of one minute she appeared much discomposed and between that and the second minute, her struggling and convulsive motions increased so much that, her head also hanging carelessly down, she seemed to be just at the point of death; from which we presently rescued her by letting the Air in upon her; So that, this Duck being reduced in our Receiver to a gasping condition within less than two minutes it did not appear that, notwithstanding the peculiar contrivance of nature to enable these water-Birds to continue without respiration for some time under water, this Duck was able to hold considerably longer than a Hen, or other Bird not-Aquatick might have done.

From the late seventeenth century through the nineteenth century, five major fields of scholarship laid the foundation for the development of ecology as a self-conscious scientific discipline in the twentieth century. These five fields were: (1) natural history and faunal exploration, (2) environmental physiology and responses, (3) evolution and natural selection, (4) population studies and demography, and (5) ecological geography and conservation.

These did not develop as independent subjects, of course. Several men played key roles in the origins of two or more areas. For example, Charles Darwin was a naturalist of the mid-nineteenth century who made outstanding contributions to natural history, faunal exploration, and evolution. Reaumur and Buffon in the early and mid-eighteenth century both made contributions in natural history and environmental physiology. Thomas Malthus provided the stimulus for population studies, and he also laid the foundation for Darwin's work in natural selection. Although these and many other interlocking forces were active in the development of ecology, it is helpful to consider them separately, to outline the major contributions of each to the subsequent rise of ecology. More complete information on any of these trends can be found in books by Allee et al. (1949), Singer (1950), and Nordenskiold (1932).

NATURAL HISTORY AND FAUNAL EXPLORATION

Man's interest in natural history has been a primary motivation and developmental source of ecology since the time of Aristotle. Some of the major naturalists of the Renaissance who contributed to this development are noted below.

Rene Reaumur (1683–1757) was a French naturalist and physicist who wrote a six volume work entitled, *Memoires pour Servir a l'Histoire des Insectes.* This work discussed the life habits and community patterns of social insects. Reaumur also investigated the natural history of crustaceans and molluscs.

Carolus Linnaeus or Karl Von Linne (1707–1778) was the first great modern systematist. His classic work entitled *Systemmae Naturae* established the binomial system of nomenclature which is the standard scientific method for naming organisms. All modern scientific names followed by a capital L were first applied by Linnaeus. He also furthered the sciences of plant and animal geography and studied plant associations in different regions.

George Louis Leclerc Buffon (1707–1788), another eminent French naturalist, wrote the forty-four volume *Histoire Naturelle, Generale et Particuliere,* a vast compendium of most of the natural history knowledge of the day. He was interested in the adaptations of plants and animals to their environments. He brought together a great deal of information about climate as a variable, compelling modification of plants and animals. In some ways his writings anticipated the evolutionary ideas of Darwin, but they were not fully expressed or developed. He definitely opposed Linnaeus' ideas of the fixity of species, and he massed considerable evidence demonstrating variation and mutability in living organisms.

In England, an outstanding naturalist of this time was Gilbert White (1720–1793). His book, *The Natural History of Selbourne,* still provides thoughtful and enjoyable reading about the natural history of southern England.

Nineteenth century England produced a golden era for natural history. One of the leading British naturalists of this period was Alfred Russell Wallace (1823–1913) who wrote *Island Life,* and several other leading books about the distribution of plants and animals of the Malay archipelago. Wallace also achieved a theory of evolution at the same time as Darwin, and this independent development of a great biological principle represents one of the fascinating chapters in the history of biology. More will be said of this later. Charles Darwin (1809–1882) was an amazingly productive scholar who wrote more than a dozen outstanding books and monographs on a variety of biological topics from 1842, when he published *Coral Reefs,* to 1881 with the publication of "Action of earthworms on the processes of soil formation." His most famous works, of course, were the *Origin of Species* (1859), and the *Descent of Man* (1871). With these, especially the former, he presented his great interpretive theory of natural selection and evolution.

Other nineteenth century English naturalists of considerable influence were H. W. Bates (1825–1892), who wrote *Naturalist on the Amazon* in 1863; Thomas Henry Huxley (1825–1895), the great physician, anatomist and zoologist, who became one of Darwin's staunchest supporters; and Edward Forbes (1815–1854), a marine biologist who was especially interested in the dynamic relations between marine organisms and their environment.

Before the time of Darwin and Huxley, James J. Audubon (1785–1851) in America was compiling his monumental *Birds of America,* and shortly thereafter Louis Agassiz (1807–1873), the Swiss-American naturalist, compiled four volumes of *Bibliographia Zoologica,* along with other books on fossil fishes, glacial activity and naturalistic expeditions to Brazil.

Toward the end of the nineteenth century, Henri Fabre (1823–1915) in France, published ten volumes on the *Life of Insects,* and A. E. Brehm (1829–1884) of Germany was compiling material for the subsequent publication (1911–1918) of the thirteen volume, *Tierleben*. Other influential scholars of the nineteenth century were C. Hart Merriam (1855–1942), who devised a classification of major life zones—a system of habitat classification still in wide use, and S. A. Forbes (1844–1930), who wrote a series of papers on the concept of ecological communities. This began the period of the self-conscious rise of ecology as a scientific discipline, and we shall return to this topic after brief consideration of other trends in the eighteenth and nineteenth centuries.

ENVIRONMENTAL PHYSIOLOGY AND RESPONSES

Rene Reaumur and George Buffon were primarily naturalists of the eighteenth century, but they were also interested in environmental responses to temperature, light and altitude, and they may be considered early students of environmental physiology. This field did not really gain momentum until the nineteenth century, however, when Justus Liebig (1803–1873) developed his analysis of limiting factors (1840). He was interested primarily in physical factors which limit the distribution and abundance of plant and animal life. We recognize his basic contribution today in Liebig's Law of the Minimum, which states that an organism will be limited by that factor most closely approaching its minimum life support requirements. Also at this time, Jacques Quetelet (1796–1874) studied dormancy and hibernation in plants and animals in relation to temperature (1846), and shortly thereafter, Alphonse de Candolle (1806–1893) studied plant germination in relation to temperature and moisture (1865).

The two men who gave environmental physiology its greatest advance in the nineteenth century were Charles Davenport (1866–1944) and K. G. Semper

(1832–1893). Semper wrote an important ecological treatise in 1881 entitled, *Animal Life as Affected by the Natural Conditions of Existence,* in which he established the principle of food chains and what we now call a "pyramid of numbers." He was interested in predator-prey interactions, protective coloration, and the spatial relations of animals. He showed that pond snails and fresh water isopods would be stunted if crowded in a small volume of water, and he postulated the existence of chemical factors passing between animals in the environment and influencing growth and development. We now recognize these as "ectocrines" or environmental hormones, and they represent a major area of current research. This was a fairly sophisticated concept in the 1800's.

Davenport continued this line of investigation and wrote two volumes in 1897 and 1899 entitled, *Experimental Morphology*. He studied animal behavior and development in a variety of artificial environments and media, and he analyzed a wide range of environmental responses: chemotaxes, hydrotaxes, geotaxes, phototaxes, thermotaxes, electrotaxes, and so forth. He also demonstrated the viability of frozen cells, and thereby opened up several lines of study and practical utilization. Semper and Davenport gave strong impetus to a branch of ecology which is still at the forefront of modern ecology, the interrelations of physiology and ecology.

EVOLUTION AND NATURAL SELECTION

The history of evolutionary thought is closely interwoven with that of natural history, and most of the leading figures in developing theories of evolution were the great naturalists of the eighteenth and nineteenth centuries.

George Buffon and Jean Baptiste de Lamarck (1744–1829) were both environmental evolutionists of the eighteenth century and early nineteenth century respectively. That is, they believed that plants and animals were changed by environmental influence. Lamarck is most famous, of course, for his *Theory of Acquired Characteristics,* which we now know contains some major misconceptions about the mechanism of evolution. Some 60 years before Lamarck, Buffon was developing somewhat similar ideas of evolutionary change in his *Histoire Naturelle* of 1749:

> If we again consider each species in different climates we shall find obvious varieties both as regards size and form; all are influenced more or less strongly by the climate. These changes only take place slowly and imperceptibly; the great workman of Nature is Time: he walks always with even strides, uniform and regular, he does nothing by leaps; but by degrees, by gradations, by succession, he does everything; and these changes, at first imperceptible, little by little become evident, and express themselves at length in results about which we cannot be mistaken.

These writings of Buffon strongly influenced Lamarck in two major books, *Philosophie Zoologique,* in 1809, and *Histoire Naturelle des Animaux sans Vertebres,* in 1815. In these works he presented his theory of evolution, based on the inheritance of acquired characteristics, sometimes referred to as the "use and disuse theory." He believed, for example, that the giraffe's neck had become long because he continually stretched higher for leaves to feed on, and thus each generation was born with a slightly longer neck as a result of stretching. Similarly, he believed that blind cave fish lost their eyes because they failed to use them. Although we can recognize elements of truth in these theories, we now know that Lamarck's understanding of the mechanism was incorrect. We know that environmental changes affecting only the somatic cells of the body (in contrast to the germinal or reproductive cells) are not transmitted to subsequent generations.

The evolutionary theories of Darwin and Wallace emerging about 40 years later were similarly ecological in emphasis, but radically different in how evolution occurred. They developed their views independently and almost concurrently, placing emphasis on population pressure and struggle for existence as the driving forces of evolutionary change. They envisaged survival of the fittest and natural selection as the mechanism of evolution, working on the foundation of inherited variations in plants and animals. They knew nothing of genetic mutation, but with this knowledge added some years later, their central ideals of natural selection remain basically intact at the present time.

Darwin's and Wallace's initial papers were published simultaneously in 1858, and this provided an unusually interesting footnote for the history of biology. This is well-described, along with some brief history of Charles Darwin, in a book by Dr. R. B. Brown (1956):

> Darwin, after an unsuccessful attempt at the study of medicine, trained for service in the Church of England. He was much interested in natural history, and was recommended by one of his botanical friends for a position as naturalist on the British ship *Beagle,* which was about to sail around the world to take soundings and study coastlines in the interests of the British Navy. Darwin's father disapproved of the idea, but to settle his son's insistent entreaty said that if Charles could find any man of common sense to recommend his going, permission would be granted. Charles had been a frequent visitor to the home of his cousin, Emma Wedgewood, a vivacious young lady with whom he was falling in love. Emma's father quickly recommended that Charles be sent off on the voyage, thus fulfilling the necessary condition, and Darwin spent almost five years on the trip. During this time he collected an immense amount of data on the kinds and geographic distribution of living organisms.
>
> After he returned, married his cousin Emma, and began to work up this material, he became more convinced than ever of the validity of the evolutionary hypothesis, and in 1842 drew up an account of his explanation of it. From then on he devoted his time to working out the details and corollaries of this theory. He discussed it with some of the scientific leaders of his country, who urged him to publish his ideas. But

Darwin insisted on waiting until he had accumulated enough evidence to convince even the most skeptical opponent.

In 1856, however, he was finally persuaded to start work on what was planned to be a monumental volume. While he was still writing on this project, one day in June, 1858, he received a latter from a young naturalist, Alfred Wallace (1823–1913), who was exploring the East Indies. Wallace had been taken ill with a tropical fever, and while he was recovering, he had been thinking over the distribution of various species, and had a flash of insight which explained some of his discoveries. This flash was a duplicate of the theory Darwin had been working on for some years. Entirely unaware of Darwin's ideas, Wallace wrote to Darwin a brief summary of his notions, and suggested that if Darwin thought well of them, they might be published.

Although somewhat stunned by this turn of events, Darwin felt honorbound to publish Wallace's paper first, and then corroborate it with his own work. His friends would not hear of this arrangement. They felt it deprived Darwin of the priority which was his due. The dilemma was settled by the simultaneous publication of Wallace's paper and Darwin's 1842 outline. Then, the following year, Darwin published an abridged version of the book he had been working on. This volume summarized its message in its title: "On the Origin of Species by Means of Natural Selection, or the Preservation of Favored Races in the Struggle for Life." The book aroused a furor of controversy. Wallace generously gave full credit for originating the theory to Darwin, and supported it faithfully by continued researches in biogeography—the distribution of plants and animals. (pp. 477–478)

POPULATION STUDIES

From the sixteenth century onward, some scholars have worried about population growth and pressures. Machiavelli (1469–1527), Botero (1540–1617), Buffon (1707–1788), and Franklin (1707–1790) were all scientists who partially anticipated Malthus by expressing concern about population growth, starvation, disease and war. Yet Malthus, of course, was primarily responsible for focusing controversial attention on these topics.

Although we often remember him as the Reverend Thomas Robert Malthus, he was, for most of his life, a Professor of History and Political Economy at Haileyburg College. While still a curate in Surrey, however, he published his controversial book, *An Essay on Population* (1798). In this expansive volume he observed that human populations tend to increase geometrically and thus exceed the means of subsistence which increase arithmetically. He felt that unless populations were controlled by voluntary constraint, they surpassed their level of support and were then decimated by starvation, disease, war and poverty.

This theory produced a storm of protest, and it has remained controversial to modern times. Generally, it fell into some scientific disrepute, for many scholars believed that Malthus failed to recognize the great potential of modern science in overcoming famine and disease. Many felt it painted too dismal a view of human potential. Thus, 30 years ago there were few scholars who actively endorsed the views of Malthus, and many books were written to refute them.

More recently Malthusian doctrine has come back into favor, and it is being taken more seriously now than in the 1930's (Hardin, 1964). Many writers who grant the potential of science and technology to solve the problems of population pressure, admit that the Malthusian fears of starvation, war and disease still haunt the earth. In future sections, more will be said of this, and we will spend considerably more time on the history of population study and demography.

Incidentally, it has been an odd quirk of scientific history that Darwin was favorably influenced by Malthus, and relied heavily on his doctrine to establish his theories of natural selection. Subsequently, Darwin's theories were accepted in scientific circles long before those of Malthus.

Population studies resumed great importance in the twentieth century with the work of Raymond Pearl (1925), Thomas Park (1933), Charles Elton (1933), and many others. These studies all form part of the modern and self-conscious development of ecology.

ECOLOGICAL GEOGRAPHY AND CONSERVATION

In the mid-nineteenth century, the field of ecological geography became established through the writings of George Perkins Marsh. One book published in 1864, *Man and Nature, or, Physical Geography as Modified by Human Action,* laid the foundation of ecological geography and conservation. As Stewart Udall pointed out in *The Quiet Crisis* (1963), this book marked "the beginning of land wisdom in this country." Lewis Mumford (1931) described *Man and Nature* as "the fountainhead of the conservation movement."

Marsh's basic theme in *Man and Nature* was that man's economic progress has often disrupted the balance of nature to his own detriment. He clarified the reciprocal interaction of man and environment. Marsh traced the history of civilizations, particularly those in the Mediterranean area, in terms of physical geography, natural resources and land changes induced by man. He first called attention to the striking effects of deforestation and land scarring on erosion, agriculture, and water resources. He forced his readers to look upon man in a different light—not as a conqueror but as a despoiler. David Lowenthal (1967) has noted that, "Few books have had more impact on the way men view and use land."

George Perkins Marsh was an amazing man. Born in Vermont in 1801, one of eight children, his personal life was marked by tragedy and business failure.

Although trained in the legal profession, he was not well-suited to the practice of law, and he became involved in a wide variety of business enterprises, including a woolen mill, a newspaper, a marble quarry, real estate, timber and a railroad. Most of these were financial failures, and he was virtually bankrupt on several occasions. In 1833 his wife and elder son died, leaving him with an infant boy to raise, and in 1839 his second wife became a bedridden invalid. During these difficult years, Marsh entered politics and was elected to the Vermont Legislative Council in 1835. In 1843 he ran for the United States Congress and was elected to four terms. His political life was not brilliant, but it won sufficient recognition for President Van Buren to appoint him Minister to Turkey in 1848. Later, President Lincoln appointed him Ambassador to Italy in 1861, primarily because Marsh had distinguished himself as a writer and scholar.

In the years preceding this appointment to Italy, Marsh's greatest achievements were in scholarly fields. He became an accomplished linguist, a master of twenty languages, and he wrote several books of historical and linguistic scholarship. His Vermont background fostered a keen interest in landscape and nature, and he became intensely aware of man's influence on his environment. During his Mediterranean travels from 1848 to 1851, and his subsequent tenure in Italy beginning in 1861, he viewed the rocky and arid landscapes of the Middle East and the Mediterranean countries with an alarming new insight. He undertook historical investigations of former vegetation and water resources, and he documented carefully his conclusions on the role of man in shaping once fertile lands into present-day deserts.

Although *Man and Nature* was a financial success and was widely read upon publication in 1864, its influence on the conservation movement in the United States was slow in coming. In fact, the period of the 1870's and 1880's in this country witnessed unprecedented environmental exploitation. Forests were cut at a tremendous rate without regard for conservation, lands were cleared, plowed and abandoned with increasing speed. Devastation of animal resources accelerated, resulting in the decline of the American bison from more than 60 million prior to 1850 to a total United States population of less than 600 by 1890. The passenger pigeon was so ruthlessly hunted and its nesting habitat so effectively destroyed that it was virtually extinct by the end of the nineteenth century and the last zoo specimen died in Cincinnati in 1914.

Not for almost 40 years, in fact, were the conservation concerns expressed in *Man and Nature* turned into political realities through the activism of Gifford Pinchot (1865–1947) and Theodore Roosevelt (1858–1919). Pinchot, later governor of Pennsylvania, was one of the first persons to advocate planned forest conservation in the United States. He became a member of the new National Forest Commission in 1896 and Chief of the Division of Forestry in 1898. Pinchot served as Chief of the Forest Service of the U.S. Department of

Agriculture from 1905 to 1910 when he became president of the National Conservation Committee. In the same year he published an influential book entitled, *The Fight for Conservation*.

The greatest political momentum to the conservation movement in the United States came through the forceful efforts of Theodore Roosevelt. As a Dakota rancher from 1884 to 1886, Roosevelt became intensely interested in wildlife and natural habitat. He wrote profusely throughout this period, centering his writings on the history of the West and the benefits of wilderness. Twenty years later, as President of the United States, he emphasized environmental conservation as one of the country's most important domestic problems. He added more acreage to the national forests than all presidents before and after. He regulated grazing and lumbering on national lands, and in 1908 he called the first White House conference on conservation. He worked closely with Pinchot in establishing the National Conservation Commission.

SUMMARY

This brief historical review of the origins of ecology prior to the twentieth century has traced certain highlights and significant trends in the scholarly fields of natural history and faunal exploration, environmental physiology, evolution, population studies, and finally ecological geography and conservation. Although these themes interweave at many points, each can be seen to have made distinctive contributions to ecological thought. All of these areas of scientific and scholarly endeavor were more or less confined to the intelligentsia. They were the activities of the elite and very little filtered down to the level of governmental concern or action until twentieth century activism which began with Gifford Pinchot and Theodore Roosevelt.

PART 3

Ecological Principles

6

Principles of the Ecosystem

ECOLOGY is a vast and encyclopedic subject. It can be confusing and discouraging at the outset because it seems so diffuse. After all, it includes the life habits of over a million different kinds of animals and plants, and it considers all manner of influences and interactions among them. Thus ecology must include not only the life sciences, but chemistry, geology, geography, meteorology, climatology, hydrology, paleontology, archaeology, anthropology and sociology as well. In fact, so extensive is the scope of ecology that it seems to have no limits at all, and ecologists could possibly claim dominion over all of the natural and social sciences. Figure 6-1 represents the cross-indexing of ecology undertaken by a scientific information retrieval service (Biosciences Information Service and Biological Abstracts), and it graphically portrays the complexity of ecology.

DEFINITIONS OF ESSENTIAL TERMS

Fortunately, the innumerable facts of ecology can be distilled into some remarkably basic and simple principles. The logical starting place is the ecosystem. An **ecosystem** is any spatial or organizational unit which includes living organisms and nonliving substances interacting to produce an exchange of materials between the living and nonliving parts.

The term ecosystem is more inclusive than the terms population and community, and is somewhat more similar in scope to the terms environment and

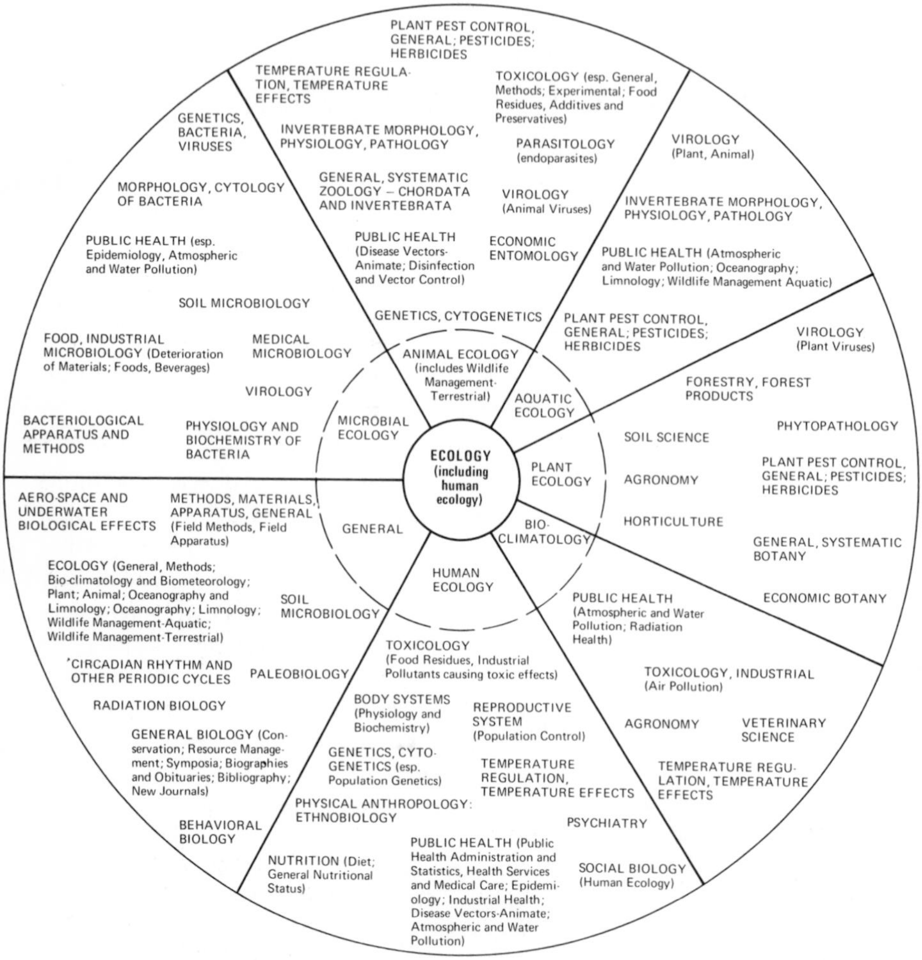

Figure 6–1 Ecology (including human ecology) subject model. (*Scope of ecology research in Biological Abstracts and Bioresearch Index.*)

habitat. For the benefit of understanding and clarity, definitions of these terms should be kept in mind.

A **population** is a group of interacting individuals, usually of the same species, in a definable space. Thus we can speak of the population of deer on an island, the population of rats in Baltimore, the population of starlings in New York City, the population of Coho salmon in Lake Michigan, etc.

A **community,** in the biologic sense, consists of the populations of plants and animals living together in a given place. Thus, we refer to the community of an oak forest, a marsh, a grassland, a coral reef, or a desert.

The terms environment and habitat refer to a definable place where an organism lives, including both the physical and biologic features of the place. **Environment** literally means "to surround" (from the French verb environner), and it therefore means surroundings or something that surrounds. It includes "all the conditions, circumstances, and influences surrounding, and affecting . . . an organism or group of organisms." (*Webster's New Twentieth Century Dictionary,* 1966).

A **habitat** is the natural abode or locality of an animal, plant or person (from the Latin, *habitare,* to dwell). Thus it also includes all features of the environment in a given locality. Frequently, the terms habitat and environment are used primarily for physical features such as topography, water supplies and climate, but the terms are not confined to physical features, for vegetation and other animals also form major components of any given habitat or environment.

An **ecosystem** includes populations, communities, habitats and environments, and it specifically refers to the dynamic interaction of all parts of the environment, focusing particularly on the exchange of materials between the living and nonliving parts.

COMPONENTS OF THE ECOSYSTEM

It is helpful to visualize ecosystems as consisting of four basic components: abiotic substances, producer organisms, consumer organisms and decomposer organisms.

Abiotic Substances

Abiotic substances are the inorganic and organic substances not momentarily present in living organisms. These include water, carbon dioxide, oxygen, nitrogen, minerals, salts, acids, bases and the entire range of elements and compounds outside living organisms at any given point in time. Many elements may be tightly bound in inorganic compounds, such as silicon in sandstone or aluminum in feldspar, and are unavailable to living organisms. Elements which are normally very active in biological processes, such as oxygen, may be in a form readily available to living organisms such as free O_2 or CO_2, or they may be in an inaccessible form such as silicon dioxide (SiO_2) in quartz, a major component of granite.

Similarly, potassium may be readily available to plants in the form of KCL in soil, but relatively unavailable in the form of $KAlSi_3O_8$ in orthoclase or monoclinic feldspar, one of the commonest of all minerals.

An important property of an ecosystem which determines its productivity is the form and composition in which bioactive elements and compounds occur. For example, an ecosystem may have a substantial abundance of vital nutrients, such as nitrates and phosphates, but if they are present in relatively insoluble particulate form as they would be if linked to ferric ions, they would not be so readily available to plants as if they were in the soluble form of potassium or calcium nitrate and phosphate. One of the most important qualities of an ecosystem is the rate of release of nutrients from solids, for this regulates the rate of function of the entire system.

Producer Organisms

Producer organisms are bacteria and plants which synthesize organic compounds. They are said to be autotrophic or self-productive, in that they take inorganic compounds and manufacture organic materials and living protoplasm from them. All green plants, including microscopic algae, are producer organisms since they exhibit photosynthesis, and some bacteria are producers since they may exhibit chemosynthesis or photosynthesis. Obviously, all life depends upon the basic productive capacity of green plants and bacteria.

Consumer Organisms

Consumer organisms are animals which utilize the organic materials directly or indirectly manufactured by plants. Consumers are unable to produce their own organic compounds for basic nutritive purposes. They are said to be heterotrophic, which means different or varied in nutritional source. Primary consumers or herbivores directly consume the organic compounds of plants. Secondary consumers may be omnivores or carnivores which depend partially or entirely on other animals for food. Tertiary and quartenary consumers may be the second or third-stage predator, for example, a hawk feeding on a weasel which in turn consumed a mouse.

Decomposer Organisms

Decomposer organisms are bacteria and fungi which degrade organic compounds. Their nutrition is said to be saprophytic, that is, associated with rotten and decaying organic material. In a sense they are the digestive organisms of an ecosystem—they reduce the complex organic molecules of dead plants and animals to simpler organic compounds which can be absorbed by green plants as vital nutrients. They provide the final essential link in the cycle of life. They are necessary for the renewal of life, for if decomposers were not active, organic compounds would become locked into complex insoluble molecules which could not be utilized as nutrients by plants.

Ecosystems involve, of course, a wide variety of life forms not specifically mentioned in the preceding paragraphs, but virtually all components of an ecosystem can be classified into producers, consumers, or decomposers. For example, parasites are merely specialized consumers. Plant parasites feed directly on plants and are thus herbivores; animal parasites derive their nutrition from other animals, and are thus carnivores differing from predators only in the fact that they normally do not kill the host. Scavengers such as vultures are also carnivores, differing from predators by the fact that they feed on an animal after it has died from some other cause. Figures 6-2 and 6-3 illustrate the essentian components of ecosystems and their relationships.

INCOMPLETE ECOSYSTEMS

Almost all ecosystems have all four basic components discussed above, though in some cases it is possible for incomplete ecosystems to exist. These are ecosystems lacking one or more basic components.

An example of an incomplete ecosystem lacking producers is the abyssal depths of the sea where only consumers and decomposers exist. In the realm of complete darkness green plants cannot survive. Scavengers and decomposers live on the fall-out of animals, plants and organic matter from the upper layers of the ocean. Predators might also be present to feed upon the scavengers. Hence, the ecosystem depends on extrinsic production, namely, the fall-out from upper levels. It might be possible, of course, for a few chemosynthetic bacteria to be present, but they would not produce a significant volume of organic material.

The same situation exists in caves where complete darkness prevents the growth of green plants. Again, a few chemosynthetic bacteria might be present, but they would not produce a significant amount of organic material. Practically all cave-dwelling-animals must depart from the cave, as do bats, or depend on extrinsically produced nutrients which enter the cave by flowing water or seepage.

The central core of the city might also be considered an incomplete ecosystem without producers, at least from the human standpoint. Some green plants obviously exist, but they would not supply meaningful production for humans and vertebrate animals living within the inner city, such as rats, pigeons, starlings,

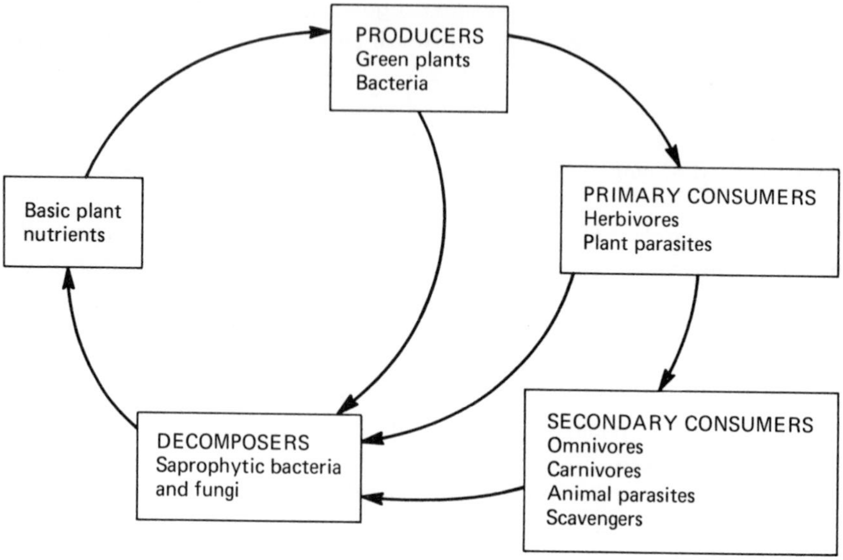

Figure 6–2 Basic components of an ecosystem. The arrows represent some of the major pathways of organic compounds.

sparrows, dogs, cats, etc. For all of these, the inner city requires extrinsic production and imported food. The only other alternative is for the inhabitants to leave the inner city and feed in peripheral areas. This undoubtedly occurs with starlings and pigeons, creating an ecological situation analogous to bats in a cave. The lack of production in an inner city is not due to a lack of light, but to a lack of soil and suitable substrate.

In other ways, cities may be considered incomplete ecosystems, ecologically parasitic upon the surrounding landscape. Not only do they import food, but they must also import fresh air and water. At the same time, they must export waste products—sewage, solid waste, carbon dioxide, sulfur dioxide, etc. If cities were encapsulated from their surrounding environments they would soon perish from thirst, starvation, asphyxiation, or the accumulation of waste products. In exchange for this life support, cities, of course, provide a great many economic and cultural benefits—jobs, housing, transportation, manufacturing, education, etc. So the relationship between city and landscape is vital in both directions, but it is particularly important to remember, as cities expand, that they cannot sustain themselves.

Incomplete ecosystems also exist in specialized cases where producers and decomposers are present without consumers. A theoretical example would be a massive bloom of some toxic algae in an aquatic ecosystem, where the algae would create toxic conditions for zooplankton and fish and all other possible

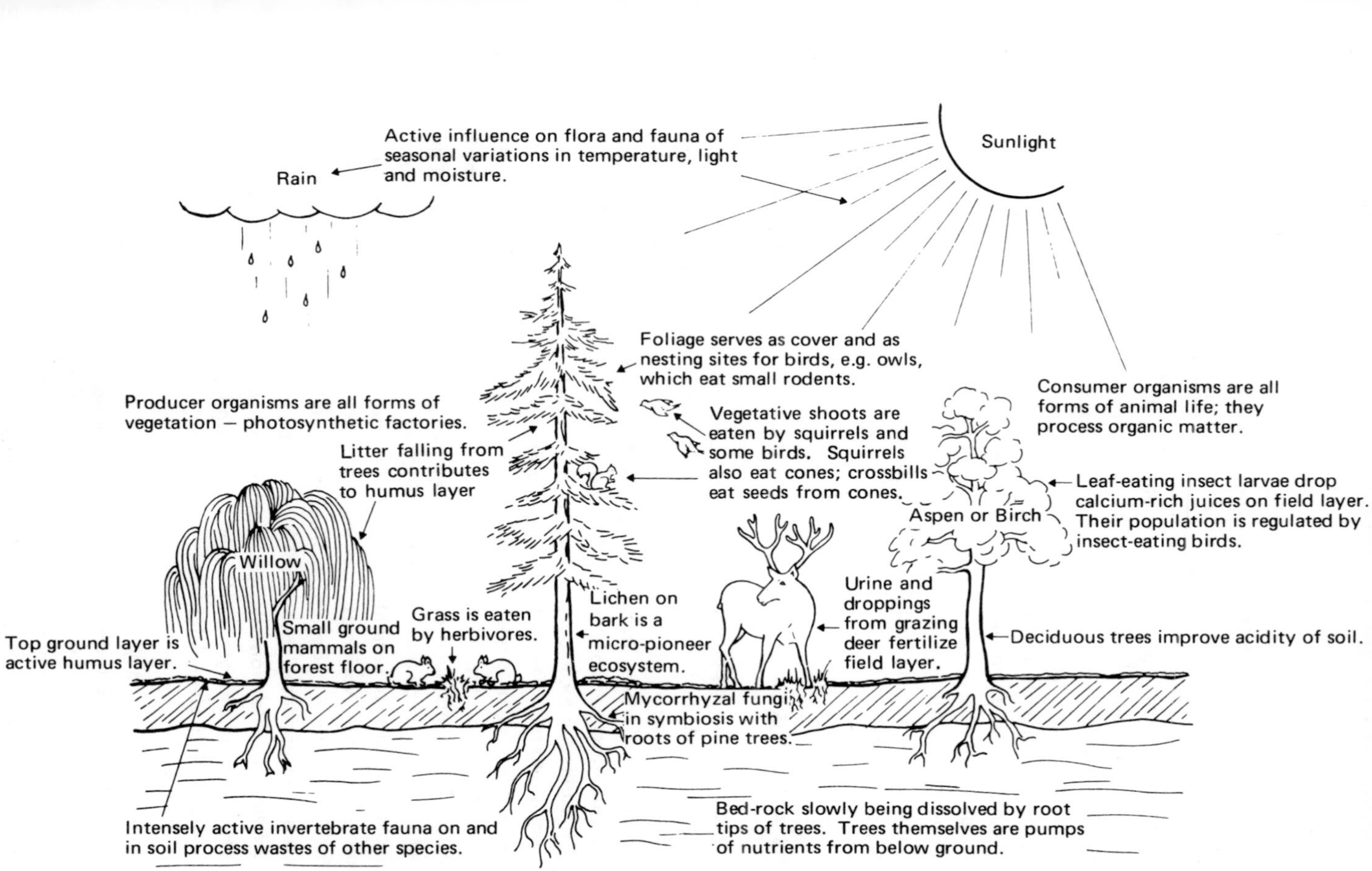
Active influence on flora and fauna of seasonal variations in temperature, light and moisture.
Rain
Sunlight
Foliage serves as cover and as nesting sites for birds, e.g. owls, which eat small rodents.
Producer organisms are all forms of vegetation – photosynthetic factories.
Consumer organisms are all forms of animal life; they process organic matter.
Vegetative shoots are eaten by squirrels and some birds. Squirrels also eat cones; crossbills eat seeds from cones.
Litter falling from trees contributes to humus layer
Leaf-eating insect larvae drop calcium-rich juices on field layer. Their population is regulated by insect-eating birds.
Aspen or Birch
Willow
Lichen on bark is a micro-pioneer ecosystem.
Urine and droppings from grazing deer fertilize field layer.
Grass is eaten by herbivores.
Small ground mammals on forest floor.
Top ground layer is active humus layer.
Deciduous trees improve acidity of soil.
Mycorrhyzal fungi in symbiosis with roots of pine trees.
Bed-rock slowly being dissolved by root tips of trees. Trees themselves are pumps of nutrients from below ground.
Intensely active invertebrate fauna on and in soil process wastes of other species.

Figure 6–3 Highly simplified illustration of integration within the comparatively simple pine forest ecosystem. (*From Darling and Dasmann in* Impact of Science on Society, *Vol. XIX, No. 2, 1969. By permission of UNESCO.*)

consumers. Then a process of excessive production and massive decomposition would go hand in hand. This would be a highly unstable and undesirable circumstance, but it has been known to occur, as, for example, in the red tides of Florida.

A third type of incomplete ecosystem might even be called an abiotic ecosystem, that is, one without living organisms, a self-contradiction in terms. They should more properly be called abiotic environments. Apollo space flights have shown so far that the moon is abiotic. Local areas on earth may be abiotic, for example, the high altitude ice plateau of Antarctica is probably devoid of living organisms over rather extensive areas. Closer home for most of us, the Copperhill basin of Tennessee is an area devoid of life, where fumes from copper smelters are toxic to all organisms within a certain downwind area. Possibly a few bacteria exist in very limited places, but for all practical purposes no plants or animals can survive.

SYSTEMS ANALYSIS IN ECOLOGY

Systems analysis is an important technique of ecosystem research and environmental management. Watt (1968) has pointed out that ecological systems are characteristically dynamic and complex, usually involving interactions between many variables, and often displaying lag effects, cumulative effects, thresholds, and nonlinear causal relationships. Such complex phenomena cannot always be studied by traditional techniques of individual research and scholarship. Computer analysis and team research are often necessary.

Systems analysis is a logical scientific method which approaches complex phenomena in several stages of operation: (1) *systems measurement,* in which the objectives of the problems are outlined and accurate data are obtained on important variables which relate to these objectives; (2) *data analysis,* in which the relationships between variables are explored through appropriate statistical procedures, and those variables most important in regulating the system are determined, (3) *systems modelling,* in which functional or mathematical models are composed to provide a theoretical basis for relating the variables, (4) *systems simulation,* in which variables are manipulated mathematically to evaluate the consequences of changes within the system, and (5) *systems optimization,* in which the best strategies for achieving the objectives are evaluated and selected.

This approach is basically similar to standard scientific procedure which involves description, classification, hypothesis formation, experimentation, and clinical tests or field trials. The particular characteristic of systems analysis, however, is that it can operate in very complex systems which require computerized data handling and do not lend themselves to direct experimentation. One cannot readily or intentionally experiment with a large ecosystem such as a river, a city, an ocean, or a range of mountains. Thus, mathematical modelling and computer simulation is sometimes the only feasible experimental approach in large-scale ecological research. Team research is also a usual component of systems analysis, for rarely can a single person possess all the skills necessary for the complete systems approach.

Systems analysis can be illustrated with the common problem of improving water quality in a badly polluted river. *Systems measurement* would first involve a statement of the objectives in water quality and river characteristics which are desired. These would be expressed in terms of physical characteristics (temperature, clarity, dissolved oxygen, nutrient levels, trace element concentrations, etc.), and biological characteristics (bacterial, algal, crustacean, insect, fish populations, etc.) which are desired in the river. Systems measurement would then involve making a detailed inventory of existing conditions, so that accurate data would be obtained on the major physical and biotic features of the river in its present polluted condition. Systems measurement would finally involve the identification of the major influences on the stream. This would require knowledge of its watershed, weather patterns, land-use patterns, geologic formations, domestic inputs, industrial operations, and stream flow characteristics. Hence, the enormity of just stage one, systems measurement, becomes apparent. This stage in a relatively small river could easily take a team of 5 to 10 scientists several years to get baseline data and understand the prevailing situation in specific terms.

Stage two, *data analysis,* would involve computer and statistical analysis of relationships between variables. How do land-use patterns, industrial operations, and domestic inputs interact with topography, weather, and stream flow characteristics to influence the physical and biotic qualities of the river? Some specific questions in this area would be: What are the sources of the nutrients? How do the nutrient levels influence the biotic communities? Are oxygen and temperature levels within satisfactory ranges? If not, why not? Are toxic chemicals present in the stream? If so, what are their origins, their effects, and their fates? A great many questions of this type can be asked, and it is apparent that multivariate analyses, as well as other statistical procedures, must be utilized.

The third stage, *systems modelling,* would be undertaken when the investigators felt they had a lead on the origins and relationships of the important variables within the system. They could then propose a mathematical relationship between various factors influencing the prevailing conditions within the river. Data analyses might suggest, for example, that pesticide run-off from agricultural operations was poisoning the river and responsible for the absence of certain organisms. The analyses might also suggest that excessive nutrients from domestic sewage were creating high bacterial populations which reduced dissolved oxygen below satisfactory levels. From these and other relationships implied by data analysis, mathematical models could be constructed to examine the effects of eliminating pesticide run-off from a certain percentage of the watershed, or of reducing domestic sewage input from a certain number of towns in the watershed.

The fourth stage, *systems simulation,* would consist of computer runs with different variables modified. The model would then predict changes in other variables. For example, systems simulation would estimate and predict changes in the river if agricultural practices were altered, if sewage treatment plants were installed, or if industrial operations were curtailed. It becomes apparent that modern computer techniques are essential at several stages in systems analysis, especially in data analysis and in systems simulation. Ideally, an integrated computer technology program should be utilized throughout the entire systems research if possible, from data collection to systems optimization.

The fifth stage, *systems optimization,* would determine the best course of action to achieve the goals. This process would relate the biological and physical data with economic feasibility studies. Alternatives and options could then be examined in realistic terms. Figure 6-4 summarizes the stages of systems analysis which can be used in any ecologic problem.

Systems analysis is also essential in working in the opposite direction from that outlined above—in determining, for example, the effects of a new industrial operation or land-use modification on existing conditions in an ecosystem. What would be the effects, for example, of constructing five or six nuclear power plants on the shores of Chesapeake Bay, and using the waters of the bay for cooling the nuclear reactors? This became an economic and ecological issue in Maryland, Virginia and Delaware in 1970 and 1971, when construction permits were issued to the Baltimore Gas and Electric Company to proceed with the construction of the first large nuclear power plant on the Bay at Calvert Cliffs below Annapolis. In numerous state and federal hearings on this issue it became clear that systems measurement data were not adequate to predict the effect of heated effluents on plankton ecology, fish spawning and growth, animal migrations, aquatic vegetation, and many other aspects of estuarine ecology. Broad spectrum systems analyses are urgently needed for a number of environmental and social problems, as well as for natural resource management problems.

Mathematical modelling and systems analysis may not provide the final answer to environmental problems, but in many situations they provide the most helpful possibilities for relating the complex array of variables which exist in ecologic problems. They provide the only reasonably objective way of evaluating cost and benefit relationships of environmental modifications, providing that accurate and complete data can be obtained. There are many problems, of course, in relating economic and ecologic values. It is not always possible to assign numerical cost or benefit figures to some ecologic considerations.

It is also essential to remember that systems analysis initially depends upon accurate descriptive data, gathered by careful field study. The most sophisticated

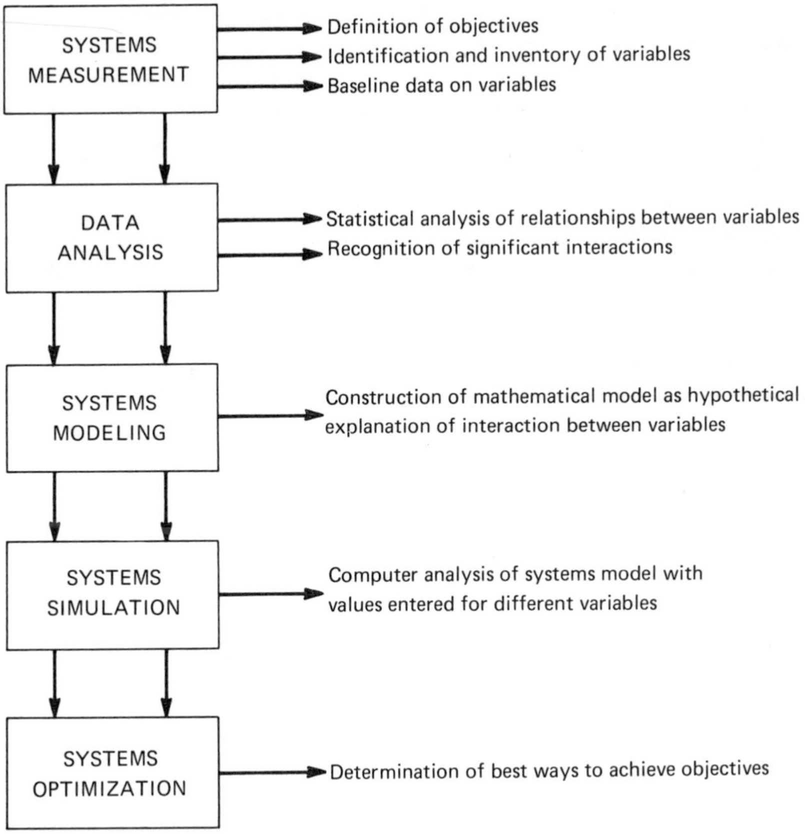

Figure 6–4 Basic stages of systems analysis.

computer technology and mathematical models are of little help if the basic ecologic data are incorrect or grossly incomplete. The field biologist still plays the essential first step in the most elaborate ecosystem analysis. In some circumstances, just as in weather prediction, a naturalist or field ecologist who has never seen a computer might have knowledge and experience of more value than an incomplete or hastily composed systems analysis program using the most advanced computer technology. Aldo Leopold never used computers, yet he provided some of the most accurate insights which have ever been achieved into the ecology and management of America's environmental problems.

Watt (1966; 1968) has supplied detailed discussions and mathematical presentations of systems analysis in ecological research. Van Dyne (1969) and Patten (1971) have compiled many important papers on systems analysis in relation to the study and management of ecosystems.

Ecosystem ecology is currently one of the most active and important areas of research in the entire field of ecology. It is increasingly important to understand the dynamics of ecosystems, their patterns of succession and evolution, and their patterns of diversity and stability. E. P. Odum (1969) has written a major review of these topics and has shown that ecosystems may be characterized as pioneer, developmental and mature. Various stages within this continuum possess certain properties of production, respiration, community structure, nutrient cycling, homeostasis, etc. For example, young developmental stages of an ecosystem tend to have greater gross production in relation to respiration, but lower species diversity and less stability than mature stages. Thus, developmental stages tend to be favored in agricultural practice, but they are also more vulnerable to ecologic damage or catastrophe. We will explore these relationships in more detail throughout subsequent chapters after various components of the ecosystem are considered separately.

7

Basic Features of Production, Consumption and Decomposition

PRODUCTION

Photosynthesis

THE most important chemical formulation in the world can be shown in simplified form as follows:

$$6\ CO_2 + 6\ H_2O + \text{light energy} + \text{enzymes of chlorophyll} \rightarrow C_6H_{12}O_6 + 6\ O_2$$

This is the *photosynthetic equation* which indicates that green plants can combine carbon dioxide with water and, using the energy of sunlight and the enzyme systems of chlorophyll, can ultimately produce sugar ($C_6H_{12}O_6$) and oxygen. This is the mainspring of life; it represents the basic productive capability of all ecosystems containing green plants; it is the means by which light energy is converted into the chemical energy of organic compounds.

This organic synthesis is not accomplished simply and in one step. It involves an elaborate series of chemical reactions, with innumerable enzymatic activities and many complex intermediate compounds. These reactions have been the subject of great amounts of research and numerous scientific papers. The fundamental pathways of carbon were worked out by Melvin Calvin and his colleagues using radioactively labeled C^{14} in carbon dioxide as a tracer. For this work Calvin received the Nobel prize in chemistry in 1961.

It is beyond the scope of this book to discuss in detail the chemistry of photosynthesis; nonetheless, a brief consideration of its major steps will elucidate certain ecologic principles.

Photosynthesis first involves the photolytic cleavage of water in which light energy absorbed by chlorophyll provides the energy for the intial separation of water. This is named the *light reaction,* and it requires some oxidant such as ferric ions. A brief representation of this stage is:

$$4\ Fe^{+++} + 2\ H_2O \rightarrow 4\ Fe^{++} + 4\ H^+ + O_2$$

This may also be portrayed by the following in which A represents any oxidant ion:

$$A + 2\ H_2O \rightarrow A\ H_2 + O_2$$

Thus, gaseous oxygen is produced early in the photosynthetic process in the same stage that produces hydrogen.

The second major stage of photosynthesis is independent of light and is called the *dark reaction*. In this, ATP (adenosine triphosphate) or NADH (nicotinamide adenine dinucleotide, formerly known as DPNH, or diphosphopyridine nucleotide) provide the energy sources. The atomic hydrogen combines with carbon dioxide to produce the first chemical union of carbon, hydrogen and oxygen. This can be represented in simplified form as follows (where A represents the oxidant ions):

$$2\ A\ H_2 + CO_2 \rightarrow 2\ A + CH_2O + H_2O$$

The carbon subsequently becomes involved in 3-phosphoglyceric acid ($C_3H_6O_3HPO_4$), and then in ribulose 1, 5 diphosphate ($C_5H_{12}O_3 \cdot 2\ PO_4$). This compound becomes the primary acceptor of CO_2; then the reactions proceed through several stages involving glycerate, glyceraldehyde, fructose, and finally glucose, 6-phosphate.

Ecologically, the most important aspects of this process are that photosynthesis requires all of the following: (1) green plants containing chlorophyll; (2) visible light energy in the wavelength of 400 to 700 millimicra (4,000 to 7,000 Angstrom units) or infrared in the wavelength of 800–850 millimicra; (3) carbon dioxide; (4) water; (5) some oxidant ion such as iron or magnesium; and finally (6) phosphorus in the form of phosphates.

Green plants have the additional capacity to synthesize higher order organic compounds, including disaccharide sugars, starches, lipids (including fats), proteins and vitamins. This, of course, is the advanced subject of biochemistry.

One type of lipids, the phospholipids, require nitrogen, primarily in the form of nitrates (NO_3). Protein synthesis also requires nitrogen and, with few exceptions, sulfur. Thus, proteins contain carbon, oxygen, hydrogen, nitrogen and usually sulfur, as their basic elemental ingredients. The nucleic acids, DNA and RNA, which form the vital and most unique organic constituents of all living cells, also contain these fundamental elements of carbon, oxygen, hydrogen, nitrogen and phosphorus. The importance of these five elements in ecology, therefore, is due to their presence in lipids, proteins and nucleic acids; hence, they form the most fundamental chemical elements of protoplasm.

Chemosynthesis

Chemosynthesis should also be mentioned as another mechanism by which organic compounds can be synthesized from inorganic materials through bacterial action, but in most ecosystems bacteria play only a very minor role in production. They are primarily decomposers. The chemosynthetic bacteria do not require energy from sunlight; rather they are able to obtain energy by chemical oxidation of inorganic compounds. For example, the oxidation of ammonia to nitrites, nitrites to nitrates, sulfides to sulfur, and ferrous to ferric ions are all oxidative processes yielding energy which can then be used in organic synthesis.

There are also a few photosynthetic bacteria which utilize sunlight energy, but differ fundamentally from green plants in that they do not produce oxygen as a by-product. The purple bacterium *Rhodospirillum* can grow anaerobically, and is an example of this latter group.

Productivity

It is interesting to compare the relative productivities of different ecosystems. Gross productivity is a measure of the total production of organic matter per unit area per unit time. Typical ecosystem productivities vary from 0.5 grams of dry organic matter per square meter per day to nearly 20 grams per square meter per day. For example, Lake Erie in winter has shown a productivity of 1.0 $g/m^2/day$, whereas in summer it has shown a productivity of 9.0 $g/m^2/day$ (Table 7-1). Coral reefs in the Pacific have produced 18.2 $g/m^2/day$. Most terrestrial ecosystems can produce 2 to 8 $g/m^2/day$, though a Hawaiian cane field has been measured as high as 23.9 $g/m^2/day$ (Table 7-2).

These figures of gross productivity do not mean actual or net productivity, nor do they mean that man can utilize all of the production. Net productivity is that amount remaining after the needs for plant respiration and metabolism have been met. Respiration is the process in all living organisms by which organic compounds are oxidized to yield energy with carbon dioxide and water as by-products. It is discussed in the following pages under the section on

TABLE 7-1 Gross Primary Productivity of Various Ecosystems as Determined by Gas Exchange Measurements of Intact Systems in Nature

Ecosystem	*Rate of Production g/m²/day*
Averages for long periods—6 months to 1 year	
Infertile open ocean, Sargasso Sea[a]	0.5
Shallow, inshore waters, Long Island Sound; yearly average[b]	3.2
Texas estuaries, Laguna Madre[c]	4.4
Clear, deep (oligotrophic) lake, Wisconsin[d]	0.7
Shallow (eutrophic) lake, Japan[e]	2.1
Bog lake, Cedar Bog Lake, Minnesota (phytoplankton only)[f]	0.3
Lake Erie, winter[g]	1.0
Lake Erie, summer[g]	9.0
Silver Springs, Florida[h]	17.5
Coral reefs, average three Pacific reefs[i]	18.2
Values obtained for short favorable periods	
Fertilized pond, North Carolina, May[j]	5.0
Pond with treated sewage wastes, Denmark, July[k]	9.0
Pond with untreated wastes, South Dakota, summer[l]	27
Silver Springs, Florida, May[h]	35
Turbid river, suspended clay, North Carolina, summer[j]	1.7
Polluted stream, Indiana, summer[m]	57
Estuaries, Texas[c]	23
Marine turtle-grass flats, Florida, August[m]	34
Mass algae culture, extra CO_2 added[n]	43

[a] Riley (1957); [b] Riley (1956); [c] H. T. Odum (unpublished); [d] Juday (1940); [e] Hogetsu and Ichimura (1954); [f] Lindeman (1942); [g] Verduin (1956) [h] H. T. Odum (1957); [i] Kohn and Helfrich (1957); [j] Hoskin, 1957 (unpublished); [k] Steeman-Nielsen (1955); [l] Bartsch and Allum (1957); [m] H. T. Odum (1957a); [n] Tamiya (1957).

decomposition. Net productivities are usually only 20 to 30 percent of gross productivities. Even total net productivities are not necessarily available to man, since much of the productivity may be in forms not usable by man. Whereas coral reefs are among the most productive of all ecosystems, most of the productivity is in the form of algae, plankton, coelenterates and other marine forms of little or no direct use to man. We will return to the subject of ecosystem productivity since it involves many topics of both theoretical and applied interest.

After primary production, the organic compounds of plants enter the dynamic pathways of the ecosystem and may have three general fates: (1) they may be metabolised within the plant itself to provide energy for growth and reproduction; (2) they may be stored within plant tissues and then consumed and assimilated by a herbivore; or (3) they may enter a cycle of decomposition in a dying plant, and return again to inorganic form.

TABLE 7-2 Estimated Gross Primary Productivity of Various Terrestrial Ecosystems[a]

Ecosystem	*Rate of Production* $g/m^2/day$
Cultivated Crops	
Wheat, world average	3.0
Wheat, high yields (Netherlands)	10.8
Corn, world average	3.0
Corn, high yields (Denmark)	5.7
Rice, world average	3.5
Rice, high yields (Japan and Italy)	10.4
Hay, United States average	3.0
Hay, high yields (California)	6.8
Sugar cane, world average	6.1
Sugar cane, average Hawaiian	12.2
Sugar cane, highest yields (Hawaii)	23.9
Noncultivated Ecosystems	
Pine Forest, England, during years of most rapid growth (20–25 yrs. old)	7.8
Deciduous forest, conditions as above	7.8
Tall grass prairies (Oklahoma and Nebraska)	3.9
Short grass, grassland (Wyoming)	0.65
Desert, 5 in. rainfall (Nevada)	0.26

[a] Based on the harvest method of measuring net productivity and converted to approximate gross productivity by the addition of 30 percent representing the average plant respiration in terrestrial ecosystems. Based on growing season only, often less than one year. (Data calculated from Odum, 1959.)

CONSUMPTION

Consumers of primary production may be conveniently classified into three categories: primary consumers or herbivores, secondary consumers or carnivores, and multilevel consumers or omnivores.

Primary Consumers

Primary consumers are those animals which feed directly upon primary producers: examples are meadow mice, deer, seed-eating birds and leaf-eating insects. In aquatic ecosystems, microscopic crustacea such as *Daphnia* which feed upon phytoplankton are primary consumers, as are some fish, such as menhaden which feed upon phytoplankton. Such fish are sometimes called "pasture" fish, since they graze upon phytoplankton and thus have the same relative position in the aquatic ecosystem as do meadow mice or cattle in the terrestrial ecosystem.

Secondary Consumers

Secondary consumers are predators or carnivores which feed upon herbivores. Thus, the hawk which eats the meadow mouse is a secondary consumer, and the bird which consumes the leaf-eating insect is also a secondary consumer. In aquatic systems, many aquatic insects and small fish which feed upon minute crustacea are secondary consumers.

There are often several levels of carnivores; that is, predators feeding upon other predators. For example, the water scorpion which feeds upon crustacea may be fed upon in turn by a frog, which is then eaten by a small fish, which is then eaten by a large fish, and the latter is finally taken by an osprey or an eagle. Thus, one could possibly speak of tertiary and quaternary consumers and so on, but the terminology becomes awkward. This brings us to the topic of "food chains" which will be discussed in a subsequent chapter.

Multilevel Consumers

Multilevel consumers or omnivores refer to those animals which feed as both herbivores and carnivores. Man is an herbivore in consuming vegetables, fruits, berries and grains, but he is a carnivore in eating meat. Bears and raccoons are also omnivores in that they feed readily upon berries and fruits, but also may prey upon fish, clams, crustacea, and other animals. Many animals are, in fact, omnivorous to a limited extent. Baboons normally consume plant material, but will occasionally eat other small animals. Shrews, which we consider carnivorous, will readily eat vegetable material and often do quite well on it in the laboratory.

A special type of consumption is that of scavengers which feed upon dead and decaying plant and animal material. Thus vultures, sea gulls and even eagles feed extensively on animals which died of other causes. They differ from true predators only in the postmortem nature of the consumption. They are similar to regular consumers in that they reconstitute the organic matter in their food, utilizing some of it for energy, some for growth and development, and excreting the metabolic by-products. Scavengers of dead plant material such as insects, millipedes and earthworms play the same role as animal scavengers, and their function becomes closely related to that of the decomposers.

DECOMPOSITION

Decomposition is the process by which complex organic materials are broken into simpler compounds that can again be utilized by plants for new growth, and it is also the process by which bacteria and fungi obtain energy and nutrients. It is, of course, an essential function, for without it, all nutrients would become tied up in dead organisms. It should also be remembered, of course, that all producers and consumers also accomplish some decomposition through their normal life processes. Thus, all organisms respire and catabolize materials releasing carbon dioxide and other waste products. It cannot be said, however, that decomposition is their primary or typical function, as it can be said of bacteria and fungi.

Some organic materials, such as sugars, lipids and proteins, are decomposed rapidly in a series of stages, whereas others, such as cellulose, lignin, hair and bones, are decomposed slowly. These latter materials may have considerable resistance to bacterial decomposition. The decomposition of cellulose may, in fact, be a limiting factor in some ecosystems. Decomposition in its most basic form may be represented by the formula for respiration. Aerobic respiration is essentially an oxidative process represented as follows:

$$C_6H_{12}O_6 + 6\ O_2 \rightarrow 6\ CO_2 + 6\ H_2O + \text{caloric energy}$$

This is not so simple as shown, of course, since it requires a complex set of enzymatic reactions of the cytochrome oxidase system, but the value of portraying it in its most basic form is to show that it is essentially the reverse reaction of photosynthesis in terms of raw materials and end products. This process occurs in living cells without bacterial action, just as it occurs in ecosystems as a result of bacterial action. Thus the metabolic process of "ecosystem respiration" may be considered analogous to cellular respiration.

Anaerobic respiration also occurs in some cells and ecosystems. It involves the decomposition of simple sugars into triosephosphates, pyruvic acid, ethyl alcohol and finally acetic acid and water with the release of energy. Oxygen is not directly involved. This sequence of biochemical reactions follows the Embden-Meyerhoff scheme, and it again can occur in living organisms without bacterial action or in dead organic material with bacterial involvement. Some organisms are capable of either aerobic or anaerobic respiration. This is true of the soil bacterium *Aerobacter,* and also of some parasitic worms which are aerobic in the free-living larval stage, and anaerobic in the adult parasitic stage (schistosome parasites are an example).

Decomposer organisms are incredibly abundant in natural ecosystems. Tepper (1969) has pointed out that a gram of soil (about 1 teaspoonful) may contain one billion bacterial cells, 5 million actinomycete fungi, 500,000 protozoa, and 200,000 molds of various kinds. The specific numbers depend upon the soil type, moisture, temperature, nutrient levels, and other environmental factors.

Although decomposers are very numerous, most of them are microscopic in size, and their total biomass is substantially less than that of producers and consumers in most ecosystems. Only some of the fungi, such as bracket fungi, puffballs and toadstools grow to large size.

No single type of bacterium or fungus performs the complete range of decomposition. A high degree of specialization occurs, with specific bacteria and fungi performing specific chemical functions through various enzymatic reactions. In the decomposition of milk, for example, *Streptococcus lactis* acts upon milk sugar, lactose, to produce lactic acid. As the pH falls, the *S. lactis* can no longer grow adequately, but the process is continued by *Lactobacilli* which can tolerate more acid conditions. Finally, as very acid conditions are approached, various species of yeasts and molds begin growth to continue the decomposition of lactic acid to carbon dioxide and water. As this is occurring, other types of bacteria, such as *Pseudomonas* begin the decomposition of proteins into ammonia and simpler nitrogen compounds. Figure 7-1 diagrams some of the population changes of bacteria, yeasts and molds in the decomposition of milk.

Although many decomposers have very specific chemical functions, quite a few of them can exist in a wide range of conditions. The soil bacterium, *Aerobacter,* already mentioned, is a good example of this. Their ubiquitous nature permits similar functions in diverse habitats.

Another important function of decomposer organisms is the production of metabolic products which have a regulatory function on other organisms. Perhaps the best known examples are the antibiotic substances produced by some molds. The mold *Penicillum,* for example, releases a substance into the environment which inhibits bacterial growth. Many of the most important antibiotics in medicine were first discovered as natural products of molds. Chemical substances produced by one group of organisms which have a regulatory influence on other organisms in the environment have been named "ectocrines" or "environmental hormones." They are not necessarily the product of decomposers

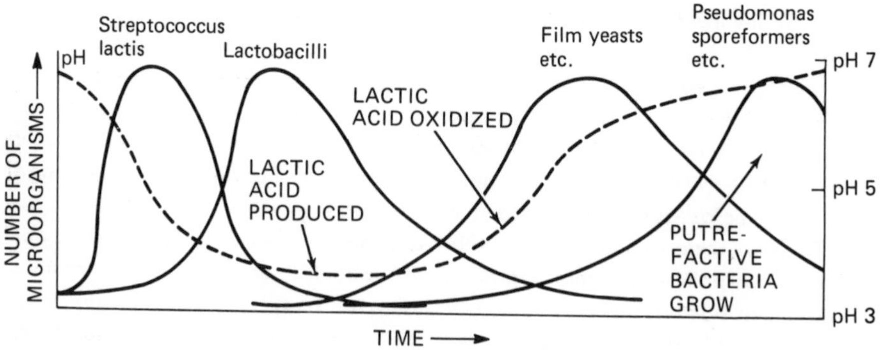

Figure 7–1 Changes in numbers and types of microorganisms in the decomposition of raw milk at room temperature. *(From Tepper, 1969, after Frobisher, 1962, and Carpenter, 1967, by permission W. B. Saunders Co.)*

alone, for it is known that higher plants and animals produce and release substances which have a regulatory influence on other organisms, but the decomposers are very important sources of ectocrines. An example of an ectocrine produced by a higher plant is a factor in wheat, corn, rice and rye, known to bestow resistance to certain diseases in experimental animals. It has been named SRF (salmonellosis resistance factor) by Schneider (1967), because it gives experimental mice resistance to intestinal pathogens of the genus *Salmonella,* and Schneider points out that it is quite distinct from antibiotics and vitamins.

An example of ectocrines in aquatic systems has been studied in rotifers. The larger rotifer *Asplanchna* is a predator of the small rotifer, *Brachionus. Brachionus* living in the presence of *Asplanchna* develop spines in subsequent generations. Detailed analysis of this system showed that *Asplanchna* releases a metabolic factor which causes the eggs of *Brachionus* to develop these spines. The parent generation of *Brachionus* remains unchanged. Thus *Brachionus* has evolved a mechanism by which it detects the presence of its predator and then adaptively modifies itself in one generation for protection. Without the presence of the ectocrine from *Asplanchna, Brachionus* develop without spines.

SUMMARY

The biochemical and ecologic cycle of life in its simplest form involves three major processes: production, in which organic compounds are synthesized; consumption, in which the basic organic compounds of production are utilized and reorganized; and decomposition, in which organic compounds are broken down to simpler substances which can again be utilized in production.

Photosynthesis is the basic productive force in the biosphere, and it is dependent upon green plants, sunlight, water, carbon dioxide and certain inorganic ions. Various carbohydrates, lipids, proteins and nucleic acids are all formed in higher-order syntheses from the sugars and starches manufactured in photosynthesis.

Consumption involves various stages of herbivore, carnivore and omnivore feeding activity. It results in the utilization and reorganization of organic materials originally formed in green plants.

Finally, decomposer organisms play three vital roles in ecosystems: the breakdown of complex organic materials into simpler forms, the production of food for other organisms, and the production of ectocrine substances which serve regulatory functions in ecosystems.

8

Biogeochemical Cycles and Ecosystem Homeostasis

SEVERAL ecosystem principles can be understood by reference to the cyclic passage of key elements, such as carbon, hydrogen, oxygen, nitrogen, phosphorus and sulfur, between the living and nonliving components of the ecosystem. Outlining these cyclic patterns in simplified terms is certainly not adequate for a full understanding of ecosystem function, for what we ultimately need is some knowledge of the quantitative and energetic relationships involved. Nevertheless, a brief consideration of these cyclic patterns is a logical starting point toward achieving this understanding.

CARBON CYCLE

Carbon, a key element in all living material, cycles relatively simply between plants, animals and the inorganic world. Figure 8-1 portrays the general pathways of carbon in an ecosystem. This diagram is oversimplified because

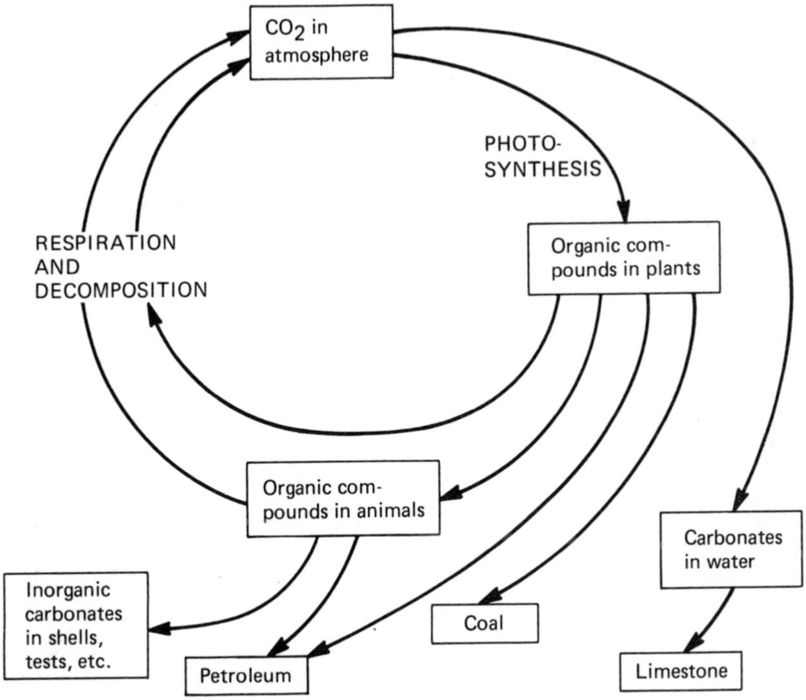

Figure 8–1 Basic aspects of the carbon cycle.

there are actually separate carbon cycles on land and in water, but they are fundamentally similar and are dynamically interrelated at the air-water interface (Bolin, 1970).

Carbon exists in the atmosphere primarily as carbon dioxide, in which form it is incorporated directly into plant protoplasm in the process of photosynthesis. For example, a tropical rain forest may incorporate between one and two kilograms of carbon per square meter per year into organic compounds (Bolin, 1970). From plants, organic carbon may go into animals, where it goes through various stages of digestion and assimilation, and from either plants or animals it may re-enter the atmosphere as CO_2 through respiration. If it remains in plants and animals, it ultimately passes into dead organic material, from which it can return to atmospheric CO_2 by oxidation or decomposition. In some animals, carbon may become tied up in hard parts, such as shells, and thus remain in the form of inorganic carbonates for a long time. Limestone can result from marine deposits of animal carbonates as well as from inorganic precipitation of carbonates in water. These carbonates in limestone can then return to the living carbon cycle only very slowly through a process of erosion and dissolution. Dissolved carbonates in water may be absorbed by plants—some aquatic plants, for example Eurasian milfoil (*Myriophyllum spicatum*), can use the carbon in dissolved carbonates as a direct carbon source in photosynthesis. Most aquatic plants, however, are more efficient when using free CO_2 in water as a carbon source.

Carbon may also become "locked" into organic deposits of coal and petroleum, remaining in this form for millions of years until released in combustion. A more detailed discussion of the carbon cycle has been presented by Bolin (1970) in an issue of *Scientific American* devoted to the ecology of the biosphere. This issue also contained review articles on the nitrogen cycle (Delwiche, 1970), oxygen cycle (Cloud, 1970), and various mineral cycles (Deevey, 1970). Some of the highlights of these cycles are considered below.

NITROGEN CYCLE

Nitrogen, another essential element of all protoplasm, has a more complicated series of cyclic pathways through the ecosystem (Figure 8-2). The atmospheric form of free nitrogen must be "fixed" or incorporated into chemical compounds such as ammonia (NH_3) which can be utilized by plants. Nitrogen fixation is accomplished by bacterial action of both free-living soil bacteria such as *Azotobacter* and *Clostridium,* and symbiotic bacteria such as *Rhizobium* living in root modules of leguminous plants. Several other bacteria can perform nitrogen fixation, and some blue-green algae such as *Anabaena* and *Nostoc* can also perform this reaction. Nitrogen fixation is also achieved as a physical process in the atmosphere by the ionizing effect of lightning and cosmic radiation, and it can be achieved industrially through the Haber and Bosch method. The greatest single source of fixed nitrogen, however, is probably terrestrial bacteria.

From the forms of ammonia and soluble nitrates, plants incorporate fixed nitrogen into protoplasm by amino acid and protein synthesis. Then the organic nitrogen compounds may follow any one of three general pathways: (1) storage or modification as proteins or nucleic acids within the plant; (2) incorporation into animal protein through consumption and assimilation by animals; or (3) decomposition to NH_3 through death and bacterial action. In animals, essentially the same three major pathways may be followed plus metabolic decomposition into urea and other excretory products.

In the decomposition cycle through death and decay, ammonia (NH_3) is produced from amino acids by the action of ammonifying bacteria such as *Pseudomonas, Proteus,* etc. Under normal circumstances it is quickly converted into nitrite form (NO_2) by nitrite bacteria such as *Nitrosomas,* and into nitrate form (NO_3) by nitrate bacteria such as *Nitrobacter*. Nitrates are then absorbed directly by plants as primary nutrients and incorporated once again into amino acids and proteins by organic synthesis within the plant.

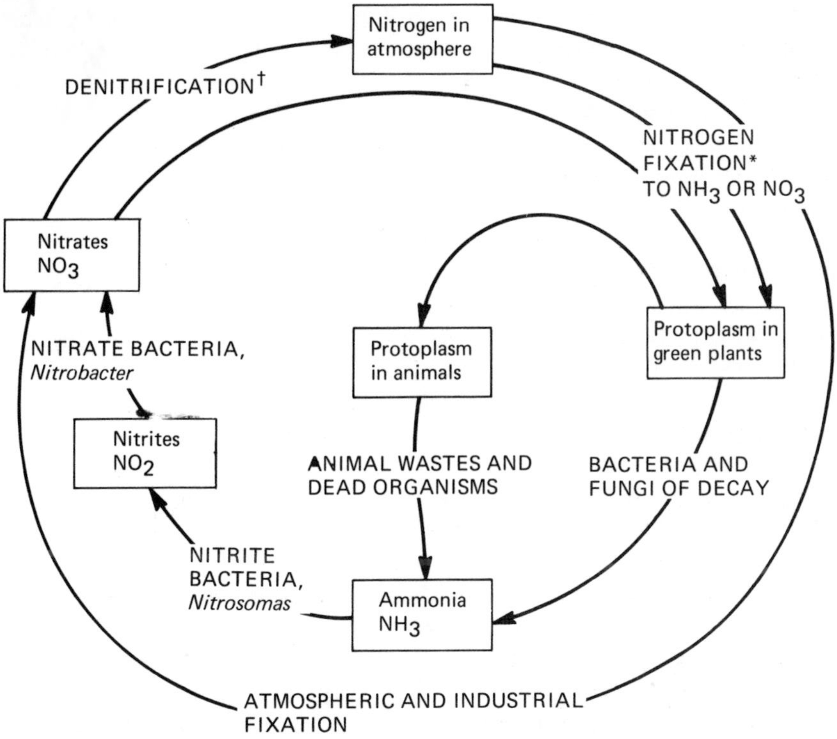

†*Dentifrifying bacteria*
Pseudomonas
Thiobacillus
Micrococcus denitrificans

**Nitrogen fixing bacteria and Algae*
Free-living: *Azotobacter, Clostridium*
Leguminous: *Rhizobium*
Blue-green: *Nostoc, Anabaena*

Figure 8–2 Nitrogen cycle.

The presence of ammonia nitrogen in ecosystems is a good measure of the balance between protein decomposition, bacterial action and plant production. High ammonia nitrogen in water is generally indicative of enrichment from fertilizer, animal wastes, or domestic sewage. Thus, field measurements of ammonia nitrogen above the levels of 1 ppm usually indicate some major source of decomposition within the system, in excess of that being utilized by bacterial action and plant growth. Field measurements of nitrites and nitrates are also good guides to the nutrient condition of lakes, streams and estuaries. The same is true of dissolved phosphates, as we shall see in subsequent paragraphs.

Nitrogen is returned to its atmospheric form by the action of denitrifying bacteria such as *Pseudomonas, Thiobacillus* and *Micrococcus denitrificans.* Obviously, the cyclic flow of nitrogen throughout the ecosystem requires balances of bacterial action involving many species, so that appropriate levels of plant nutrients are maintained without excessive accumulation of decomposition products like ammonia. A 1969 report on environmental problems by a committee of the American Chemical Society pointed out that all life could be extinguished on earth by the extinction of perhaps just a dozen species of bacteria involved in the nitrogen cycle.

PHOSPHORUS CYCLE

Phosphorus is also a key element in all living organisms, and it plays an essential role in almost every step of organic synthesis. It is much less abundant in the abiotic ecosystem than nitrogen, existing in natural levels at a ratio of 1 to 23 in relation to nitrogen (Hutchinson, 1944), but it is relatively more abundant in plants and animals. It is more likely than almost any other element to limit productivity in many of the earth's ecosystems. Phosphorus in the form of adenosine triphosphate (ATP) is a "universal fuel of living organisms" (Deevey, 1970).

Phosphorus in the protoplasm of plants and animals is broken down by cellular metabolism or the action of phosphatizing bacteria to dissolved phosphates (e.g., $CaHPO_4$) (Figure 8-3). These dissolved phosphates may be utilized directly in protein synthesis in plants as primary nutrients, or they may enter marine deposits and become fixed in relatively insoluble forms of phosphate rocks, $Ca_3(PO_4)_2$. Bone and guano deposits may also lock up phosphates for considerable periods of time until artificially recovered. In general, the loss of phosphorus to the ocean has been greater than the gain to land, and this has become a practical problem in many countries such as India where there is a shortage of artificial fertilizers. Fifty years of cultivation in temperate zones can readily reduce phosphate levels of the soil by more than one-third (Odum, 1959). A much greater loss in less time can occur in tropical regions. India has also mistreated and defeated natural ecosystem principles by burning animal wastes instead of using them as natural fertilizer. Cow dung is dried and used as a major fuel source—a process which has short-circuited natural nitrogen and phosphorus cycles and thus robbed the soil of basic nutrients. This is now being corrected by an expensive crash program of artificial fertilizer production.

The greatest reservoir of phosphates in the world lies in the relatively insoluble ferric and calcium phosphates in rocks. In this form, phosphorus may be released slowly to soluble forms by the action of dilute nitric acid formed during nitrification.

A modern source of phosphorus is in the common household detergents which now enter waste water systems and are then released into streams, lakes and estuaries. Waste detergents are often sufficiently abundant in streams

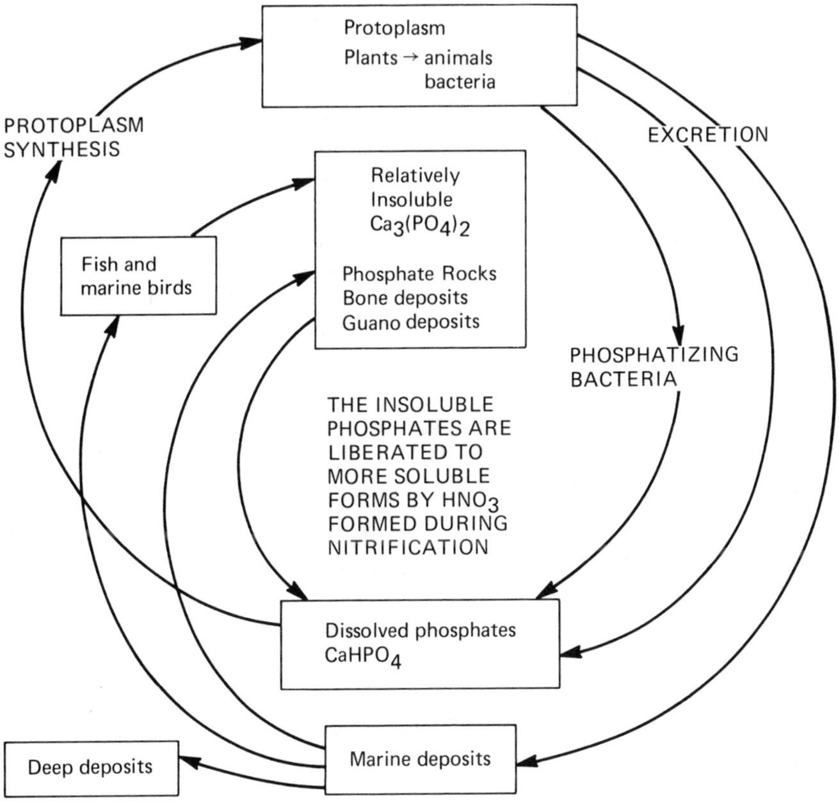

Figure 8–3 Phosphorus cycle.

to cause foaming and sudsing at waterfalls. The high phosphate content of these detergents can also stimulate undesirable plankton blooms, and hence these detergents are sometimes a major component of pollution and eutrophication.

SULFUR CYCLE

Sulfur is an essential element in protein synthesis, since it provides a linkage between polypeptide chains in protein molecules. Life, as known on earth, could not exist without sulfur. It is less likely, however, to be limiting of ecosystem productivity than phosphorus. In nature, sulfur exists in the elemental form and in several oxidation states, including hydrogen sulfide (H_2S), sulfites (SO_2) and sulfates (SO_4). Organic sulfur in plants and animals is decomposed to H_2S by bacterial action, and the H_2S is further oxidized to sulfates such as

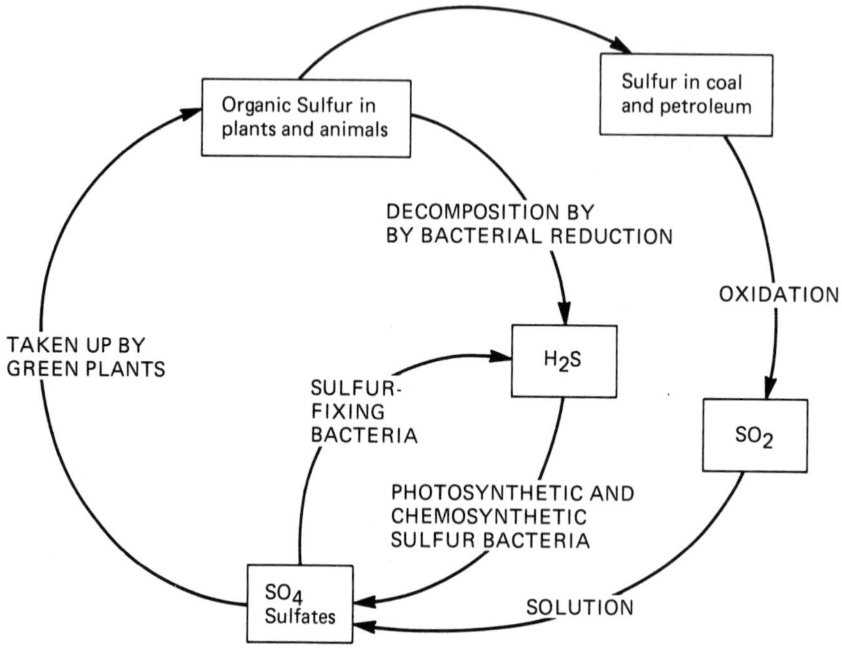

Figure 8–4 Sulfur Cycle.

NH_3SO_4 by sulfur-oxidizing bacteria (Figure 8-4). These sulfates are then taken up by plants as primary nutrients. Hydrogen sulfide can occasionally accumulate by rapid protein breakdown. In the Black Sea, below 150 meters, the concentrations of H_2S and H_2SO_4 are so great as to exclude all forms of life other than sulfur bacteria (Brock, 1966). In polluted estuaries, large accumulations of algal sea lettuce may undergo primary decomposition to produce obnoxious quantities of H_2S. In some coastal areas, this has been known to be sufficiently severe to cause paint blistering on nearby houses. Sulfur is also locked into coal and petroleum and is released as sulfur dioxide when these products are burned.

The foregoing examples of biogeochemical cycles have been greatly oversimplified, and more substantial accounts of these cycles are given by Brock (1966), Bolin (1970), Delwiche (1970) and Deevey (1970). These discussions illustrate the cyclic flow of essential elements between the living and nonliving components of the ecosystems. Similar flow-charts could be constructed for all the other elements which play a vital role in living systems, such as iron, magnesium, sodium, potassium, manganese, cobalt, and many others. Our knowledge of all of these cycles serves to emphasize the interrelations of the living and nonliving world, and to show us that the biosphere is a very complex biochemical system which depends on the proper functioning of many organisms.

ECOSYSTEM HOMEOSTASIS

Ecosystem homeostasis is a technical term for the balance of nature. It involves a good deal more than the average person may envisage in that popular term, however. In its broadest sense it refers not only to a balance of species, as for example, a balance between predator and prey or host and parasite, but also a balance of basic nutrient cycles and energetic pathways within an ecosystem. A homeostatic condition within an ecosystem implies that all aspects of ecosystem function are in balance. Thus, there would be a balance between production, consumption and decomposition, as well as between all species within the system.

The concept of homeostasis within an individual has been valuable in physiology. As developed by the eminent French scientist, Claude Bernard, in the nineteenth century, and later expanded by the American physiologist, Walter Cannon, in the twentieth century, it has led to vastly increased understanding of the regulation of body processes through nervous and endocrine control. It stimulated the discovery of vasomotor and metabolic responses in the regulation of heartrate, respiration and body temperature. It has also shown the interplay of nervous and hormonal regulation in growth, reproduction and behavior, and has contributed much to our understanding of health and disease.

In many ways, a comparable concept of homeostasis at the ecosystem level helps us to understand processes of regulation within plant and animal communities. It can pinpoint areas where more research is needed to clarify control mechanisms and routes of interaction between components of ecosystems.

The concept of intrinsic regulation or feedback is essential to our understanding of homeostasis. In physiology, for example, we know that muscular activity increases carbon dioxide concentration and decreases oxygen levels in the blood. This stimulates faster heart rate and breathing rates which serve to expel carbon dioxide and bring in more oxygen. When the carbon dioxide and oxygen levels return to normal, heart rate and breathing rate also return to normal. Thus the system remains in balance to meet the metabolic needs of the individual.

In a balanced aquatic ecosystem, there is an analogous, though less accurately controlled, homeostasis involving carbon dioxide and oxygen. For example, an increase in water temperature in the springtime which increases metabolic rate and respiration in aquatic plants and animals, results in an increase in carbon dioxide and a decrease of oxygen. The higher level of free CO_2 and increasing water temperatures stimulate more rapid photosynthesis and plant growth which utilizes the CO_2 and produces oxygen. Thus both O_2 and CO_2 tend to return to normal limits. If the temperature and metabolic rate declines, and all available free CO_2 is utilized in the water, then plant growth is limited until decomposition adds more CO_2 to the water. One can readily envisage the complexity of the system when one considers hundreds of species of plants and animals interacting over the common interface of oxygen, carbon dioxide, light, nutrients and many other resources.

Ecosystems have the ability to perform a certain amount of self-regulation within limits, but if these limits are exceeded, they may no longer be able to function and may experience various patterns of change, injury, or breakdown. Thus, if temperature levels exceed certain natural limits, the metabolic homeostasis of the system may be thrown out of balance so far that its natural regulatory capacities are impaired. It should be a major task of modern science to study ecosystems with the object of understanding natural homeostatic mechanisms and limitations. It is particularly important that we recognize the tolerance levels of different systems to disturbance. It is obvious in some areas such as Lake Erie, Delaware Bay, and many parts of the Middle East and Asia, that man has exceeded ecosystem tolerances—that he pushed the system too far, either through direct damage or negligence—and the result was irreparable damage.

It must be pointed out that man does not always desire a homeostatic ecosystem. In fact, all agriculture is based on systems in which production exceeds consumption so that surpluses of organic products result for human consumption. This can be considered either a nonhomeostatic system or an artificial homeostasis if man successfully utilizes the surplus.

In ecologic theory, any nonhomeostatic or artificial ecosystem has intrinsic instabilities that must be controlled by direct action. Thus in agriculture, weeds and pests must be controlled by cultivation, pesticides, or some other form of control. Only by constant attention can the agricultural ecosystem be maintained in a productive state.

Many of the world's ecologic difficulties arise from upsets in natural homeostatic mechanisms in ecosystems. In polluted waters, for example, excessive nutrients from sewage may cause excessive production of algae. If this production grossly exceeds consumption by herbivores, it leads to unsightly and obnoxious plankton blooms in which excessive decomposition suddenly becomes prevalent. This may produce toxic products, or it may deplete the oxygen supply so that fish and other aquatic animals die. Thus a simple imbalance of production may cause far-reaching damage to the entire system.

Imbalances often occur initially in only a few components of the system, but they may have the result of affecting the entire system. Locust plagues may result from an unusually favorable combination of factors which stimulate locust reproduction, but these plagues may then devastate the entire system which spawned them.

ECOSYSTEM MANAGEMENT

Practical ecosystem management usually involves maintaining a natural or artificial homeostasis in which the end product is of benefit to man, and in which violent or pathologic imbalances can be controlled. An example of a natural ecosystem which man might wish to maintain for esthetic or recreational value would be a wilderness area or National Park. Here the most important consideration is to insure that natural processes are permitted to continue. In a forest this would mean, for example, that fire, if already a natural part of the ecosystem, be permitted to occur at natural frequency and intensity. Several recent studies have shown that certain forests depend upon intermittent burning to recycle nutrients and minerals (Oberle, 1969). In the lowland conifer forests of Alaska, if fire is excluded a thick carpet of moss may accumulate and raise the permafrost level. This encourages the growth of black spruce, a species with little food or timber value. Browsing animals such as deer, moose and bear cannot survive on black spruce. They depend on natural fires to maintain a suitable cycle of vegetation. Many foresters now feel that the underbrush and leaf litter in the Ponderosa pine forests of western United States should be removed by light ground fires every few years to maintain soil fertility, to prevent excessive accumulation of organic matter in the form of cellulose and also to prevent more devastating fires that destroy the entire forest. In the pine forests of the South, very efficient fire suppression techniques allowed dangerous accumulation of combustible materials, and a pine fungus disease was favored in the accumulated litter. In all of these forests, proper management should not exclude all fires, but should prevent totally devastating fires. These examples illustrate the need for understanding natural processes in ecosystems in order to maintain them properly.

The Everglades of Florida provide another example of the importance of understanding natural ecosystem processes before the system can be properly managed or maintained. The unique flora and fauna of the Everglades depends upon the proper balance of fresh and saline water flowing through its channels. If a significant decline in fresh water input occurs, as has been taking place due to the diversion of fresh water to the growing megalopolis of Miami, numerous important species of plants and animals may die and the characteristic features of the system may be destroyed. In times of drought, many animals depend upon alligator wallows as a source of water. These are pools hollowed out by alligators, and they tend to retain water longer than other parts of the swamp. If alligators decline through shooting or any other mortality force, many other animals also decline. Now that some of the key factors necessary for the maintenance of the Everglades are known, its survival becomes, in simplest terms, a competitive contest between the natural ecosystem and the expanding human population pressures.

We must recognize that most of the world's ecosystems are already significantly altered by man, and we can no longer apply a simple "hands off" conservation policy to insure their preservation. Thus, programs of active management become necessary, as in agriculture, and we can often apply agricultural principles of cropping and harvest to natural populations of fish, wild game and forests to the mutual advantage of man and the ecosystem. This approach was brilliantly elucidated by Aldo Leopold in his classic book, *Game Management* (1933), and it was more recently expressed in modern mathematical terms by Watt in, *Ecology and Resource Management* (1968). We shall return to these concepts in future chapters.

It is equally important to recognize that ecosystems have a certain level of "insult tolerance," but if we continue to assault them with bad management or noxious influences, they will most certainly become irreversibly injured. If we continue to restrict fresh water from the Everglades, or surround its watershed with suburbia, it will undoubtedly suffer. If we continue to build concrete jungles around our lakes, rivers and estuaries, and fill their waters with waste materials, we can no longer expect them to produce clams, oysters, gamefish, or support recreational pleasure. It has been estimated that if we could instantly stop all pollution entering Lake Erie, and supply an unpolluted input, it would still take over a hundred years to clean out the lake and restore it to anything approaching its former value and productivity.

In upsetting the homeostatic conditions of ecological systems, we are inflicting changes that may persist for centuries. Once the scientific facts are known, proper ecosystem management becomes a matter of ethics and value judgments. As pointed out by Odum (1959), Leopold (1949), and Hutchinson (1948), the understanding and proper management of ecosystems must be recognized by mankind as a moral responsibility, and the old concept of unlimited exploitation must be replaced by one of restoration and enlightened conservation.

9

Energy Flow and Trophic Structure

ECOLOGY is intimately concerned with the sources of energy in ecologic systems and the transformation of this energy in living organisms. It is appropriate to think in terms of energy "flow" and not "cycles" as in minerals and nutrients. Energy may move in various directions, but it does not spontaneously return to its original state; hence, it does not really cycle.

Two basic physical laws which relate to energy are the first and second laws of thermodynamics. These laws, first enunciated and established by the work of Carnot and Joule in the 1830's and 1840's, are fundamental to our understanding of ecosystem energetics.

The first law of thermodynamics states that energy on earth is neither created nor destroyed; it is transformed from one type to another, as, for example, from light to heat to motion. The second law of thermodynamics states that no energy process occurs spontaneously unless it is a degradation or dissipation from a concentrated form to a dispersed form. An important corollary of this is that since some energy is always dispersed at each transformation, no transformation of energy is 100 percent efficient.

An important aspect of ecology, therefore, is to try to measure the pathways and efficiencies of energy transfer. How does energy enter and pass through an ecosystem? How much energy is lost when solar energy is transformed into the chemical energy of plant protoplasm through photosynthesis? How much is lost when plants are consumed by animals? How can man best utilize energy flow patterns to improve the quantity and quality of his own food supply? These questions are central to understanding the structure and function of ecologic systems.

THE SOURCE OF ENERGY

The ultimate source of energy for living organisms is the sun, of course. It supplies an incredible amount of energy to the earth's surface. At sea level, the intensity of solar radiation averages 15,000 calories (g-cal)[1] per square meter per minute (Clarke, 1954). This totals 9,000,000 calories per square meter per day assuming 10 hours of sunshine, or more than 36 billion calories per acre per day. The total amount of solar energy striking the earth's surface each day is equivalent to the energy in 684 billion tons of coal (6.84×10^{11} tons). This is sufficient energy to melt a layer of ice 35 feet thick over the surface of the earth if the energy were applied evenly over the globe (Clarke, 1954). It would also produce light energy equivalent to that supplied by over 1,000,000 watts of light bulbs for each acre of ground. The solar energy striking the surface of the United States every 20 minutes is sufficient to meet the country's entire power needs for one year, if it could be harnessed.

Most of this solar energy striking the earth's surface is reflected or absorbed, and thus scattered or transformed into heat. Of the light striking green plants, 98 percent is reflected and approximately 2 percent is absorbed (Rabinowitch, 1951). Of this 2 percent which is absorbed, only about one-half is in wavelengths utilized by chlorophyll in photosynthesis. Thus the ecologic efficiency of green plants is usually 1 percent or less. In large bodies of water, an even higher percentage of light is reflected or scattered, so that the ecologic efficiency of green plants in the ocean averages only 0.18 percent (Riley, 1944). The topic of ecologic efficiencies of plants and animals will be discussed later in the chapter.

[1] A gram-calorie (g-cal) is the same as a small calorie; that is, the amount of heat energy needed to heat 1 gram of water 1 degree Centigrade. A kilogram-calorie is a large calorie; that is, the amount of heat energy needed to heat 1 kilogram of water 1 degree Centigrade.

FOOD CHAINS AND TROPHIC STRUCTURE

The transfer of energy from plants through a series of other organisms constitutes food chains. The term *trophic level* refers to the parts of a food chain or nutritive series in which a group of organisms secures food in the same general way. Thus all animals which obtain their energy by directly eating grass, such as grasshoppers, meadow mice and cattle, would be a part of the same trophic level. The particular assemblage of trophic levels within an ecosystem is known as the trophic structure. Typically, ecosystems have 3 to 6 trophic levels through which energy and organic materials pass. In more vernacular terms, food chains usually have 3 to 6 "links" or groups of organisms which derive their nutrition in the same general way. An example of a short, practical food chain would be grass → cattle → man, which is the agricultural equivalent of grass → bison → man of Indian and frontier times.

In aquatic systems, algae, phytoplankton and other aquatic plants occupy the same trophic level as grass, and herbivorous animals including some crustacea, some insect larvae, and a few herbivorous fish occupy the same trophic level as the bison or cattle. The menhaden is such a fish, and it is sometimes described as a "pasture" fish capable of "grazing" on phytoplankton in the water.

The shorter the food chain, the greater the biomass which can be produced from a given amount of energy. The reason for this is that some energy is lost at each transfer, thus a 5 link food chain (such as algae–crustacea–insect–minnow–bass) is considerably less efficient than a 3 link food chain (such as algae–minnow–bass). In practical terms, this means that more pounds of bass could be produced in a given pond if it contained algae-eating minnows than if it contained insect-eating minnows, provided all other conditions were similar.

These considerations help us to understand why the antarctic seas, during the polar summer, are among the most productive oceans in the world. They typically have short, simple food chains; for example, from phytoplankton to baleen whales in a two-link food chain. During the polar summer they also have a 24 hour energy input from sunlight, and a sharp turnover of water strata with an upwelling of nutrients from the bottom to the surface. This favors the growth of plankton. Finally, the organisms in the cold polar water have a low respiratory rate, and therefore relatively little energy is lost through respiration. In other words, net productivity is high in relation to gross productivity.

Figure 9-1 provides a simplified diagram of an arctic terrestrial food chain, and Figure 9-2 provides a diagram of an antarctic aquatic food chain. In both cases, one can see the relative simplicity of ecologic systems in the

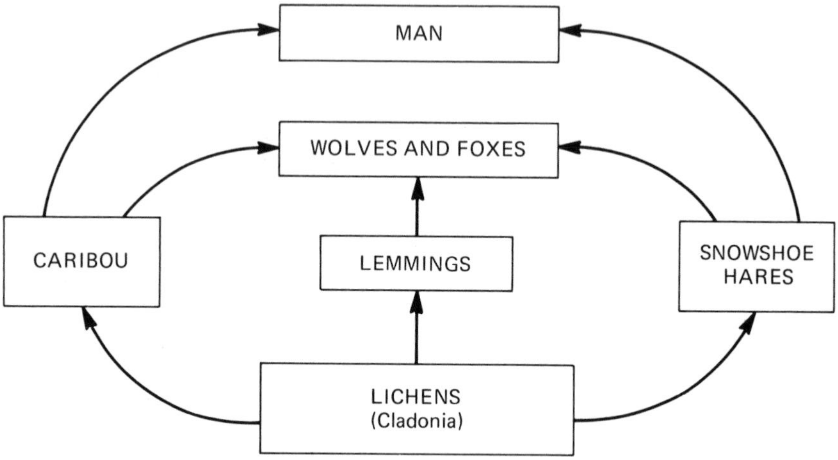

Figure 9–1 A simplified food chain for an Arctic terrestrial ecosystem.

polar regions. This is one reason why ecologists like to work in polar regions—ecologic relationships are relatively simple and amenable to study.

In the terrestrial ecosystem of the high arctic, the primary producer trophic level is represented mainly by lichens, a commensal combination of algae and fungi. Lichens form the main food for the primary herbivores, caribou, lemmings and snowshoe hares, and these in turn are the major foods for the predators—foxes, wolves and man. Obviously, this is not the complete story of life in the arctic, for other plants and animals are present as well, including some flowering plants, musk oxen in certain regions, birds, fish, polar bears, etc., but they all fit into a relatively simple trophic structure.

In the antarctic seas, there is also a relatively simple trophic structure in which all primary productivity depends upon phytoplankton. The primary herbivore is a small shrimp, *Euphausia superba,* also known as "krill." These shrimp form the primary food source for fish, birds, such as penguins, and baleen whales. Baleen whales may also directly consume and utilize some of the phytoplankton, thus they are both herbivorous and carnivorous. The major predators at the top of the trophic structure are seals, squid, sperm whales, smaller toothed whales and man.

Trophic structures tend to be simple in the polar regions, and they become more complex in progressing through the temperate regions into the equatorial tropics. Some of the reasons behind this important ecological principle will become evident in subsequent chapters.

For our consideration now, we can illustrate this principle by reference to a temperate food chain. Figure 9-3 shows the simplified diagram of some of the important food chains in a temperate stream in England. Even though it is

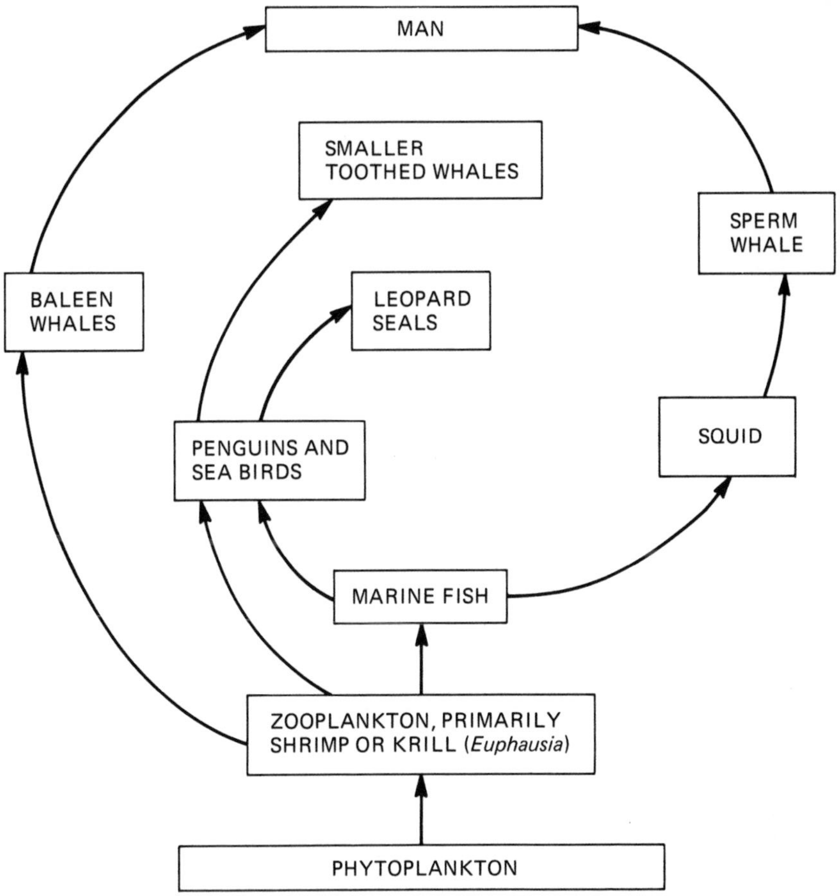

Figure 9–2 A simplified food chain for an Antarctic aquatic ecosystem.

purposefully simplified, it remains very complicated in comparison to polar food chains. In fact, it is more appropriate to call such trophic structures "food webs" rather than "food chains." The interlocking nature of these relationships is so typical of ecology, that a prominent ecologist, John Storer, chose the title, *The Web of Life* to present some of his basic concepts of ecology in book form (1953).

In most temperate food webs, the patterns of energy flow become so complicated that it is difficult or impossible to diagram all of the possible relationships. In a typical temperate forest, for example, we might have 40 or 50 species of insectivorous birds feeding on several hundred species of insects. In a tropical rain forest, even greater complexities occur, and we may have several hundred species of insectivorous birds feeding on several thousand

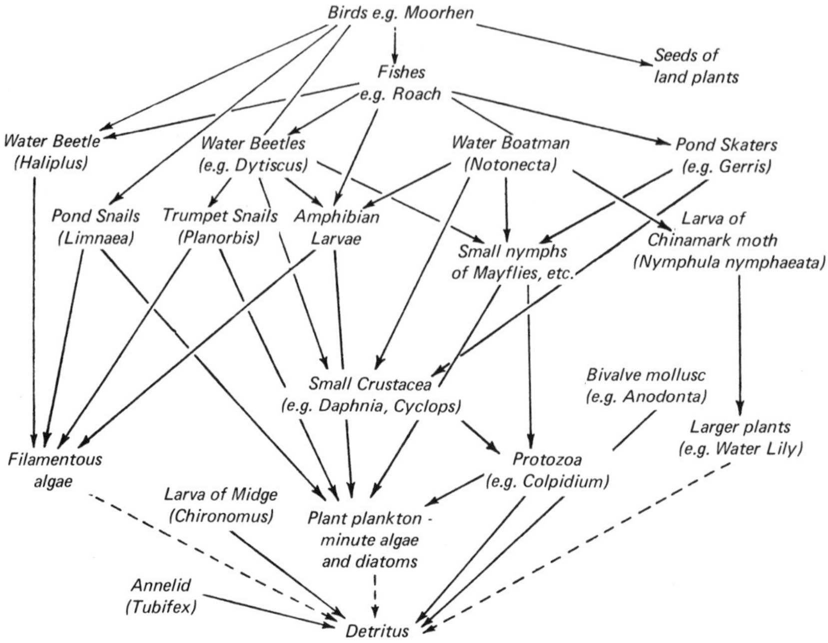

Figure 9–3 Some food chains of animals inhabiting static fresh water. (*From Dowdeswell, 1961*).

species of insects. This increasing complexity of the tropical forest will be discussed when we consider characteristics of major biomes.

TROPHIC STRUCTURE AND ECOSYSTEM STABILITY

One major consequence of the varying complexities of trophic structure is the relative stability of different ecosystems. Ecosystems with simple trophic structures are usually more vulnerable to drastic ecologic change than are ecosystems with complex trophic structures. In the arctic ecosystem, for example, if the production of lichens was impaired, the entire system would collapse, since all life depends upon lichens. Similarly, in the antarctic seas, if krill were eliminated by some ecologic accident, there would be a catastrophic decline of virtually all marine mammals, birds, and fish which depend upon krill for food.

In temperate or tropical systems, on the other hand, where alternate food supplies are available, the temporary loss of any one species does not necessarily endanger the entire system. There are exceptions to this, of course, for if a dominant species of grass in a grassland community were eliminated then all herbivore species depending upon this would be adversely affected, but natural grasslands usually consist of dozens of species of grasses and herbs. The original American prairies, in fact, were richly structured with hundreds of species of plants.

Thus there is ecologic strength and security in complex trophic structures that simplified systems do not have. It is no surprise, therefore, to note that complex ecosystems are usually more stable than simplified systems. Instability increases as a direct function of trophic simplicity.

INFLUENCE OF MAN ON ECOSYSTEM COMPLEXITY AND STABILITY

One of the major ecologic influences of man has been to simplify the world's ecosystems. Agriculture, which is the applied management of food chains, fosters simplified systems and reduces food chains to their simplest terms. Thus, man plows the prairie, eliminating a hundred species of native prairie herbs and grasses, which he replaces with pure stands of wheat, corn or alfalfa. This increases efficiency and productivity, but it also increases ecologic vulnerability and instability. If we have a pure stand of wheat, then it amplifies the possibility of ecologic catastrophe by having some pathogen (such as wheat rust) or herbivore (such as grasshoppers or locusts) sweep in suddenly to decimate the entire system.

In animal populations man also tends to oversimplify. He has taken the complex ungulate faunas of Africa and India, each consisting of many species of wild antelope, wild buffalo and other ungulates, and he has tended to replace them with a single species of domestic cattle. This has often had a drastically damaging effect on the native grasslands and forests, and it has tended to produce exaggerated epidemics of diseases such as rindepest and foot and mouth disease. Man clearly pays a price for oversimplifying ecosystems and thereby reducing their natural stability.

Laymen often ask what is the good of this species or that—why do we need so many different kinds of plants and animals? So what if a few of them do become extinct? Questions like this can seldom be answered in very specific terms, but the answer lies in the ecologic principles we have been discussing. There may be no immediate ecologic catastrophe if a certain species does disappear, but the system is thereby somewhat impaired from its naturally evolved state, and is more vulnerable to ecologic instability. We will have many examples throughout this book of how a minor change in one part of the system had major consequences for other parts of the system.

In these terms, the eminent ecologist Charles Elton (1958) built a strong case for the "conservation of diversity." By conserving diverse habitats we help to insure a more natural, more complex and more stable flora and fauna. In the English hedgerows, for example, Elton noted a variety of beneficial species that could not survive in areas with barbed wire fences. This included many insectivorous birds and mammals. In these areas more natural means of insect control were operative, and less dependence on massive insecticide application was necessary.

ECOLOGIC PYRAMIDS

Elton was also one of the first ecologists to emphasize the value of analyzing trophic structures in terms of ecologic pyramids. Ecologic pyramids are diagrams of data representing the standing crops at each trophic level. They may be expressed in terms of numbers of organisms, total biomass, or total energy flow at each trophic level.

Figure 9-4 represents a numbers pyramid for one acre of grassland. It is obviously not drawn to scale, but it helps portray the relative numbers of organisms at each of 4 trophic levels—producers, herbivores and two levels of predators.

Figure 9-5 illustrates two biomass pyramids for Weber Lake, Wisconsin,

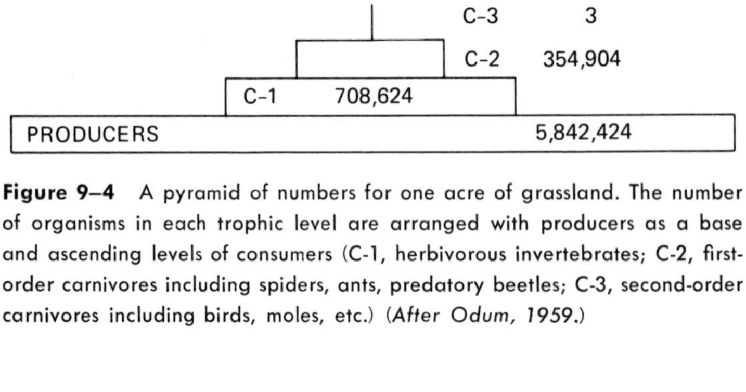

Figure 9–4 A pyramid of numbers for one acre of grassland. The number of organisms in each trophic level are arranged with producers as a base and ascending levels of consumers (C-1, herbivorous invertebrates; C-2, first-order carnivores including spiders, ants, predatory beetles; C-3, second-order carnivores including birds, moles, etc.) (*After Odum, 1959.*)

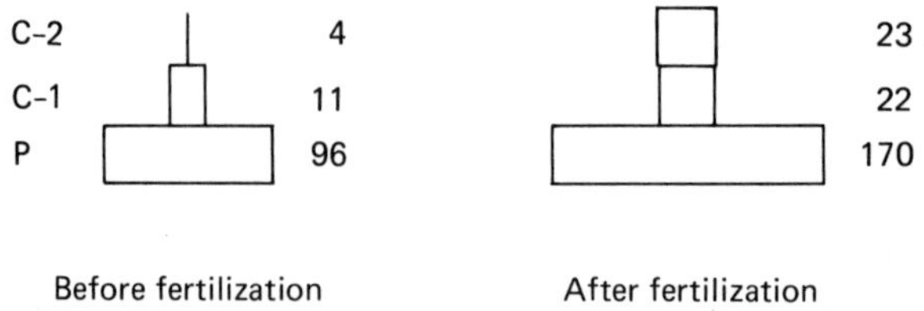

Figure 9–5 Biomass pyramids for an aquatic ecosystem in Weber Lake, Wisconsin. Figures represent grams of dry biomass per square meter. (*Data from Juday, 1942, after Odum, 1959.*)

before and after artificial fertilization of the lake. The numbers represent grams of living organisms per square meter. The data show that fertilization of the lake resulted in a doubling of producers and herbivores, and five fold increase in C-2 consumers.

Figure 9-6 shows all three types of pyramids for a hypothetical alfalfa → calf → boy food chain based on 10 acres over the course of a year. These help us to visualize both the numerical and energetic relationships within the system. If data like these were available for agricultural crops it would assist greatly in agricultural management. We could compare the efficiencies of various crops; we could precisely analyze the loss of energy to undesirable herbivores such as insects; we could evaluate the effectiveness of fertilizers, cultivation and watershed management on energy flow and productivity.

Since modern agriculture actually makes many of these evaluations, at least in terms of net productivity and profit yield, there is an even greater need for such data on natural ecosystems. With ecologic pyramids like these on estuaries, wetlands, forests and natural grasslands, we would be in a much better

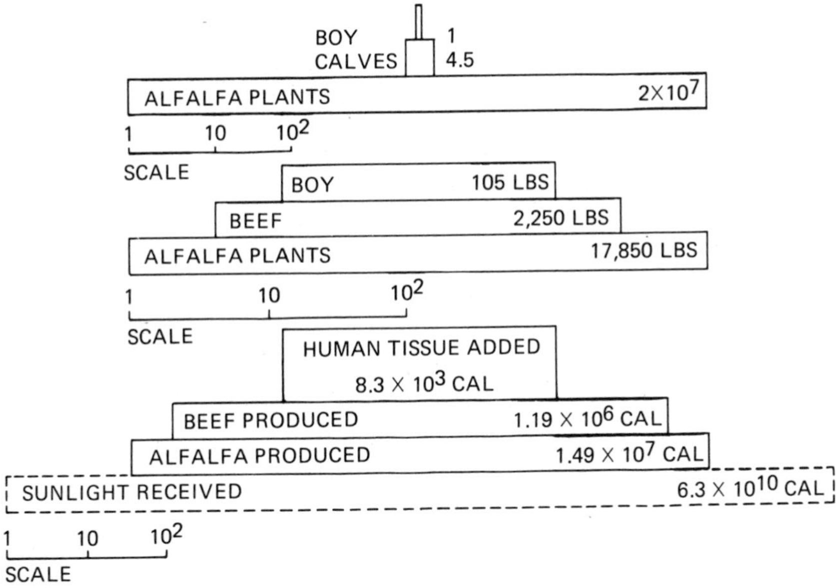

Figure 9–6 The three types of ecological pyramids illustrated for a hypothetical alfalfa-calf-boy food chain computed on the basis of 10 acres and one year and plotted on a log scale. Compiled from data obtained as follows: Sunlight: Haurwitz and Austin (1944), "Climatology." Alfalfa: "USDA Statistics, 1951"; "USDA Yearbook 1948"; Morrison (1947), "Feeds and Feeding." Beef calf: Brody (1945), "Bioenergetics and Growth." Growing boy: Fulton (1950), "Physiology"; Dearborn and Rothney (1941), "Predicting the Child's Development." (*After Odum, 1971.*)

position to estimate their value in economic terms. When the developer wants to drain and fill an estuary, such as San Francisco Bay, the ecologist is hard pressed to express the value of the estuary except in terms of esthetics and recreational value. It would be helpful to know the energy budget and total ecologic function of the estuary. What is its potential production in terms of shellfish? How vital is it as a spawning and rearing ground for commercial and game fish? What role does it play in ecosystem balance in regard to adjacent land forms? Does it provide essential habitat for the natural predators of mosquito larvae? Do its patterns of energy utilization offer productive opportunities for sea farming? Numerous questions like these should be asked before any natural ecosystem falls heir to modern development. Ecologic pyramids offer one important way of organizing data to answer some of these questions.

ECOLOGIC EFFICIENCIES

It was previously mentioned that the ecologic efficiency of green plants is usually 1 percent or less; that is, green plants utilize for photosynthesis only about 1 percent of the total solar energy striking their surface. Approximately 98 percent of this surface energy is reflected or scattered, and only about one-half of the remainder is in wavelengths suitable for photosynthesis.

With subsequent energy transformations through each trophic level, usually 80 to 95 percent is again lost through dispersion, heat loss, motion, etc. Another way of saying this is that only 5 to 20 percent of the energy in green plants is incorporated into the herbivores, and only 5 to 20 percent of that in herbivores is incorporated into primary carnivores. Table 9-1 shows representative data on energy transfer at various trophic levels in aquatic ecosystems. This table shows that all ecologic efficiencies from herbivores to large

TABLE 9-1 Efficiency of Energy Transfer at Various Trophic Levels in Aquatic Ecosystems[a]

	PERCENT ENERGY REACHING SURFACE OR TROPHIC LEVEL WHICH IS CONVERTED INTO ORGANIC MATERIAL		
Trophic Level	*Cedar Bog Lake, Minnesota*	*Lake Mendota, Wisconsin*	*Silver Springs, Florida*
Plants	0.10	0.40	1.20
Herbivores	13.3	8.7	16.0
Small carnivores	22.3	5.5	11.0
Large carnivores	absent	13.0	6.0

[a] (from Odum, 1959.)

carnivores in these aquatic systems fall within the range of 5.5 to 22.3 percent, meaning that each trophic level incorporates only this much of the energy it consumes into its own organic structure.

Engineers may point out that these efficiencies are relatively poor, and considerably less than that achieved by many machines. Machines vary in their efficiency (ratio between energy produced by the machine to energy supplied to it) from 5 percent to nearly 100 percent. No machine operates with 100 percent efficiency because of friction, and hence a perpetual motion machine has never been achieved. Nonetheless an automobile engine achieves an efficiency of 25 percent, and some machines that transmit mechanical energy only may have an efficiency of over 90 percent. By these comparisons, the 1 percent efficiencies of plants and the 5 to 20 percent efficiencies of animals seem quite low, but the comparison is not entirely valid. Living organisms must also achieve growth, reproduction, dispersion and self-maintenance—processes that machines do not undergo—and these items all enter into the energy budget.

There was considerable interest several years ago in algal culture as a way of achieving maximum efficiency in primary production and thereby helping to solve the world's food problem. In artificial algal cultures, mainly of the single celled algae *Chlorella,* with the proper combination of temperature, nutrients, harvest rate, and light of exactly the right wave lengths, efficiencies of 20 percent were obtained, and even 50 percent on certain limited occasions. These methods have not proven feasible for mass production, however. In daylight, artificially managed algal cultures usually achieve no more than 2 to 8 percent efficiency. These figures can probably be considered practical norms for photosynthesis at the present time.

In summary, we have noted that the analysis of energy flow in ecologic systems is a topic of great practical and theoretical importance. It directs attention to both the efficiency and stability of natural systems. Man has had the world-wide effect of simplifying ecosystems and trophic structures. In so doing, he has often increased efficiency but always at the risk of instability. Ecosystems with simple trophic structures are more productive for man, but they are also more vulnerable to sudden change and ecologic catastrophes.

10

Limiting Factors and Tolerance Levels

A central problem of ecology is to understand the abundance and distribution of plants and animals. Why are tent caterpillars abundant one year and not the next? Why are ring-necked pheasants common in Ohio and absent from Alabama? Why are tigers in India and not in Africa? Why has the lake trout disappeared from Lake Erie? Why is ragweed common in New England, but not in England? Why is yellow fever virus present in Africa and Latin America, but not in India? Why is water-milfoil a serious weed pest in the United States, but not in its native Eurasian habitats? Why do the river basins of the Nile, Congo and Niger in Africa have schistosome parasites, while those of the Ganges and Brahmaputra in India do not?

These questions represent many different types of problems, but it is questions like these, and countless others, concerning the abundance and distribution of plants and animals, which represent some of the core questions of ecology. From a practical standpoint, they represent some of its most vexing problems.

If we start with the obvious, it is not too difficult to explain, at least superficially, why polar bears are not found in the Caribbean, and why howler monkeys are not found in Canada. We say that polar bears are adapted to cold climates, and by virtue of their insulation and metabolism, they cannot tolerate heat; they are limited to cold and icy waters. Conversely, howler monkeys are animals of the hot tropical rain forest. They cannot tolerate temperatures below 65° F, and they are thus limited to perennially warm tropical regions.

SHELFORD'S AND LIEBIG'S LAWS

One of the best systematic expressions of these principles was provided about 50 years ago by the American ecologist, Victor Shelford, and we recognize his statement today as Shelford's Law of Tolerance. This states that the *abundance or distribution of an organism can be controlled by factors exceeding the maximum or minimum levels of tolerance for that organism.* This focuses upon the ecologic requirements of plants and animals in terms of climatic, topographic and biologic factors. Thus an organism might be limited by numerous factors which exist at levels above or below that which the organism requires. There might be too much light, or not enough; too much moisture, or not enough; too many dissolved minerals in the soil, or not enough; too many predators, or too little protective cover; too little food, or improper balances in the types of food; too many pathogenic diseases, etc. In specific instances, one or more factors might be critical, or various factors might work in combination. For example, there might be an adequate amount of food in the summer time to support a feral house mouse (*Mus musculus*) population in the forests of Maryland, and they can certainly withstand temperatures colder than those of the Maryland winter if they have adequate food. But the combination of winter temperatures and reduced winter food act together to eliminate forest-dwelling house mice in Maryland. On the other hand, the native white-footed wood mouse, *Peromyscus leucopus,* fares well in the woods throughout the year, but the house mouse is usually limited to barns and human dwellings in the winter in Maryland.

As a point of historical interest, approximately 70 years before Shelford, the German botanist and physiologist, Justin Liebig, expressed a similar but more limited concept that we now recognize as Liebig's Law of the Minimum. Liebig noted that *the essential material available in amounts most closely approaching the critical minimum would tend to be limiting.* Since Liebig was a botanist, he thought primarily in terms of light, temperature, nutrients and essential elements, and he sought to explain such common observations as the lack of vegetation above certain altitudes in the alps, or the lack of some plants in shaded areas. Thus, he found explanations of plant distribution in terms of not enough light, not enough warmth, or not enough nutrients, and he did not develop at length the corollary that there might be too much of these factors as well as too little.

The principles elaborated by Liebig and Shelford are valuable guidelines in analyzing factors which limit plant and animal abundance. The practical search for limiting factors often becomes a fascinating and difficult mystery and one of the most important aspects of applied ecology. It may take two major directions. One in which the management objective is to increase productivity or extend the range of some desirable plant or animal, and a second one in which the objective is to find and implement some limiting factor which can reduce the abundance of a pest. Both types of activity are common aspects of applied systems research in agriculture, wildlife management and public health. Thus, man may wish to encourage the numbers and spread of wild turkeys (*Meleagris gallopavo*) throughout eastern United States, since they are desirable native game birds, but he may also wish to reduce the numbers and extent of starlings (*Sturnus vulgaris*), since they are agricultural pests. In both cases we need to know more about natural limiting factors, so that they may be alleviated in the case of the turkey, and enhanced in the case of the starling.

As another example, man may wish to increase the number of impala on the East African savannahs, since they are attractive and desirable animals, but he wishes to reduce the number of ticks and tsetse flies which transmit disease between wild animals and domestic ungulates.

Animals differ greatly, of course, in their tolerance ranges to different environmental factors. Some, such as the starling, are exceedingly adaptable and are able to exist in a wide range of habitats and climates. Since their introduction in the United States in 1880, starlings have successfully spread from New York to California and from Alaska to Mexico. They are found in cities, farmlands, conifer forests of the far north and deserts of the southwest.

A native species with wide tolerance ranges to many ecologic factors is the white-tailed deer (*Odocoileus virgianus*). It occurs in eastern deciduous forests, western mountains as far as British Columbia, and south to the tropical rain forests of Panama. The cougar or mountain lion (*Felis concolor*) is another very adaptable species, ranging from the east coast to west and from Alaska to Panama. The same species inhabits cold mountain forests in Canada and hot rain forests in Central America. The deer mouse (*Peromyscus maniculatus*) occurs from Nova Scotia to Alaska and south to southern Mexico (Figure 10-1).

Other species are more restricted in range and ecologic tolerance. The Oldfield mouse (*Peromyscus polionotus*) occurs only in four southeastern states (Figure 10-1). The North American mountain goat (*Oreamnos americanus*)

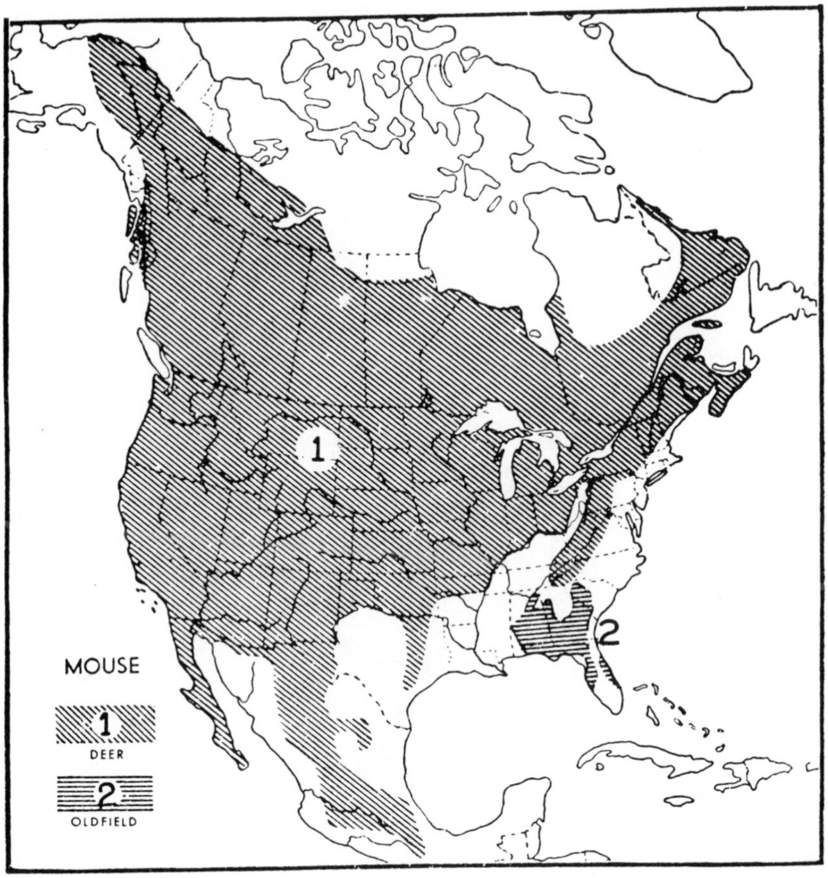

Figure 10–1 Distribution of the deer mouse (*Peromyscus maniculatus*) and Oldfield mouse (*Peromyscus polionotus*) in North America. (*From Burt and Grossenheider,* Field Guide to the Mammals, *Houghton, Mifflin, 1952. By permission.*)

occurs only from Alaska to Washington, Idaho and Montana. The Columbian ground squirrel (*Citellus columbianus*) is limited to a relatively small section of the northwest from Idaho and Montana to southern Alberta and British Columbia (Figure 10-2). The Mt. Lyell Shrew (*Sorex lyelli*) is limited to a few small areas of the high Sierras above 6,900 feet in California. Kirkland's warbler (*Dendroica kirtlandii*), a small insectivorous song bird, nests only in a few restricted forests of young Jack pine in Michigan, and the golden cheeked warbler (*Dendroica chrysoparia*) nests only in a very small forested area in the Edward's plateau of Texas—an area now planned for commercial development, which may eliminate this species. The koala bear (*Phascolarctos cinereus*), or familiar "teddy bear" is very sparsely distributed in eastern Australia, and it is limited in food to about 12 species of eucalyptus trees.

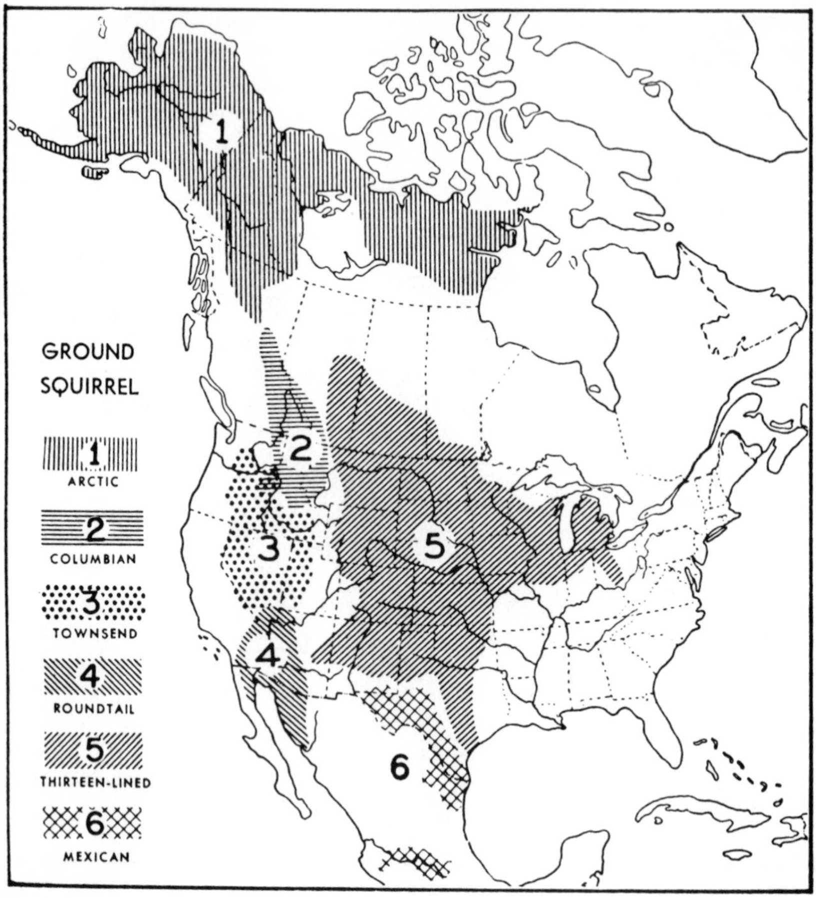

Figure 10–2 Distribution of six species of ground squirrels (*Citellus* sp.) in North America. (*From Burt and Grossenheider,* Field Guide to the Mammals, *Houghton, Mifflin, 1952. By permission.*)

Among primates, the rhesus monkey (*Macaca mulatta*) is a species with wide tolerance limits in climate and habitat. It occurs in crowded cities of India, hot lowland mangrove forests of the Gangetic delta, arid deserts of Rajasthan and Punjab, and the cool pine forests of the Himalayan foothills as high as 7,000 feet. On the other hand, the lion-tailed macaque (*Macaca silenus*) in India is very limited in habitat, restricted to the monsoon forests of the western ghat mountains south of Bombay, and occurring now in just a few isolated pockets. Like the golden-cheeked warbler, the lion-tailed macaque may be extinct in 10 or 15 years.

A species with wide ecologic tolerance levels is said to be euryoecious, and one with narrow ecologic limits is said to be stenoecious. The prefixes, eury- and steno- may be applied to more specific stem words to refer to specific traits. Thus, a plant or animal with wide tolerances to temperature would be eurythermal; one with narrow tolerances to temperature would be stenothermal. An organism with wide tolerance to salt concentration, such as the gangetic dolphin (*Platanista gangetica*), which passes readily from salt water to entirely fresh water would be described as euryhaline, whereas one limited to salt water, such as most starfish would be described as stenohaline. A species with a wide food tolerance such as the Virginia opossum (*Didelphis virginiana*) would be described as euryphagic, whereas the koala bear would be stenophagic. Many other compound terms of this type may be assembled to refer to the tolerance ranges of animals to a great variety of physical or biotic factors.

The systematic study of tolerance levels and limiting factors forms a logical interdisciplinary bridge between physiology, ecology and biogeography. We often do not know the tolerance levels of many species to the wide range of environmental influences impinging upon them, and we are just beginning to analyze some of the key factors involved in the adaptability or lack of adaptability of different species.

CLIMAGRAPHS

A relatively simple but very effective way of comparing habitat suitability in regard to tolerance levels of organisms is by means of climagraphs. *Climagraphs are graphic representations of important ecologic features in a species' natural range*. These may then be compared with other areas in which the species does not occur, or in which the species has considerable difficulty in developing successfully. Climagraph analysis has been valuable, for example, in interpreting the relative success of the Hungarian partridge (*Perdix perdix*) in the United States. The Hungarian partridge is a native of eastern Europe, but it was brought to the United States in 1908 in attempts to establish it as a game bird. It has become successfully established in many northern midwestern States, but has failed in more southern states. Figure 10-3 is a climagraph of temperature and rainfall conditions in Montana where the Hungarian partridge has been successful, and Missouri where it has failed. When temperature and rainfall of these states are compared with the native European habitat, it is immediately seen that the summers in Missouri are hotter and wetter than in its European environment. One would immediately suspect that the Hungarian partridge cannot tolerate heat and moisture during its period of nesting, incubation and early chick survival. One's attention would be directed, therefore, to more specific studies of the bird's nesting requirements and the patterns of egg survival, hatching success and chick survival. On the other hand, in Montana, it is seen that temperature and rainfall of the entire nesting and fledgling periods are within the European ranges. In the winter, Montana becomes colder than the European habitats but apparently the birds

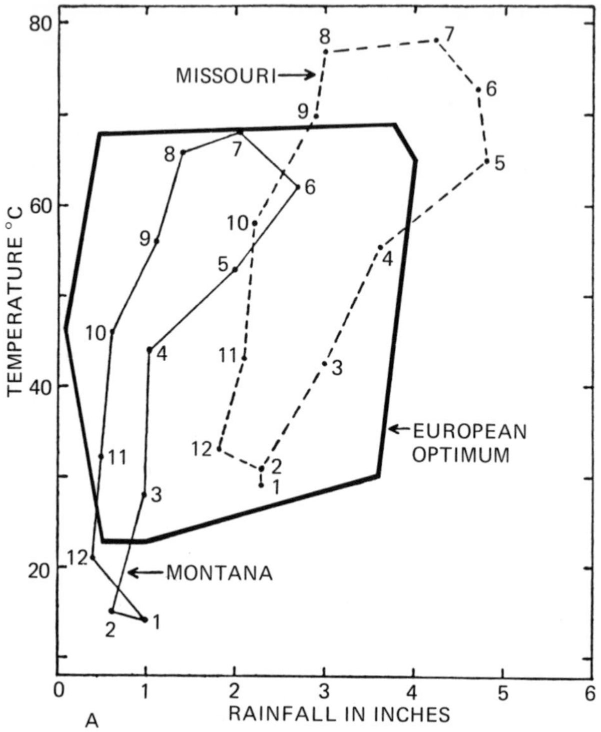

Figure 10–3 Temperature-moisture climagraph for Havre, Montana, where the Hungarian partridge (*Perdix perdix*) has been successfully introduced, and Columbia, Missouri, where it has failed, compared with average conditions in its European breeding range. (*After Odum, 1971.*)

can tolerate this. If population failures did occur in Montana, one's attention would initially be directed to winter food supply, cover availability and patterns of winter mortality.

Climagraphs can be constructed for a variety of factors relative to specific problems, and they can often be powerful tools in the analysis of specific problems.

TOLERANCE LEVELS AND POLLUTION

Tolerance levels become key questions in our evaluation of pollution. What are the tolerance levels of different aquatic organisms to pollutants in the water? How severe a shortage of dissolved oxygen can different species of fish tolerate? How much acidity or alkalinity can zooplankton tolerate? How susceptible are clams and oysters to copper, chromium, mercury, lead and other trace metals in water? In human terms, how many contaminants and pollutants in our air and water can we tolerate before harmful effects begin to appear?

The investigation of these major questions is just beginning. We are surrounded by controversial points of view on the relative dangers of pesticides, radioisotopes, excessive nutrients, food additives and trace toxicants because in so many cases we just do not know what our tolerance levels to these substances are, not to mention the tolerance levels of all the diverse organisms within an ecosystem.

Thus, the study of tolerance levels becomes a matter of survival as well as basic ecological science. At no time in the history of the earth has man elaborated and expelled such a wide variety and volume of new chemicals and new influences on himself and his environment as at the present time. It is no wonder that strange chemicals such as thalidomide, cyclohexamate, and monosodium glutamate spring into widespread use before potential dangers are realized. When potential dangers are recognized, there is often inadequate knowledge to evaluate them accurately, and the public can react with either excessive alarm or total indifference.

We know so little about the tolerance levels of ecosystems to various types of insults and stimuli, that we may not discover the damage until it is too late. A recent television program asked the question, "Who Killed Lake Erie?" It is a misleading question and one answerable only in terms of a long description. Lake Erie is not completely dead in the biological sense, but it has been seriously polluted and damaged. Much of its former value and beauty has been destroyed. Its beaches are dirty; its shoreline waters are often foul and odorous. It still produces large populations of fish, but its valuable gamefish such as lake trout (*Salvelinus namaycush*), whitefish (*Coregonus clupeiformis*), and northern pike (*Esox lucious*) have been displaced by less desirable rough fish, including carp (*Cyprinus carpio*), sheepshead (*Aplodinotus grunniens*), and gizzard shad (*Dorosoma cepedianum*). Certainly no single industry, no single city, or no single populace changed Lake Erie. But the collective pollution of hundreds of industries, towns and cities, gradually overloaded the system until its ability for self-renewal was destroyed. Its tolerance to pollutional insult was exceeded, and the damage became virtually irreparable.

It is important, therefore, to view tolerance levels as properties of ecosystems as well as individuals. The ecosystem is more complex, possibly more resilient, but its alteration and death can be just as real as senility and death in an individual.

INDICATOR ORGANISMS

The term "indicator organisms" refers to plants, animals or microbes which are indicators of ecologic conditions. That is, by their presence or absence, or specific appearance they give clues to various ecologic conditions. In western United States, for example, plants of the genus *Astragalus*, grow in association with selenium, a soil mineral commonly found in or near uranium depositions. Thus these plants are practical indicators to the possible presence of uranium ore. In Maryland, the growth of Eastern colombine (*Aquilegia canadensis*) is often indicative of high limestone content (calcium carbonate) in the substrate, and redbud (*Cercis canadensis*) is often associated with dolomite (calcium and magnesium carbonate).

In public health practice, the presence of coliform bacteria, especially *Escherichia coli*, a common commensal inhabitant of the intestinal tract of man and animals, is taken as an indicator of fecal pollution. If coliform bacteria counts exceed certain standards in a lake or stream the area is closed to public swimming. Certain algae are also indicators of high nutrient content in water, and blooms of the green algae (*Chlorella*) often occur in waters enriched by domestic pollution.

Organisms with relatively narrow tolerance levels to specific environmental factors make the best indicator organisms. Their reliability as indicators is more or less in inverse proportion to the extent of their tolerance. Thus, the narrower the tolerance, the greater the accuracy in indicating specific ecologic conditions. For example, if an aquatic organism grows only in a pH from 8.0 to 9.0, its presence would be an accurate field indicator of alkalinity within this range.

Another type of ecologic indicator is represented by a specific symptom on a plant or animal associated with specific conditions. Some plants develop lesions or spots in response to specific air pollutants. For example, terminal and marginal scorching of tulip leaves is indicative of gaseous fluorides; intercostal markings on violet leaves indicate high sulfur dioxide concentrations; banding on snowstorm petunia leaves indicate aldehydes in the air; and small white flecks on tobacco plants indicate high ozone levels in the air (Altshuller, 1966; Waggoner, 1971).

Ecologic indicators may also provide evidence of former land use practices. Some plants such as bull thistle (*Cirsium lanceolatum*) are graze resistant—that is, they are not eaten by cattle, and often flourish in pastures where competitors are removed by grazing cattle. Large numbers of bull thistle scattered across an old field are usually firm evidence of former overgrazing on this field. Red cedar (*Juniperus virginiana*) or arbor vitae (*Thuja occidentalis*) are similar indicators of overgrazing in pastures that may have been abandoned years ago. In the East African savannah, an excessive number of thorn bushes indicate overgrazing coupled with a lack of periodic burning which is necessary to maintain grass. A skilled plant ecologist can frequently "read the landscape" with remarkable accuracy, describing past land use events by a critical analysis of the present-day plant community.

BIOGEOGRAPHY

Some of the questions raised in the opening paragraph of this chapter, such as why tigers are in India and not in Africa, are related to the geologic history of continents and the facts of evolution. They obviously go far beyond the question of limiting factors and tolerance levels. Each major continental area has developed a characteristic assemblage of plants and animals, and the relative uniqueness of this is a function of the timing and extent of geographic isolation. For example, Australia became a distinctly separate land mass about 50 million years ago, just before the main origin of the placental mammals. It developed its own unique group of marsupial mammals which often paralleled placental mammals in their overt appearances but had a basically different pattern of reproductive physiology.

Africa and South America became separated at an earlier date, probably 150 million years ago. The theory of continental drift which envisions many of today's continents as one large land area millions of years ago, has gained strength and support in light of recent research. According to Sir Edward Bullard (1969), "It is virtually certain that the Atlantic ocean did not exist 150 million years ago." At that time, Europe and North America and Africa and South America were thought to have been joined as one major land mass, and they have slowly drifted apart in intervening years. All of these continents have had a fairly complete representation of placental mammals of similar types, but of quite distinct species and often distinct genera. The study of this, as well as the fascinating ecology and faunal history of islands, is really the subject of biogeography and evolution, and cannot be treated adequately in this book. Those interested in further reading will profit from books such as *Biogeography* by Dansereau (1957), *Ecological Animal Geography* by Hesse, Allee and Schmidt (1951), and *Zoogeography* by Darlington (1964).

ALIEN ORGANISMS

Whereas in the course of geologic history and evolution each continent evolved its own characteristic flora and fauna, man over the last 500 years has done much to intermix these biotic communities. He has purposefully or inadvertently transported plants and animals throughout the world. The early European colonizers of New Zealand found no native mammals on this oceanic island, and in the course of two centuries from 1700 to 1900 they imported more than 200 species of vertebrates. Many of these imported animals found virgin niches, and without predators, disease, or other natural checks and balances, they exhibited explosive population growth and became serious problems. Rabbits and deer from Europe proliferated beyond all normal limits and began to devastate pastures, orchards and forests. California quail (*Lophortyx californicus*) exploded in population numbers and became agricultural pests. Durwood Allen (1954) once described New Zealand as a land where, "A herbivorous opossum from Australia is eating the forest and orchard from the top down, while deer, goats and rabbits from Europe and North America are denuding the land from the ground up."

In other countries, similar problems have prevailed to a lesser extent. In Australia, the European rabbit (*Oryctolagus cuniculus*) spread throughout the country, becoming a serious pest on pasture and rangeland. At one point, the Australian government went to unprecedented expense to build a rabbit-proof fence across hundreds of miles of rangeland to contain the rabbit population, only to discover later that the prolific rabbits were already on the other side. In the United States some of our conspicuous vertebrate pests are imports from other continents. The domestic rat (*Rattus norvegicus* and *R. rattus*) and house mouse (*Mus musculus*) were early immigrants, probably arriving on the ships of colonists. The English sparrow (*Passer domesticus*) and starling, came much later as purposeful introductions. The Asian carp (*Cyprinus carpio*), a deliberate introduction in 1870, has spread throughout most of the United States and now dominates many lakes and rivers to the detriment of native fish.

Many agricultural pests throughout the world are also immigrant animals, showing the familiar pattern of increasing markedly in areas where they are introduced by purposeful or accidental means. The Japanese beetle (*Popilla japonica*), European corn borer (*Pyrausta nubialis*), Mexican bean beetle (*Epilachna varivestis*), giant African snail (*Achatina fulica*), Mediterranean fruit fly (*Ceratitis capitala*), Oriental fruit moth (*Laspeyresia molesta*), cattle horn fly (*Haematobia irritans*) and the imported cabbage worm (*Pieris rapae*) are all alien organisms which have spread accidentally in luggage or cargo from continent to continent, and all have become economic problems in their newly invaded countries. With the accelerating spread and frequency of travel, there have been increasing accidental transportation of plants, animals and microbes between continents. It is now a common occurrence for a fly or mosquito to rest upon a piece of luggage or even a person's clothing and make a non-stop flight between continents in a matter of hours. Not long ago I watched a small beetle on a canvas carry-on bag enter a jet plane in Calcutta, India, and exit from the same plane in San Francisco 19 hours later. At that point, I swatted the beetle, but I had convinced myself that this type of thing could easily occur. This entire problem has been discussed at some length by Charles Elton in the brilliant book, *The Ecology of Invasions by Plants and Animals* (1958), and more recently in a popular book by George Laycock, *The Alien Animals* (1967). These books document, with many examples, the common phenomenon of irruptive population growth in alien organisms in areas where they are freed from their normal limiting factors.

It would be misleading to imply that all plant and animal introductions are invariably harmful. There are certainly some desirable and beneficial ones as well. The ring-necked pheasant (*Phasianus colchicus*), a prized game bird in the United States, was deliberately introduced from Asia in 1880, and the Hungarian partridge, also a desirable game bird, was introduced from eastern Europe in 1908. The striped bass (*Roccus saxatilis*) of the Atlantic coast was introduced to the Pacific coast, and is now an economically important species in the West. A ladybird beetle (*Vedalia cardinalis*) from Australia was purposefully introduced to California to control the Citrus scale insect (*Icerya purchasi*), and it did an admirable job without becoming a pest in itself. A predaceous bug (*Cyrtorhinus mundulus*) from Australia was successfully introduced into Hawaii to control the sugar cane leafhopper which was threatening to destroy the sugar industry. Throughout the world many predaceous and parasitic insects have been used in the biological control of other insect pests, and this type of control offers great promise for future pest control. Thus, there is a positive side to the ledger, but by and large, the history of alien animals has not been favorable.

The rapid spread of living organisms from continent to continent has become a major concern of public health as well as agriculture. The importation of plants and animals must be watched with particular care in order to reduce the possibilities of introducing pathogenic organisms. Containerized air cargo becomes a likely infestation source for insects and small mammals which could carry disease organisms, and special fumigation and decontamination procedures are often necessary. Air passengers themselves may become inadvertent carriers of pathogenic organisms, and there exists a renewed danger of diseases such as smallpox, malaria, dengue fever, plague, cholera and typhus traveling from Asia to Europe or the United States within a matter of 12 hours in the form of an inapparent infection in man or animals. As Slobodkin has pointed out (1961): "Not only is man the most shocking innovation since the first appearance of the terrestrial vertebrates, but his activities are proceeding at an accelerating rate. The importance of man continues to increase and the possibility of the biologic world ever being as stable as it was in prehuman times becomes more and more remote."

These, of course, are the realities of modern life, and we have been able, on the whole, to cope with these problems satisfactorily with constant surveillance, skilled technology, and a rather massive cost, but the cost of alien organisms is high—much higher than we usually realize if we could compile the accumulated price of agricultural crop protection, pesticide expenditures, quarantine measures, and public health machinery necessary to prevent global epidemics.

SUMMARY

We have emphasized that the study of the abundance and distribution of plant and animal life is central to the concerns and interests of ecology. In many cases, patterns of abundance and distribution are understandable in terms of tolerance levels of each species to various ecologic factors such as temperature, light, moisture, nutrients, predators, disease, etc. Some species have evolved with wide tolerance levels to many factors, and may therefore have a wide distribution both geographically and ecologically. The starling, house sparrow, Norway rat, and house mouse are examples of such animals. Other animals may have very narrow tolerance levels to one or more factors, and be quite limited in distribution. The golden-cheeked warbler, koala bear and lion-tailed macaque are all examples of such species.

Over the course of geologic time, each major land mass has evolved characteristic assemblages of plants and animals, which had developed patterns of balance and homeostasis. Man has done much in recent centuries to intermix species throughout the world, by both purposeful and accidental introductions. These exchanges have often had drastic and far-reaching consequences. Exotic plants and animals from one continent have often found virgin niches in other continents, where, freed from their naturally evolved checks and balances, they have shown explosive population behavior and have created serious dislocations in nature with resultant pest problems in agriculture, forestry, range management and public health.

The study of tolerance levels in plants and animals, including man, becomes increasingly important in assessing the effects of pollution and environmental change. Ecosystems as well as individuals and populations have tolerance limits to chemical and physical alterations. With increasing dissemination of new chemicals and pollutants, and increasing rates of change in our physical, biotic, and social environments, the accurate identification of tolerance levels becomes a matter of survival. Man seems disposed to push his own tolerance limits to the extreme, and this very philosophy implies that we can safely recognize the precise boundaries of these tolerance levels. If we can accept the lessons of history, there seems to be abundant reason for a more humble attitude.

PART 4

Population Ecology

Photo by Stephen Frantz

II

Structure and Dynamics of Single Species Populations

POPULATIONS are fundamental units in ecology, as important to the ecologist as are tissues and organs to the anatomist and physiologist. They may be thought of as major components of communities and ecosystems, just as tissues and organs are major components of individual organisms. Whereas tissues and organs are composed of cells in functional groups, populations are composed of individuals in functional groups.

A population is a group of living individuals set in a frame that is limited and defined in respect to both time and space (Pearl, 1937; Sladen and Bang, 1969). Another important qualification is that a population is an interacting or potentially interacting group of individuals, usually of the same species. We speak of the population of ants in an ant hill, the population of salmon in a river, the population of starlings in a city, the population of deer in a state, and so forth. Biologically, a population should have some natural boundaries, as in the ants in an ant hill, but the term is often used with artificial boundaries in mind, as in referring to the population of deer in a state. A state might have several biological populations of deer, but for economic or political considerations it is sometimes convenient to group them together. The human population of a nation is a convenient figure for political and economic purposes, but biologically most nations consist of many different human populations representing various ethnic, religious or economic groups. These populations within one nation might have quite different characteristics.

In the initial study of populations it is helpful to recognize certain attributes of populations which distinguish them from individual organisms. Like individuals, they have a structural organization, a functional unity, and a pattern of growth and development. Unlike individuals, they have group attributes and statistical properties that no single individual possesses. Thus, the structure of a population is definable in terms of numbers, density, spatial distribution, age groups, sex ratios and breeding organization. Its physiology is definable in terms of birth rates or natality, death rates or mortality, and changes through emigration or immigration. In both structural and functional terms these are properties which individual organisms do not have, and thus it can be said that populations are more than the sum of their parts. They represent a higher level of biological organization than that of an individual—a level which has its own unique qualities not found at the level of the individual organism. An understanding of these qualities is essential for an appreciation of what ecology is all about.

POPULATION SIZES

Some of the most important aspects of population structure are population size, density, spatial distribution, age and sex ratios, breeding structure and social organization. Population sizes may vary from a few individuals to millions of individuals. The population of whooping cranes (*Grus americana*) numbers about 50 individuals—a single group of birds barely clinging to existence. Other rare and endangered species around the world also represent very small populations. The world's population of the Indian lion (*Panthera leo*) numbers less than 200 individuals living in the Gir forest of western India. The California condor (*Gymnogyps californicus*), golden marmoset (*Leontideus rosalia*), Javan rhinoceros (*Didermocerus sumatrensis*), orangutan (*Pongo pygmaeus*), Gray jungle fowl (*Gallus sonnerati*), and many other species are now represented by small remnant populations wavering on the edge between life and death (Fisher et al., 1969; Guggisberg, 1970).

At the other extreme, the world contains many species of tremendous abundance. A population of starlings in a single winter flocking roost in eastern United States may number 5 million birds. The white-tailed deer population of Pennsylvania was estimated in 1946 to contain 1 million deer; the rhesus monkey population of one province in northern India was estimated in 1960 to contain between 800,000 and 1 million monkeys (it now probably contains less than half that number); the Norway rat population of Baltimore was estimated in World War II at 400,000; and the bison population of the Great Plains was estimated to contain 60 million individuals in the early nineteenth century, though by the 1890's, the bison population had crashed to less than 600 through the greed and disruptive influence of man.

POPULATION ESTIMATES

These figures raise the question, of course, as to how such data are obtained, and this is a large and complicated subject in itself. Censusing wild animal populations is fraught with many difficulties of sampling error and statistical bias. Occasionally, direct counts are possible, as in aerial photographs of big game herds or sea bird colonies, but for most animals this is impossible and various kinds of sampling methods must be employed. Most animals are not readily visible because of their behavior and habitat, or because they exist in such abundance or scarcity that they cannot be readily counted. It therefore becomes necessary to estimate numbers through programs of capture and resampling. With many animals a systematic program of capture-mark-recapture has been helpful in obtaining estimates of abundance. The ratio of marked to unmarked animals in subsequent trapping runs provides a population estimate known as the *Lincoln Index*. If, for example, one is studying a resident bird or rodent population in a forest, and succeeds in capturing and banding 100 individuals all of which are released, and then if on a second trapping program, one again catches 100 animals of which 20 were previously banded, one would estimate the total original population at 500 animals. This is based on the Lincoln Index ratio which may be written:

$$\frac{P}{M_1} = \frac{T_2}{M_2}$$

where P equals the unknown population, M_1 equals the total number of individuals marked in the first capture period, T_2 equals the total number of individuals captured in the second capture period, and M_2 equals the number of those in the second capture period which were marked. The validity of this method involves several major assumptions: (1) that the marked animals mix randomly in the population, and (2) that the probability of recapture is the same for each individual regardless of whether or not it is marked, and (3) that there be no immigration or emigration, death or births between the sampling periods. Actually, these conditions are rarely attained in natural populations. Many studies on birds and mammals have shown that some individuals are easily captured, whereas others are rarely captured. There are thus "trap-happy" individuals and "trap-shy" individuals. In some species, one experience with trapping on the part of an individual animal may reduce the probability of subsequent recapture, whereas in others it may enhance it. Such behavior patterns bias the data and reduce the value of the population estimates made with this method. Also, birth, death, immigration and emigration are often continuous processes within populations, so their effect can rarely be eliminated entirely.

Nonetheless, the Lincoln Index approaches validity in certain cases. It can be a helpful tool for population estimates, particularly in the following cases: (1) if a capture method is available which is relatively nonselective and random in its action; (2) if a relatively confined population is available, so that emigration and immigration do not occur, as an island population, a woodland population surrounded by grassland, a grassland population surrounded by forest, an oasis, etc.; (3) if the population can be sampled at a time when it is relatively stable, not reproducing during the sampling period, and not migrating. Under these circumstances, the capture-mark-recapture technique can produce the best available results on certain animal populations.

An example of a population estimate by the Lincoln Index on a relatively stable population of terrestrial snails (*Achatina fulica,* the giant African land snail) in Hawaii, is given in Table 11-1. The snails were captured by careful

TABLE 11-1 Population Estimates of the Giant African Land Snail, *A. fulica*, on Oahu

Census Trial	*Date*	*Total Marked Population*	*No. Found in Census*		*Percent Marked in Census Sample*	*Population Estimate*	*Population per 1000 Square Meters*
			Total	*Marked*			
1	July 22, 1965	50	88	14	15.9	314	419
2	July 23, 1965	137	87	33	37.9	361	481
3	July 24, 1965	207	127	78	61.4	336	448
4	July 25, 1965	269	134	98	73.1	368	491
5	July 26, 1965	305	142	124	87.3	349	465

and thorough searching of the habitat. They were marked by quick-drying enamel numbers painted on the shell. Presumably there was no capture or recapture bias. On July 21, 1965, 50 snails were marked and released on a lawn and garden habitat of approximately 700 square meters. On July 22, 88 snails were found of which 14 were marked, giving a population estimate of 314 snails. On the next four days, capture samples of 305 snails were marked and released, and the Lincoln Index for each successive day provided population estimates ranging from 314 to 368. The mean population estimate was 346 snails, which indicated a population of 461 snails per 1,000 square meters, or 1,865 snails per acre. These were large snails with an average live body weight of 32.3 grams per snail, thus the estimated population represented a biomass of 60 kilograms of snails per acre.

Another method of estimating animal population size is based on a change in sex ratio before and after a known harvest of one sex has been removed. This has been called the *Kelker Ratio,* and has been used in the estimation of deer, pheasants, and other wildlife populations where hunting statistics are available (Davis, 1963). In a study of California deer by Dasmann (1952), the sex ratio observed before the hunting season was 74 bucks to 139 does, and after the hunting season in the same area it was 50 bucks to 139 does. The harvest in the hunting season was known to be 246 bucks. Hence, the removal of 246 bucks produced the observed change in the sex ratio from 74/139 to 50/139. The total buck population should be in the same proportion to the original sex ratio as the total buck harvest was to the change in sex ratios produced by the harvest. Thus, the following Kelker formula was applied to calculate the total population of bucks:

$$\frac{S_1}{P_1} = \frac{S_1 - S_2}{H}$$

where S_1 equals the pre-hunting sex ratio, P_1 equals the pre-hunting buck population, S_2 equals the post-hunting sex ratio (therefore $S_1 - S_2$ equals the change in sex ratio during the hunt), and H equals the total harvest.

In the Dasmann population, the calculations were as follows:

$$\frac{(74/139)}{P_1} = \frac{(74/139) - (50/139)}{246}$$

$$P_1 = 759$$

Hence, the total buck population before the hunting season was estimated to be 759, and the post-hunting buck population was estimated to be 513. Dasmann also knew that the ratio of fawns and does to bucks was 6.1 to 1, and he therefore estimated the total post-hunting deer population to be approximately 3,100 deer.

One can readily see some of the observational problems and potential biases in the Kelker Ratio as in the Lincoln Index. The Kelker Ratio assumes that the observations of sex ratios before and after harvest are equivalent. If the bucks become shyer than the does and are harder to observe after the hunt, this enters a major bias. It is essential that the field ecologist be fully experienced and knowledgeable in the habits of the animal he is studying before he applies these population estimation techniques. He can then recognize the sampling problems and possible sources of error. If reasonably unbiased sample estimates are obtained, it is then important to subject them to statistical treatments, such as standard error calculations, or analyses of variance, to evaluate the sampling errors which might occur by chance alone. It is thus important for ecologists and statisticians to work closely together in population studies.

Population density is an important aspect of both population size and distribution. In many populations, the precise boundaries of the population are unknown. Though the distributional limits of the species may be recognized, each species usually consists of several populations. The essential feature which distinguishes a population is whether or not the individuals are potentially interactive, or potentially interbreeding. If a species consists of two or more discrete aggregations, which do not interact or influence each other, these aggregations can then be considered populations. If this condition persists over an extended period of time, different races, subspecies and even species may emerge as gene frequencies change and evolution occurs. Since populational boundaries are often unknown, the most practical description of population abundance is frequently in terms of densities. As an example, many ecologic studies of small rodents have shown population densities usually in the range of 0.5 to 20 animals per acre. Burt (1940) showed population densities of the deer mouse (*Peromyscus leucopus*) in southern Michigan woodlots to range from 3.08 to 10.32 individuals per acre, whereas the population density of chipmunks varied from 0.8 to 1.63 per acre. Blair (1951) estimated the population density of beach mice or oldfield mice (*Peromyscus polionotus*) to range from 0.83 to 1.41 individuals per acre. Ashby (1967) estimated population densities of the bank vole (*Cleithrionomys glareolus*), a relative of the meadow mouse, to run as high as 17 to 18 individuals per acre in central England.

An important consideration, which will be emphasized later, is that many populations fluctuate widely in density and size in the course of natural events. Occasionally small mammal populations reach amazingly high densities. In California, Krebs (1966) described a meadow mouse or vole population (*Microtus californicus*) which achieved a density of 150 to 300 individuals per acre in a relatively small area, and Pearson (1963) described a population of house mice (*Mus musculus*) in California which increased from a density of less than 5 mice per acre to 300 mice per acre within 6 months. At these high levels, of course, rodents can become agricultural problems and can precipitate economic losses of crops or stored food products. Spillett (1968) has estimated that wild rats (*Bandicota bengalensis*) in Calcutta grain storage warehouses may achieve densities of one rat per square meter, equivalent to 4,000 per acre. These represent extreme concentrations, of course, but they indicate the population potential of some species.

SPATIAL DISTRIBUTION

Another important consideration in concepts of both population size and density is the spatial distribution of individuals. Spatial distribution involves a fascinating interplay of behavioral and ecologic factors, and is a prime example of the inseparable tie between behavior and ecology. We have been speaking of population densities as though animals were randomly distributed. This is rarely true, and it is much more common for populations to be clumped and nonrandomly distributed.

A typical pattern of spatial distribution within a population is shown in Figure 11-1. This might accurately portray the distribution of insects in an old field, isopods or millipedes in a forest, nesting birds or prairie dogs in a grassland, baboons in a savannah, fish in a lake, or even plankton in the sea. The reasons for such clumping or nonrandom distribution might be numerous. In some cases clumping would be attributable to the heterogeneity of the environment—the distribution of food, cover, and shelter. For example, dung beetles in an old field would be clumped around the dung, monarch butter-

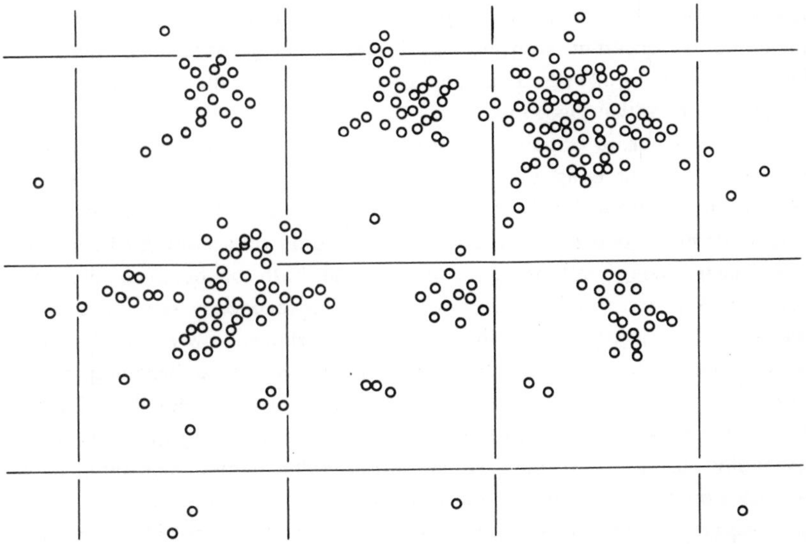

Figure 11–1 Typical pattern of spatial distribution of individuals within a population. (The dimensional units of this diagram are entirely hypothetical. They might be millimeters or micra in the case of microorganisms; inches or feet in the case of soil invertebrates; miles in the case of birds or mammals; hundreds of miles in the case of whales.)

flies around the milkweed, isopods and millipedes around the fallen logs, etc. In other cases, the clumping might be due to the social behavior of the species. Thus, schools of fish, flocks and coveys of birds, and herds of mammals are all clumped as a function of social behavior. This clumping may be related to environmental resources, but not necessarily so.

This is not to say that random dispersion of animals never occurs. Park (1933) pointed out that flour beetles (*Tribolium confusum*) are distributed in flour according to a Poisson distribution, indicating that the individuals are spaced at random. Cole (1946) found that certain spiders were distributed randomly under boards placed on the forest floor. These examples are probably exceptions to the general rule, and they are most likely attributable to either the homogeneous or artificial nature of the environment in each case. For purposes of generalization it can be stated with reasonable assurance that most animals and many plants in natural environments show a clumped or nonrandom pattern of distribution.

SEX RATIOS AND AGE STRUCTURES

It is obvious that animals within a population differ in a number of ways: sex, age, breeding condition, health, physical condition and social status. Demography, or the numerical analysis of populations, is concerned with the sex, age and breeding condition of the population. Medical science and epidemiology are primarily concerned with the health and physical condition of the population, and behavioral science is primarily concerned with the social and behavioral states within the population.

In most vertebrate populations primary sex ratios at hatching or birth approach 50 percent male and 50 percent female. In some forms, there is a slight deviation from this; in man, for example, sex ratio at birth is closer to 52 percent male, and this is true for rabbits, cattle and many birds. In domestic chickens, sheep and horses, sex ratios at birth or hatching are usually closer to 49 percent males (Leopold, 1933). The secondary sex ratios or adult sex ratios of many vertebrates show greater deviations, however. In the Alaskan fur seal (Peterson, 1968), one adult bull may dominate a breeding group of 30 females. Whereas the total adult male sex ratio is greater than 3 percent, the excess adult males are on the periphery of the breeding grounds, they are essentially surplus, and they suffer substantially higher mortality. In most primate populations, adult females outnumber adult males two or three to one. Apparently males have higher mortality rates throughout life than females. In some bird populations, the reverse is true, and females have higher mortality rates than males, particularly during the nesting season. In waterfowl and gallinaceous birds, for example, males often outnumber females. The sex ratio of blue-winged teal (*Querquedula discors*) in fall and winter flocks averaged 59 percent males in a sample of 5,090 birds (Bennett, 1938), and in the early summer breeding populations of mallards (*Anas platyrhynchos*) and pintail (*Anas acuta*) in Manitoba, males may outnumber females 4 or 5 to one (Hochbaum, 1944).

Age ratios represent another important element of population structure, and one that is also particularly valuable in analyzing population dynamics. Figure 11-2 shows age structure of populations in terms of three basic types: (A) a declining population, with a low percentage of young in the population; (B) a stable population, with a larger percentage of young than adults, and (C) an increasing population, with a very large percentage of young. The exact shape and proportions of these age pyramids are a function of natality, mortality and population turnover. Some populations characteristically have tremendous production of sex cells and young, with massive mortality of young. The Pacific herring, for example, produces 8,000 eggs per female per season, of which 95 percent hatch but only 0.1 percent survive to maturity (Allee et al., 1949). The female shad lays from 30,000 to 100,000 eggs per female per season, and less than 0.1 percent grow into mature fishes. Populations such as this would have age pyramids with very broad bases with tiny segments ascending into adult levels. At the other extreme, elephants and whales usually produce only one young per female every few years, and they would have age pyramids with relatively narrow bases and expanded adult proportions. That is, the majority of animals in the population would be adults.

Figure 11-3 shows population age structure data on rhesus monkey populations. Rhesus populations in India show different age structures in different habitats. In roadside habitats and villages, where juvenile rhesus monkeys (1 to 3 years of age, postweaned but preadult) are trapped for export for use in biomedical research and pharmaceutical production throughout the world, there

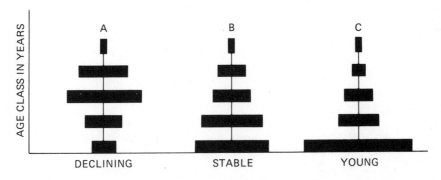

Figure 11–2 Theoretical age pyramids representing populations with low, medium, and high percentages of young individuals. (*Reprinted by permission of The Macmillan Company from* Ecology of Populations, *by Arthur S. Boughey. Copyright © by Arthur S. Boughey, 1968.*)

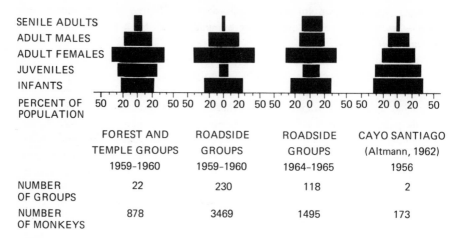

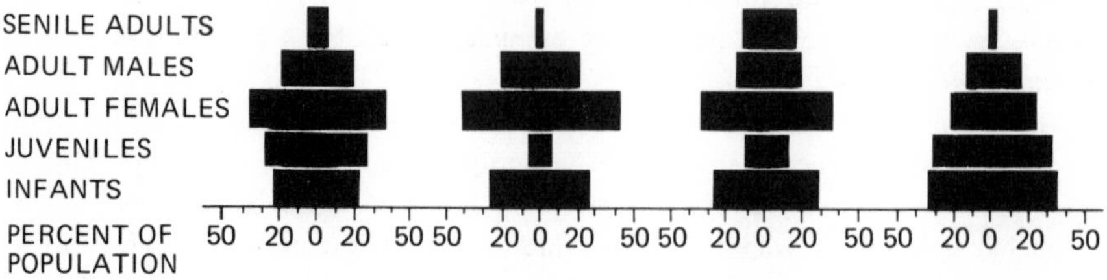

	FOREST AND TEMPLE GROUPS 1959-1960	ROADSIDE GROUPS 1959-1960	ROADSIDE GROUPS 1964-1965	CAYO SANTIAGO (Altmann, 1962) 1956
NUMBER OF GROUPS	22	230	118	2
NUMBER OF MONKEYS	878	3469	1495	173

Figure 11–3 Age structures of rhesus populations.

is a conspicuous juvenile "gap," that is, an unnatural shortage of these individuals. These populations have been declining throughout the 1950's and 1960's at the rate of approximately 5 percent per year, as India has exported anywhere from 40,000 to over 100,000 rhesus monkeys per year, primarily juveniles. In temples, where rhesus monkeys are protected to a certain extent for religious reasons, and in forests where they can escape trapping more successfully, the age structure shows a better balance of juveniles. These populations are relatively stable. By comparison, the age structure of an island population of rhesus monkeys on Cayo Santiago Island off the eastern coast of Puerto Rico, is also shown. This population has been fed and protected by a governmental research program of the National Institutes of Health, and it shows a broad base of infants and juveniles. This population was increasing at the rate of 16 percent per year in the early 1960's, and gave a good indication of the natural tendency of primate populations to increase if protected in a satisfactory habitat.

LIFE TABLES

Life tables represent tabular data on age structures of populations, and they also provide valuable information on mortality rates and longevity patterns. The essential data of a life table may be of two basic types: (1) census data on a population with accurate counts of the numbers of individuals in each age category, (2) mortality data on the number of individuals of each age group dying in a given period of years. From such data, survivorship and longevity can be calculated. Insurance companies use life table statistics on human populations to estimate the probability of death at each age group, and from these data they establish rates for life insurance.

Table 11-2 shows life table data on a population of Dall mountain sheep in Alaska. These are wild sheep which may be aged by tooth and skull char-

TABLE 11-2 Life Table for Dall Mountain Sheep, Alaska[a]

Age (years)	Age as Percent Deviation from Mean Length of Life	Number Dying in Age Interval Per 1,000 Born	Number Surviving at Beginning of Age Interval Per 1,000 Born	Mortality Rate Per 1,000 Alive at Beginning of Age Interval[b]	Expectation of Life, or Mean Lifetime Remaining to those Attaining Age Interval (years)
0–0.5	−100.0	54	1,000	54.0	7.06
0.5–1	−93.0	145	946	153.0	—
1–2	−85.9	12	801	15.0	7.7
2–3	−71.8	13	789	16.5	6.8
3–4	−57.7	12	776	15.5	5.9
4–5	−43.5	30	764	39.3	5.0
5–6	−29.5	46	734	62.6	4.2
6–7	−15.4	48	688	69.9	3.4
7–8	−1.1	69	640	108.0	2.6
8–9	+13.0	132	571	231.0	1.9
9–10	+27.0	187	439	426.0	1.3
10–11	+41.0	156	252	619.0	0.9
11–12	+55.0	90	96	937.0	0.6
12–13	+69.0	3	6	500.0	1.2
13–14	+84.0	3	3	1,000.0	0.7

[a] After E. S. Deevey, in **The Quarterly Review of Biology,** 22:283-314, 1947.
[b] Divide by 10 to convert to annual mortality on a percentage basis.

acteristics, and thus estimates of the number dying in each age group may be obtained in the field by examination of the skulls of dead animals. The data are arranged to show the number dying in each yearly age interval per 1,000 animals born. It can be seen that the total longevity is 14 years, and by 8 to 9 years of age approximately 50 percent of the animals have died. Thus, if life expectancy at birth is defined as the age by which there is a 50 percent chance of life or death, then the life expectancy at birth is 8.5 years for the sheep in this population. The table also shows that there is approximately 20 percent mortality within the first year of life, but then from one to five years of age, the annual mortality is only 1.5 percent to 3.9 percent. After five years of age, annual mortality rates begin to increase from 6.26 percent in the fifth to sixth year, to 61.9 percent in the tenth to eleventh year.

Life tables provide ecologists with essential information for the management of animal populations. They pinpoint mortality patterns, indicating when deaths occur and how many. If one finds abnormally high mortality occurring in infancy, adolescence, or early adulthood, it may suggest specific management procedures to alleviate this mortality. Excessive infant mortality, for example, may be associated with inadequate food for young animals, or inadequate cover by which they can escape predators, or improper patterns of parental care due to behavioral disturbances within the adult population. Excessive adult mortality might indicate a shortage of adult food supplies, infectious diseases or parasitic burdens which become more severe in adulthood, or possibly excessive crowding during the reproductive period so that sexual fighting becomes violent. Thus, life table data provide important clues for more detailed study. They seldom give the final answers on why or how certain events occur, but they form the essential starting point of what is happening in a population.

In summary, we have noted that populations may be studied in both anatomical and physiological terms. That is, we may analyze the structure of a population in terms of numbers, density, spatial distribution, sex and age ratios, and other descriptive qualities. Similarly, we may analyze patterns of change in populations in terms of birth rates, mortality rates, longevity patterns, movements, and other dynamic attributes. The ultimate objectives are to understand the organization of populations and their vital processes. As we see more and more populations getting out of balance, and going toward extremes of great abundance or great rarity and extinction, such understanding has now become critical to human affairs and the quality of life on earth.

12

Population Growth

POPULATIONS are not static entities. They are constantly changing, and at any given point in time they may be exhibiting growth and expansion, or decline and contraction, or some neutral phase representing the net result of these dynamic processes. Our object in the next two chapters will be to consider some of these processes, particularly as they relate to population growth and fluctuation.

Virtually all organisms have a reproductive capacity for substantial population growth. Reproductive potential is almost always considerably greater than that actually achieved. A single female insect or fish may lay hundreds or thousands of eggs at a time, of which only a few will survive. A female mouse has the potential to ovulate 10 to 12 egg cells every five days, but only a small percentage of these can become fertilized and develop into living young. If fertilization occurred at its optimal rate, one pair of house mice and their offspring could produce over 3,000 progeny in a single year. This great potential results from the fact that gestation in the house mouse is only 21 days (21 days from conception to birth), offspring are weaned at 21 days of age, and they mature sexually within 21 days after weaning. Thus at 42 days of age, they are ready to breed themselves. Furthermore, the female house mouse has a postpartum heat, which means that she comes into heat or estrus within 24 hours after the birth of her litter, and at this time she ovulates a new set of eggs and can mate again. Thus, the female mouse has the capacity to support both gestation and lactation at the same time.

Many animals do not have this great reproductive potential, of course, and some of the largest mammals with long life spans have a relatively small reproductive potential. Elephants and whales have long gestation periods (12–18 months for whales, 20–22 months for elephants), and they usually give birth to only one young at a time. However, with a life span of 30 to 60 years, a female may still give birth to a dozen or more young throughout her lifetime.

The human female normally ovulates between 300 and 400 ova in the course of her lifetime, any one of which can potentially develop into a young. If conceptions occur at maximum rate, however, a human female can give birth to about 1 young every 12 to 15 months, or a total of 20 to 25 in the course of her lifetime. This still means a reproductive potential 8 to 10 times greater than the average family size in most countries.

PATTERNS OF POPULATION GROWTH

The great reproductive potential of most plants and animals gives them a capacity for rapid population growth in favorable habitats. Many newly developing populations reveal exponential increases in the early stages of population growth. An exponential increase is one in which the base number or starting number is repeatedly multiplied by itself. Thus an exponential increase beginning with a base number of 6 would be represented by the series 6. . . . 36. . . .216. . . .1,296. . . .7,776. . . .46,656. . . . and so on. Obviously exponential growth cannot continue for long. It must and it does come to an end, and the pattern in which this is achieved results in a population growth curve.

MALTHUSIAN GROWTH

At the close of the eighteenth century, Thomas Malthus, an English economist, noted this tendency for populations, including human populations, to increase rapidly. In 1798, he published his famous work, *Essay on the Principle of Population,* in which he put forth the idea that populations tend to increase faster than their means of subsistence. In general, he felt that populations tended to increase geometrically or exponentially, whereas their food supplies and means of subsistence tended to increase only arithmetically. A geometric increase is a series of numbers having a common ratio, such as 2. . . .4. . . .8. . . .16. . . .32 . . . .64, in which each succeeding number is twice the former. An arithmetic increase is a series of numbers having a common difference, such as 2. . . .4 . . . .6. . . .8. . . .10. . . .12, in which each number differs from the other by 2. Thus, Malthus thought that populations constantly tended to outstrip their food supplies and were then decimated by poverty, starvation, disease, warfare, or other catastrophes. Malthus was primarily concerned with human populations, and he felt that moral restraint was the only solution to human population pressures.

The concepts of Malthus produced a storm of protest, of course, from both the church and the scientific community. The church saw Malthus' thesis as a threat to man's natural procreation, and the scientific community viewed it as an insult to the capabilities of science and technology. Thus, there was an increasing feeling throughout the nineteenth century that Science could grow more food, conquer disease, and provide more satisfactorily for expanding populations. The prevailing beliefs in the Baconian Creed (Science shall conquer all problems), and the Abrahamic concept of land (go ye forth and subdue all nature) were clearly anti-Malthusian. The ideas of Malthus generally fell in disrepute and were considered little more than historical oddities.

In more recent years, however, Malthusian doctrine has become far more respected. His name is reappearing in the literature of economics, sociology and biology, more often as an honored scholar and less as an historical heretic. Modern man finds himself enmeshed in many of the Malthusian predictions. There is, throughout the world, war, starvation and disease. Poverty still remains the major problem of world affairs despite unprecedented prosperity and technologic expansion. Even in the modern "Green Revolutions" of Asia, in which dramatic increases of food production occurred throughout the late 1960's, there remain war, poverty and famine in many areas. In the fall of 1969, for example, when India and southeast Asia experienced bumper harvests of wheat and rice, starvation conditions still prevailed in India's Rajasthan province and Thailand's southern states. In the spring of 1971, the tragic events in East Pakistan followed a predictable Malthusian pattern of war, starvation, disease and mass migration.

Disease also continues to be a prominent problem of modern man, along with warfare and starvation. Whereas we have tended to emphasize the accomplishments of modern medicine and public health in controlling infectious disease, we often forget that many problems of infectious disease have not been solved. Some infectious diseases show signs of substantial resurgence. Malaria, for example, is rebounding in many parts of the world. In Ceylon, where malaria was considered eradicated in the late 1950's and early 1960's, it has skyrocketed upward again from just a few cases in 1965 to several hundred in 1966, to over 1 million new cases in 1968. This resurgence is due in part to the worldwide problem of DDT resistance in the vector mosquitos. In Africa there have been renewed increases of trypanosomiasis (Sleeping sickness), as the expensive measures against tsetse flies have been dropped by newly independent governments. In both Africa and Asia, cholera erupted in epidemic proportions in 1971, even though modern medicine has provided both a vaccine and a very effective clinical treatment once the disease is contracted. Neither of these were adequate to prevent thousands of cholera deaths in India, Pakistan, Chad, and Uganda in May and June of 1971.

Perhaps the world's fastest growing disease is schistosomiasis, sometimes called bilharziasis or snail fever. It is caused by a small blood fluke which spends part of its life cycle in freshwater snails. It has followed the paths of irrigation throughout many parts of Asia, the Middle East and Africa. More than 200,000,000 people throughout the world are affected by this debilitating disease. Modern medicine and public health authorities are unable to solve this increasing problem at the present time.

The threat of world-wide infectious epidemics is still very much with us. In fact, their probability is increased by the speed and frequency of modern jet travel. In the late 1960's, several waves of viral influenza including the "Hong Kong flu" swept around the world. Modern medicine and international public health measures were unable to stop these pandemic waves. Fortunately, Hong Kong flu was a relatively mild disease, but Dr. Joshua Lederberg has pointed out that it was only a minor accident that this virus was not more lethal. By the quirks of viral mutation a considerably more potent virus could arise. In referring to these global epidemics, Dr. Lederberg noted (1969): "there is a considerable amount of self-delusion . . . that the antibiotics will take care of any bacteriological infection; . . . that the plague has been conquered by medicine; that virus infections will somehow be taken care of." But "when you see a pandemic like the Hong Kong flu, you have a foretaste of what really can happen. That was a world-wide epidemic. The attack rate was something like 20–30 percent of the world's population. . . . It was not a particularly lethal one, but it is only a minor accident that it was not. Such events are undoubtedly going to occur in the future that will be very much nastier."

Thus, the possibility of infectious disease acting as a major controlling factor on human populations still exists. It is these realities of war, starvation, disease and poverty which led Spengler (1969) to conclude: "Malthus' fears may at last be irremediably confirmed."

If the ideas of Malthus were to be condensed into a single diagram of population growth, they would depict a curve like that in Figure 12-1. This curve shows an exponential or geometric increase in the early stages of population growth, with population control achieved by a series of catastrophic events. These would limit further population growth abruptly through the

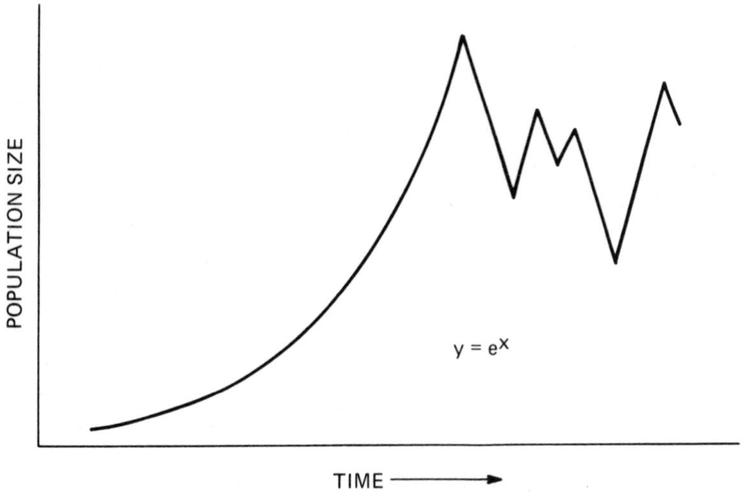

Figure 12–1 Population growth pattern typical of Malthusian or irruptive growth.

mechanisms of disease, starvation and violence. Thus, the upper limits of population growth would be characterized by sudden, often drastic mortality.

LOGISTIC GROWTH

The first major scientific challenge to Malthusian doctrine as a pattern of population growth appeared in the 1830's in the work of Pierre Verhulst. In 1839 Verhulst proposed that populations normally grow in a much more orderly fashion than that proposed by Malthus, describing in their growth a curve of S-shaped proportions. This became known as the Logistic Theory of population growth. The logistic theory asserted that populations have a slow initial growth rate, which increases exponentially until it reaches a maximum, and then becomes progressively less as the population approaches an upper limit of its growth. The upper limit is approached gradually and in an orderly, predictable manner. The resulting curve is the familiar S-shaped curve of population growth (Figure 12-2).

It should be noted that the logistic curve and the Malthusian or irruptive curve do not differ in the early stages of population growth—both show a slow start followed by a period of exponential or geometric growth—but they do differ fundamentally in the upper or controlling stages of growth. The Malthusian curve is characterized by an erratic, often catastrophic, pattern of limiting growth, whereas the logistic curve is characterized by a smooth, orderly and gradual pattern.

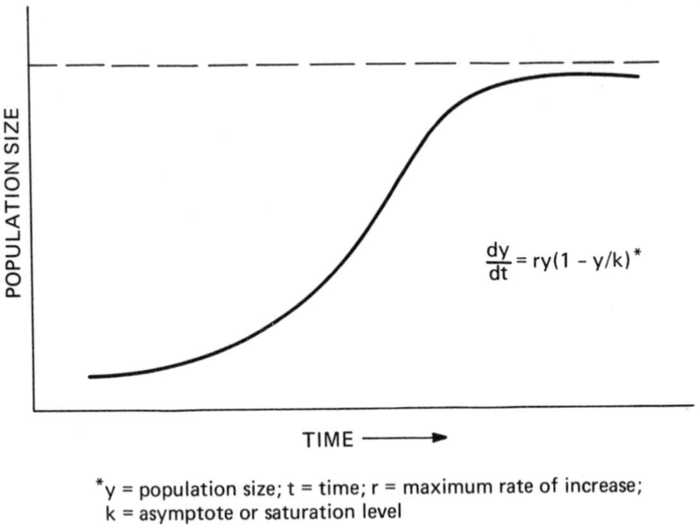

Figure 12–2 Population growth pattern typical of logistic or S-shaped growth.

Following the work of Verhulst, the logistic theory lay dormant until it was independently derived and popularized by Raymond Pearl in the 1920's. Pearl applied the logistic curve to the population growth of yeast, *Drosophila* and man (Pearl, 1925). This ushered forth a burst of activity resulting in the applica-

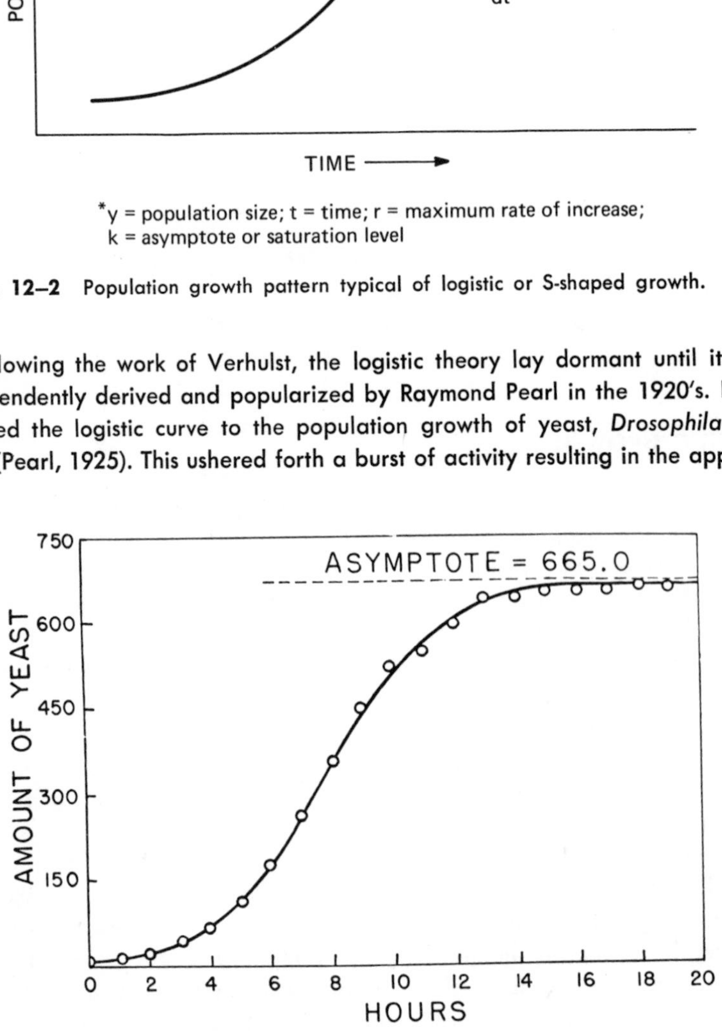

Figure 12–3 The logistic growth of a laboratory population of yeast cells. *(From Raymond Pearl,* The Biology of Population Growth, *Knopf, 1925. By permission.)*

tion of the logistic curve to the population growth of protozoa, water fleas, pond snails, thrips, ants, bees and other organisms by many investigators (Allee et al., 1949). Figures 12-3, 12-4, 12-5, and 12-6 show population growth curves for yeast cells, fruit flies (*Drosophila*), water fleas (*Moina*) and worker ants (*Atta*).

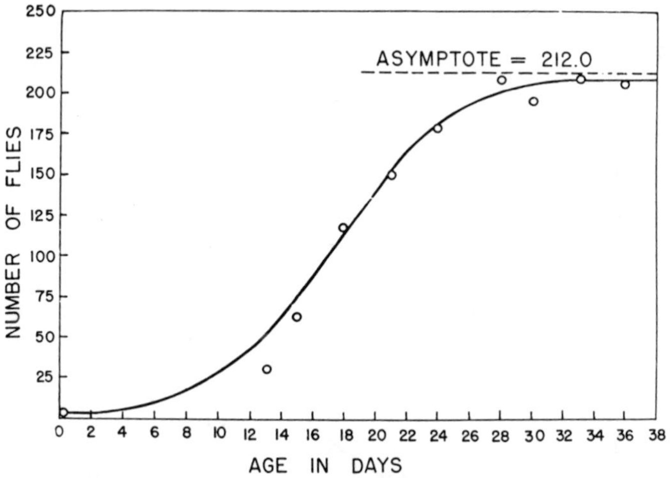

Figure 12–4 The logistic growth of a laboratory population of *Drosophila melangaster* (*From Raymond Pearl,* The Biology of Population Growth, *Knopf, 1925. By permission.*)

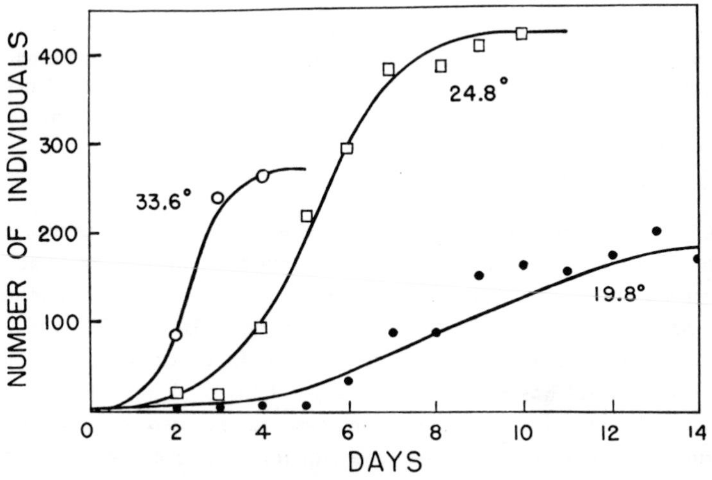

Figure 12–5 The logistic growth of three laboratory populations of the water flea, *Moina macrocopa,* at three temperatures. (*From Terao and Tanada, 1928.*)

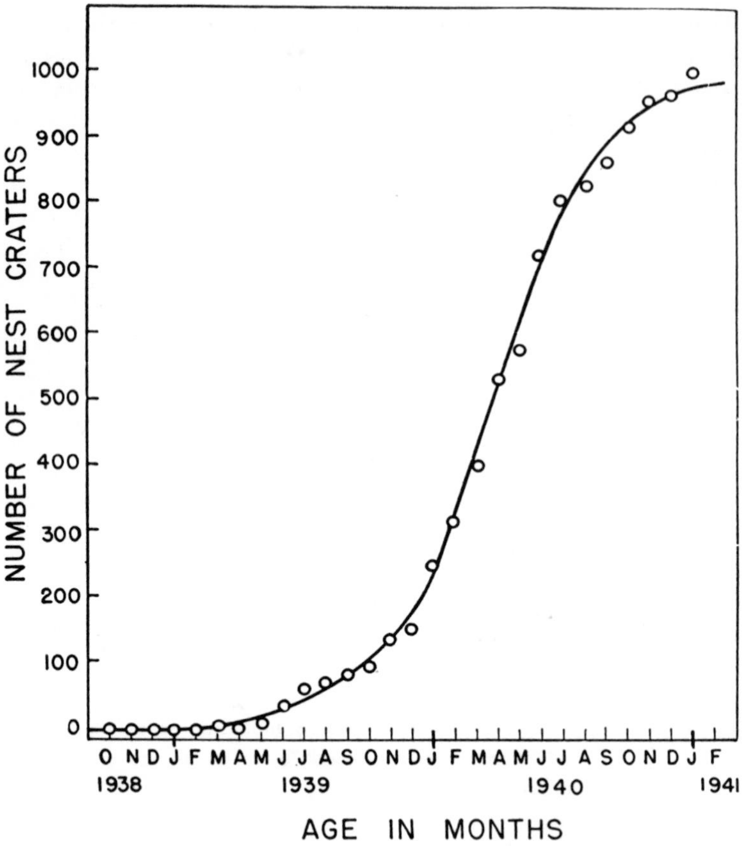

Figure 12–6 The logistic growth of the population of workers (an estimate) within a nest of the ant, *Atta sexdens rubropilosa*. The ordinate plots the number of crater openings, which are roughly proportional to the number of ants. The circles are observed counts of craters; the curve is the fitted functions. (*From Bitancourt, 1941*.)

With this wealth of supporting evidence, the logistic theory attained wide acceptance. It was proclaimed as a law of population growth by some biologists and was used to predict future population levels of experimental and natural populations (Pearl, 1930; Pearl, Reed and Kish, 1940; Reed, 1936; Udny-Yule, 1925). Pearl's investigations led him to conclude (1930): "It has been shown in what has preceded that populations of living things, from the simplest, represented by yeast, to the highest, represented by man, grow in accordance with the logistic curve. One can now feel more certain that this curve is a first and tolerably close approximation to a real law of growth for human populations. . . . We can, for example, upon a more adequate scientific basis than mere guess-work, predict future populations, or estimate past populations, outside the range of known census counts."

That some biologists accepted Pearl's interpretation rather freely is indicated by statements of Clarke (1954): "Populations of a wide variety of organisms, ranging from bacteria to whales, have been found to follow the logistic curve in their growth form. . . . The growth of man's population follows a similar pattern whether examined in individual regions or in the world as a whole." Clarke presented a graph of population growth in the United States fitted with a logistic curve, and observed an upper asymptote of 184 million in the year 2100 A.D.

Many other biologists and mathematicians, however, severely criticized the significance of the logistic theory. In fact, the theory met stern opposition since its first expression. Kavanagh and Richards (1934) emphasized that the generalized logistic equation could be made to fit a wide variety of curves, often of very different natures. Gray (1929), Hogben (1931), and Wilson and Puffer (1933) registered doubt as to the universal validity of logistic theory and commented on its inadequacies as a description. Feller (1940) pointed out that the actual fit of the logistic curve to much of the published data was poor and better agreement could be obtained with other formulas.

Andrewartha and Birch (1954) and Sang (1950) accurately pointed to limitations of the logistic equation. Sang (1950) in a careful reexamination of population growth in *Drosophila,* emphasized that, "The data we have summarized show that only in very exceptional circumstances would logistic growth occur in a *Drosophila* culture. . . . The ecological situation is too complex to be adequately described as the Pearl-Verhulst law." Sang continued to discuss general aspects of the logistic theory by stating, "This acceptance of a plausible formula, in spite of its inaccuracy, may be justified; provided this attitude does not inhibit further research, particularly into fundamental conceptions. In the case reviewed, it seems that wide acceptance of the logistic law has led to just this kind of inhibition of further work, and that many interesting ecological and physiological processes have been ignored as a result."

The study of vertebrate populations in both natural and confined circumstances has produced some population growth curves deviating considerably from logistic growth and tending toward irruptive or Malthusian growth. The sheep population of southern Australia showed a pattern of logistic-like growth from 1840 to 1890, but its upper level was reached abruptly and was followed by substantial mortality (Figure 12-7). The deer of Kaibab National Forest in Arizona grew from an estimated population of 4,000 animals in 1918 after a period of extensive predator control to an estimated population of 100,000 in 1924 (Allen, 1954), but in the winter of 1924, they died by the thousands after the population had destroyed its food supply. Their growth and decline described a classic Malthusian curve. The pheasant population of Protection Island, Washington, showed a phenomenal increase from 8 birds on the 397-

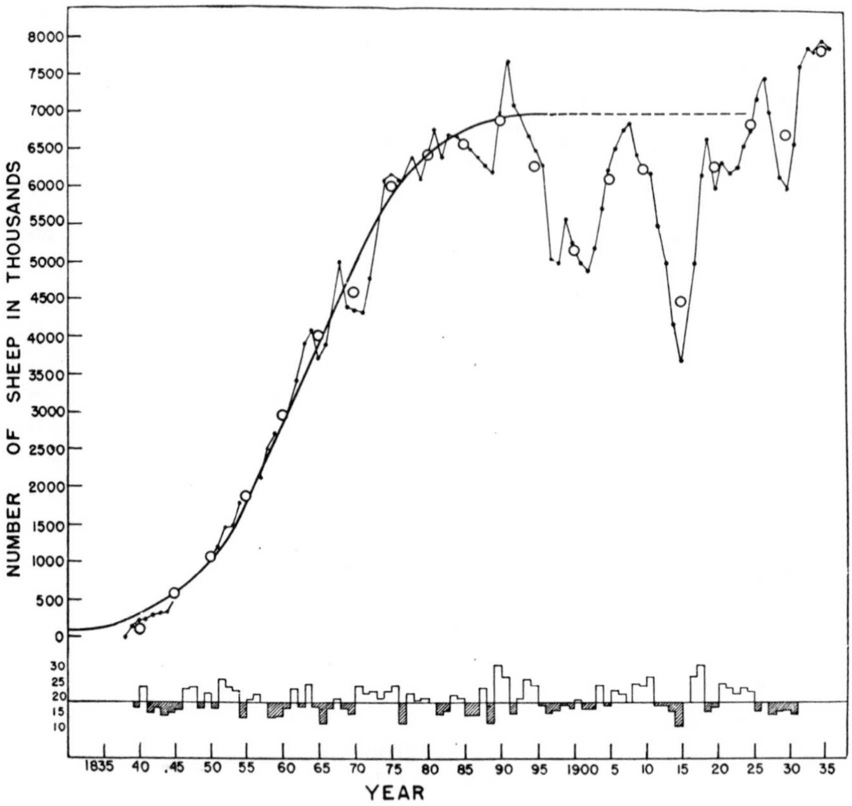

Figure 12–7 The logistic growth of the sheep population of South Australia. Annual rainfall in inches appears as the lower chart. (*After Davidson, from Allee et al., 1949.*)

acre island in 1937 to nearly 2,000 birds by 1942. In one year the population showed an increase of 300 percent (Einarsen, 1945). Unfortunately, the field research on this population was terminated before natural control occurred, but the population seemed to be describing a typical irruptive population curve (Figure 12-8).

Several studies of seminatural populations of wild house mice have also shown irruptive population curves (Figure 12-9). Seminatural populations have been maintained in confined, but free-ranging conditions, with a natural habitat for this species, except for the element of confinement. Various factors such as space, food supply, nesting cover and the structure of the physical environment have been experimentally controlled to elucidate their roles in population growth and limitation. Work of this type at the University of Wisconsin and the Institute of Ecology in Warsaw, Poland, has shown that population growth in house mice is highly variable, and often dependent upon patterns of social

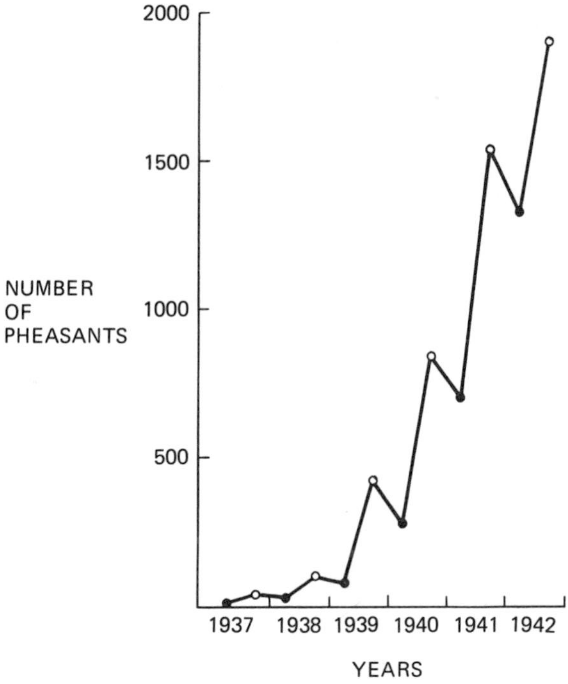

Figure 12–8 Population growth of pheasants on Protection Island, Washington. (*From Einarson, 1945.*)

behavior within the population. For example, populations in identical environments showed widely variable patterns of population growth (Figure 12-10). The variability was related to social behavior within the population in terms of aggressive behavior, dominance hierarchies, sexual behavior and parental care. All populations became controlled even though abundant food and nesting space was still available when aggressive behavior increased to an abnormal degree, when the nesting territories of females were no longer properly maintained, and when parental care of the nestlings became aberrant. Infant mortality increased through nest desertion and cannibalism. The densities at which these behavioral pathologies occurred were highly variable between populations, and they were a function of the crowding tolerance and behavioral stability of individuals. This complicated interplay of ecology and social behavior was not adequately accounted for by the logistic theory, and the resulting population curves were totally individualistic and unpredictable.

A fairly complete evaluation of logistic theory has been presented by Andrewartha and Birch (1954). They have analyzed the inherent assumptions of the logistic equation, and have reviewed many studies in the light of these assumptions. This analysis has led them to conclude that the logistic curve

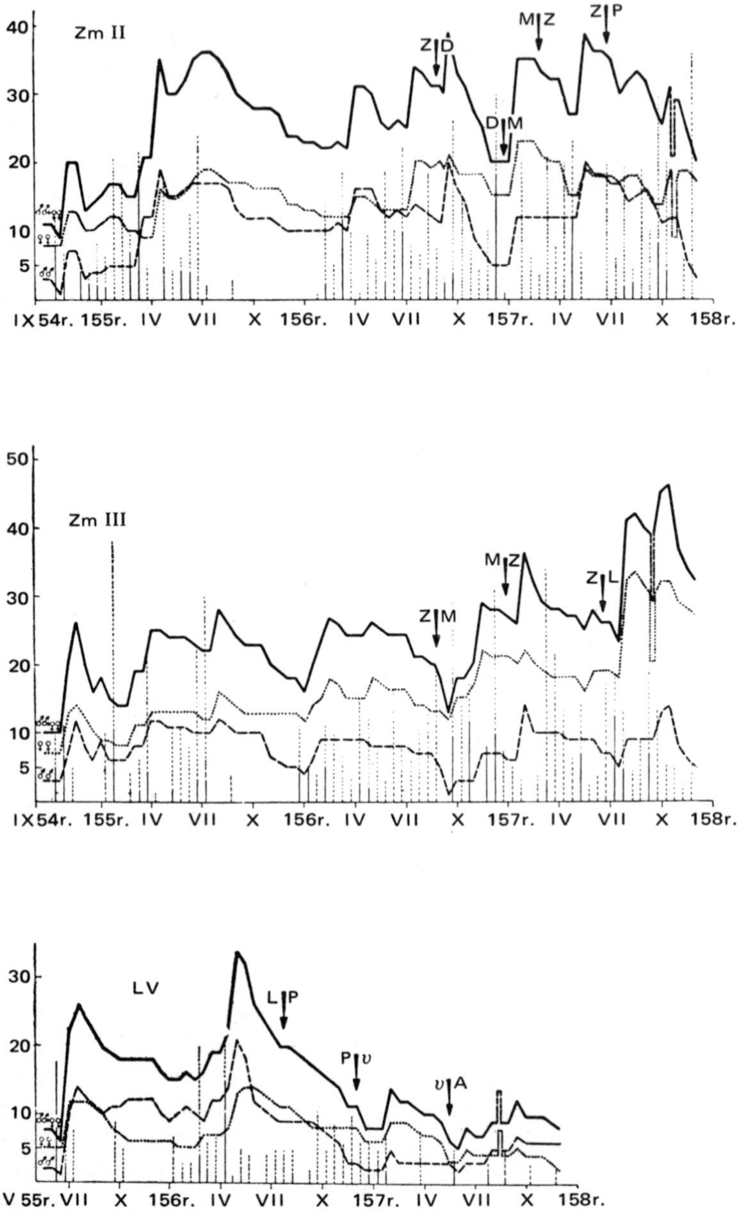

Figure 12–9 Irruptive population growth of house mice (*Mus musculus*) in confined enclosures. (*From Petrusewicz, 1957.*)

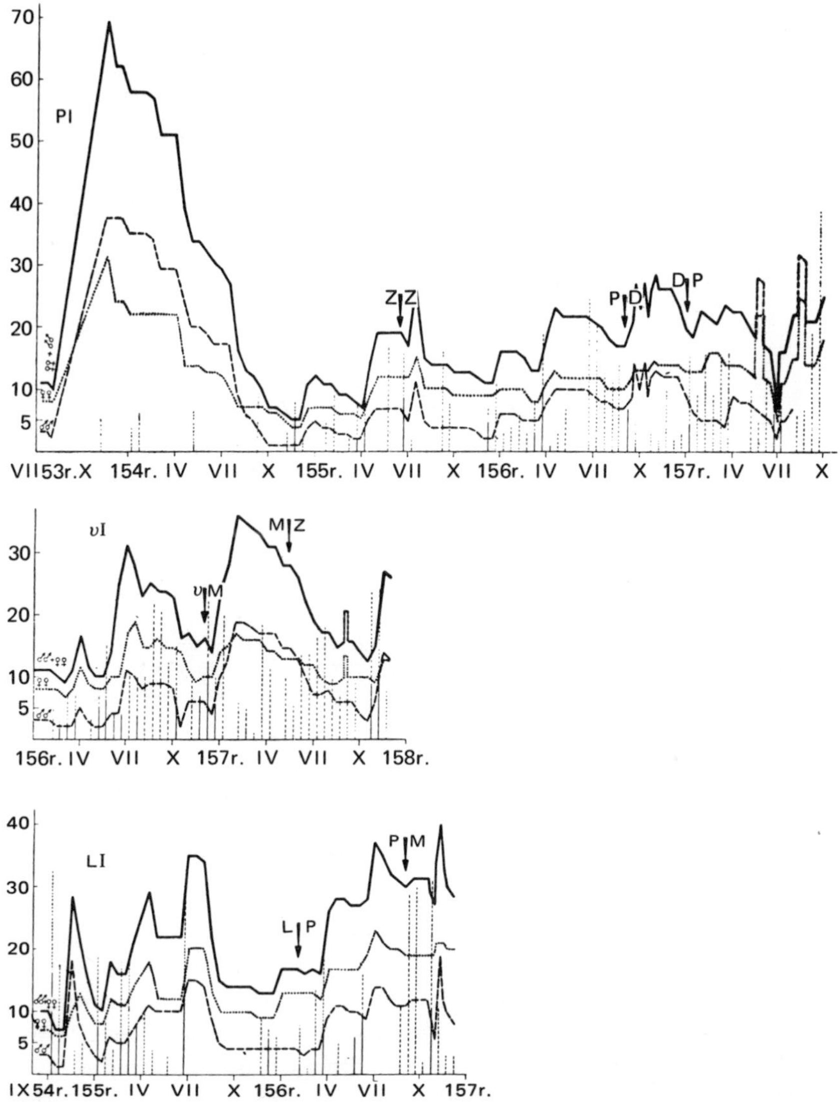

provides an imperfect, yet useful, description of the population growth of organisms with simple life histories (such as yeast and *Paramecium*); but for animals with complex life histories (such as insects and vertebrates) the population growth usually does not conform closely to the logistic curve except for small parts of the data.

The concluding attitude of Andrewartha and Birch is well expressed in the following: "despite its theoretical limitations, the logistic curve remains a useful tool for the ecologist. Because of its limitations, too much reliance should

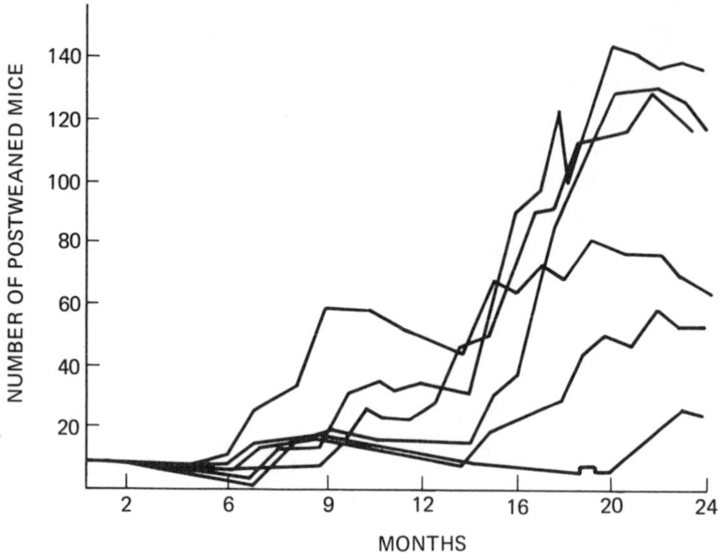

[1]Six different populations were each started with four pairs of wild house mice in population pens of 150 square feet. Each population was provided with nest boxes, nesting material, and unlimited food and water (after Southwick, 1955a and 1955b).

Figure 12–10 Growth of confined seminatural populations of house mice. (Six different populations were each started with four pairs of wild house mice in population pens of 150 square feet. Each population was provided with nest boxes, nesting material, and unlimited food and water. (*After Southwick, 1955a and 1955b.*)

not be placed on it in particular cases until it has been verified empirically for each case." Hence, it appears fallacious to utilize the logistic equation for a prognostic purpose in any population without experimental precedence.

The evidence is clear, therefore, to support the early contention of Wilson and Puffer (1933) that the logistic formula should not be considered a law of population growth to such an extent that it permits extrapolation of the curve for forecasting purposes or the interpretation that the constants of the equation are the constants of nature. Rather, it does provide a convenient description of growth in certain well-studied populations.

The major question in human terms is, of course, what pattern of population growth characterizes man? From the early seventeenth century to the present time, the world's population has displayed exponential growth. There seems to be widespread faith that human populations shall ultimately reveal a logistic pattern, but certainly all attempts to predict human population levels by means of the logistic model have been very far from the mark. Pearl predicted a world population of 2.6 billion by the year 2100. This population level was exceeded more than 10 years ago, and any economic planning based on Pearl's prediction would indeed have been disastrous. Although we continue to hope that human populations throughout the world will achieve a pattern of gradual logistic balance through orderly and intentional process, there is no sound scientific evidence that this will in fact occur. The possibilities of catastrophic mortality through war, disease and starvation are still very much with us. So also are other patterns of mortality which are not necessarily catastrophic in their total effect, but are clearly related to density, crowding and modern population pressures. In this category are automobile accident mortalities, which are responsible for 50,000 deaths per year in the United States alone, various stress-related diseases, including coronary artery disease and high blood pressure, and various pollution-related diseases including emphysema and chronic bronchitis. If deaths attributable to these causes continue to show present trends, these mortality factors will begin to play a more significant role in population regulation, especially in technologically advanced countries.

SUMMARY

The growth patterns of many populations often follow one of two general patterns, irruptive or logistic. There are, however, many variations of each. Both types of population growth have a similar pattern of slow initial growth followed by a period of rapid geometric or exponential increase. In the upper stages, however, the growth curves differ markedly. Irruptive population curves are limited by sudden, often catastrophic mortality. Logistic curves achieve a gradual leveling off to a more stable asymptote.

Neither pattern of population growth can be considered a law of population growth. Neither pattern has reliable predictive value unless empirically verified in individual cases for many replicates.

In general, irruptive growth patterns are commonly found in many insect and vertebrate populations in unstable environments. They are also characteristic of disease epidemics. Logistic growth patterns are more common in organisms with simple life histories, particularly in stable or well-controlled environments. In higher animals, both types of growth are influenced by behavior and social interaction.

It is not possible to predict which pattern will be ultimately characteristic of the human population of the world. At present, the world's population is still in an exponential phase, and it could potentially display either a logistic or a continued irruptive pattern in its later stages of growth. Individual nations may independently exhibit logistic or Malthusian tendencies, but these are not necessarily predictive of the world's total population.

13

Population Fluctuations

DIFFERENT patterns of population growth discussed in the previous chapter may be exhibited by populations recovering from major depletions, or by populations invading new habitats. Since most populations are already established, however, and are not growing *de novo,* they are instead showing varying degrees of fluctuations. Our familiar concept of the balance of nature implies stability, but the fact is that most animal populations undergo continual change. Natural populations often exhibit a dynamic resiliency, rising and falling in response to many factors. In one sense, a pattern of population fluctuation, such as that normally exhibited by most species, may be considered a continuous series of growth and decline sequences. Thus, the study of population fluctuations becomes closely allied with the study of population growth and natural control.

SEASONAL FLUCTUATIONS

At the outset it is convenient to separate population fluctuations which are seasonal (i.e., obviously related to seasonal weather), from those which are nonseasonal (i.e., not obviously related to season). In the temperate latitudes, most animal populations have breeding seasons so that population growth typically occurs in the spring and summer seasons. Most arthropods and vertebrates begin producing young in the spring and summer seasons, so that these seasons are characterized by population growth, and they cease the production of young in late autumn and winter. Figure 13-1 shows the seasonal fluctuations in a population of California quail, with the high point of the population in each year reached in midsummer. This would be typical of most of the familiar animals in the United States, including insects, spiders, fish, amphibia, reptiles, birds and mammals.

In aquatic ecosystems, many populations also show marked seasonal fluctuations. Both phytoplankton (microscopic algae), and zooplankton (protozoa, rotifers, small crustacea, etc.) usually show spring and fall increases in population. These sharp increases, often 10 to 20 fold, are known as "pulses." They are sometimes correlated with temperature changes or with turnovers in the water strata which recirculate nutrients. Figure 13-2 shows characteristic plankton pulses in Lake Erie, with peaks in early spring and fall.

In the tropical regions of the world, where there are not sharply defined seasons on a temperature basis, reproductive seasonality still occurs in many

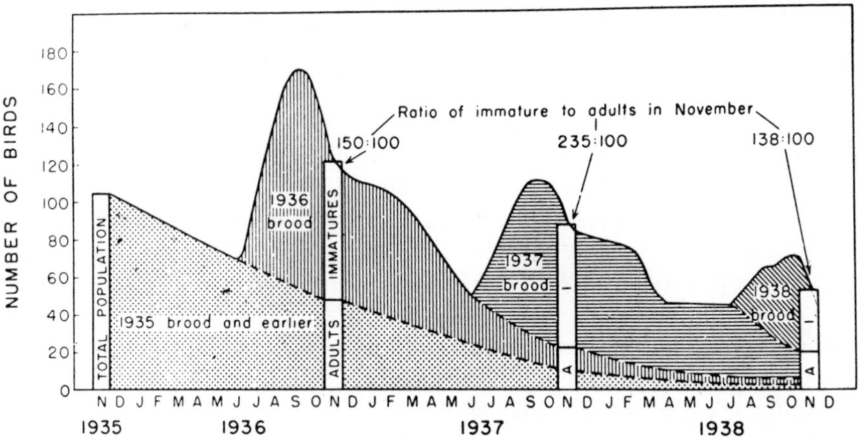

Figure 13–1 Changes in the population of quail on the University of California Farm at Davis, 1935–1938. Figures at the left represent the actual number of birds. Columns show the population and age composition for November of each year. (Estimated normal ratio of immature to adult for this locality in November—200:100.) (*After Emlen, in the* Journal of Wildlife Management, 1940.)

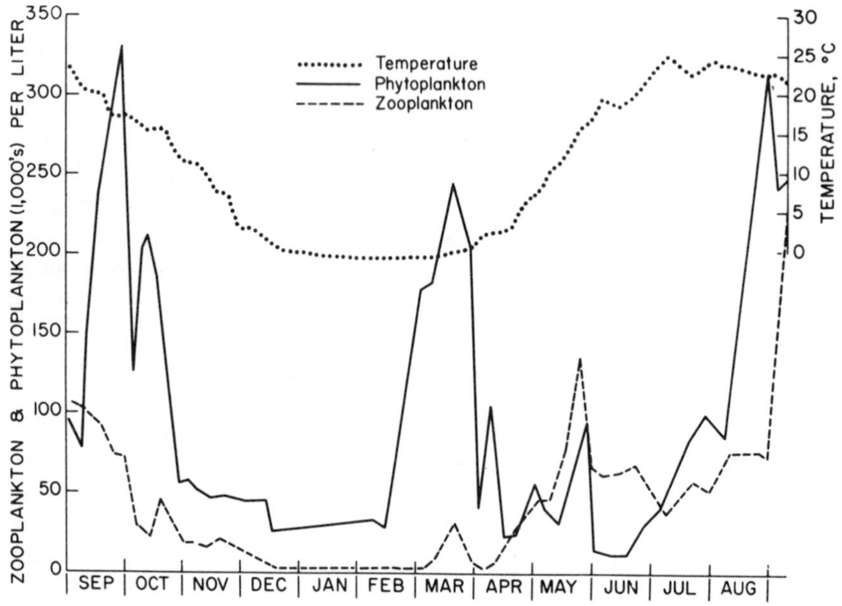

Figure 13–2 Seasonal plankton populations in western Lake Erie through a year (*After Chandler, 1940.*)

plants and animals, and it is often related to the cycle of wet and dry seasons. Many tropical insects, for example, have sharp peaks of abundance related to the monsoon seasons. Even vertebrate animals, including primates, have been found in nature to have definite periods of breeding and birth seasons, often correlated with the monsoons.

In both tropic and temperate zones an important principle seems to be that the young appear at the most favorable time of the year in terms of food and climate. Thus young deer and antelope are born when grass and browse are starting growth to provide new young shoots for food. Birds hatch when insect food is most readily available. There are exceptions to this general principle; for example, young rhesus monkeys are born in the midst of the hot dry season in northern India, when temperatures are excessively high (often over 100° F), water is extremely scarce, and plant growth is virtually nil. In this case, however, the infant feeds exclusively on the mother's milk for two or three months until the monsoon season starts. It is possible that natural selection has acted in such a way as to place breeding, conception and the first trimester of pregnancy at the most favorable time of the year, namely, the immediate postmonsoon months when food is most abundant. In other words, the events of conception and early embryonic growth may have been more critical in the life cycle of the rhesus monkey in terms of environmental influences than the period of early infancy which is well protected by the mother and the social group. This is speculation and not established fact at this time. It also fits with the observation, however, that the monsoon season with its luxuriant growth again arrives when the infant is two or three months old and is relying less on maternal milk and starts taking its own food.

Population fluctuations which are relatively independent of seasons are of two general types: (1) random, or (2) cyclic. Cyclic fluctuations are those which have a definite periodicity, or rhythmicity; that is, the population reaches a peak at fairly regular intervals. In vertebrate populations, certain species reach population peaks every 4 years, while others have a 10-year cycle. These will be discussed later as specialized cases of more generalized random fluctuations.

RANDOM FLUCTUATIONS

Random fluctuations may be minor perturbations of fairly stable populations, producing relatively flat curves, or they may be major changes in abundance, producing erratic, unstable curves.

Examples of stable populations are increasingly difficult to find in the modern world. They were probably characteristic of many organisms in undisturbed ecosystems; that is, ecosystems in balance. They were also probably characteristic of complex systems, such as forests and particularly tropical forests where great species diversity produced a complicated web of natural checks and balances on each species. Within such a system most vertebrate populations were probably stable and exhibited only their typical seasonal fluctuations.

Probably some of the best modern examples of stable populations may be found among the birds and larger mammals. In birds where considerable species diversity occurs and fairly extensive food resources are available, considerable stability in population size may be present. John Gibb (1961) reviewed a number of bird population studies several years ago and noted: "Notwithstanding the considerable annual fluctuations in bird populations that have been recorded in almost every species investigated, it is their comparative stability over long periods that is their most remarkable attribute." Thus, although wide variations in number occur from year to year in many bird populations, these populations often show a tendency to return to typical levels. Figure 13-3 shows population studies on a small insectivorous bird (the Great Tit, *Parus major*) in four different woodland habitats in England (Lack, 1966). In two of these woodlands (Marley Wood and the Forest of Dean), the populations fluctuated widely from year to year, but tended to return to a population density of 1 to 2 pairs of birds per hectare (a hectare is 2.471 acres). In two other woodlands (Breckland and Veluwe), the populations of this species were less dense and considerably more stable, varying only from 0.1 to 0.5 pairs per acre.

An example of a relatively stable vertebrate population is afforded by a

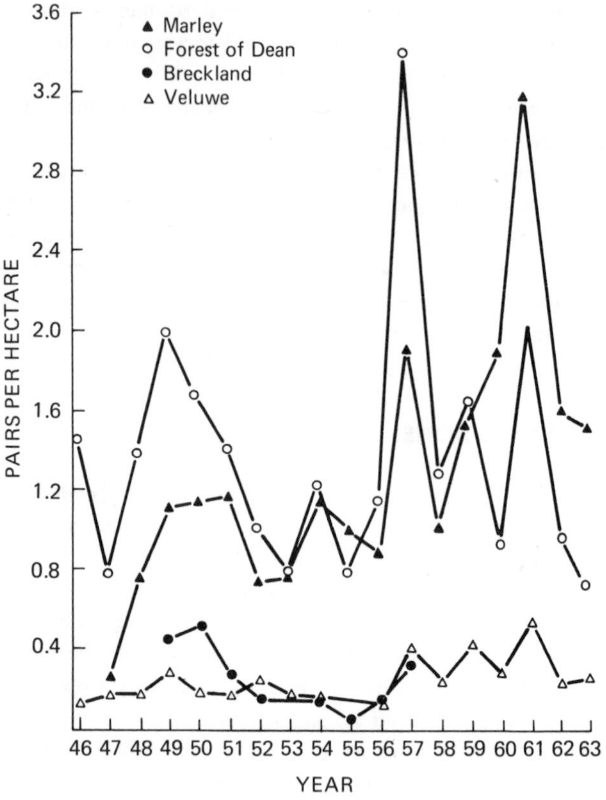

Figure 13–3 Density of breeding pairs of Great Tit in Marley wood, the Forest of Dean, the Breckland, and the Veluwe (1946–1962). *(From Lack, 1966.)*

census study of rhesus monkey populations (*Macaca mulatta*) in northern India. With complete protection, good health and abundant food supplies, rhesus populations can increase in experimental colonies at the rate of 15 percent per year (Koford, 1965). But in India, where monkeys experience limited food supplies, varying degrees of protection from the human population, and a considerable number of infectious diseases, populations do not show this pattern of increase. Figure 13-4 shows a population sample of 17 groups of rhesus monkeys in northern India censused at regular intervals from 1959 to 1970. Fifteen of these groups had not been protected by nearby villagers, and they were subject to trapping. The population in these groups has been declining irregularly since 1962. Two of the rhesus groups were rigorously protected by villagers in neighboring villages, and the population in these two groups was remarkably stable throughout the 10-year period. This stability raises the question of what is the natural limiting factor on this population. Is it a social property of the monkeys themselves; that is, is there some social mechanism by which the groups expel members above a certain level? Or is it a property of the human population; that is, do the villagers permit only so many monkeys in their area? Unfortunately, these questions cannot be fully answered. It is known that the birth rate of these groups was high (approximately 85 percent of the adult females give birth to one young per year, a pattern of fertility as good or better than that achieved under ideal conditions), but the infant mortality rate was also high (at least 25 percent per year, substantially higher than that in well-managed laboratory colonies).

In general, throughout India as in most of the world, primate populations are declining due to habitat deterioration (cutting of forests, overgrazing of savannah lands, etc.), human competition (where native peoples shoot monkeys for food), and high rates of trapping for commercial trade. As an example of the latter, in the late 1950's India exported more than 100,000 monkeys per year for use in biomedical research and pharmaceutical production, though the current figure is only about one-half of this. These monkeys were essential in the development of polio vaccine, measles vaccine, and many types of medical and psychological research.

The long term population trend of many animals, especially wild vertebrate animals, is downward. Since 1600 A.D., 120 species of birds and mammals have become extinct, a number several times greater than the natural rate of extinction. At the present time, approximately one vertebrate animal becomes extinct on the face of the earth every year, and within the next 30 to 50 years, 100

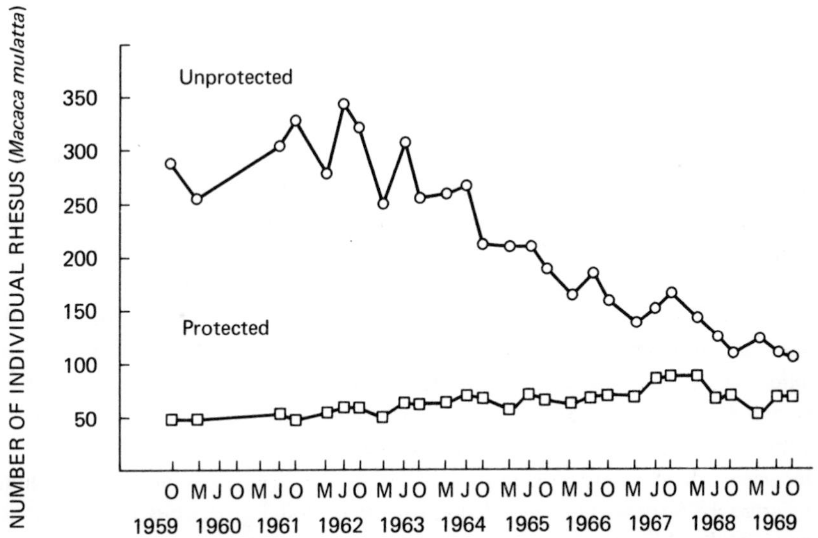

Figure 13–4 Rhesus population, Aligarh district, India, 1959–1969: total individuals.

more species will probably be lost. The International Union for the Conservation of Nature periodically publishes a list of rare and endangered species (the publication is known as the *Red Data Book*), and it now lists nearly 300 species of birds and 275 species of mammals as endangered. Within our own country in the last century, the passenger pigeon (*Ectopistes migratorius*), Heath hen (*Tympanachus cupido*), Carolina parakeet (*Conuropsis carolinensis*), and several other species have vanished. The peregrine falcon (*Falco peregrinus*), osprey (*Pandion haliaetus*), and bald eagle (*Haliaeetus leucocephalus*) are presently among the vanishing animals of North America. Others have come close to extinction, but were saved by intensive conservation movements at the last moment.

Throughout the world, the basic causes of extinction are habitat deterioration, poaching, poisoning, and direct or indirect competition with man. From the ecological standpoint, some extinction is an inevitable evolutionary process, but when it becomes excessive as in the last century, it is a serious disruption to biotic communities and ecosystems. Each extinction lessens species diversity, and thereby reduces ecosystem stability. Each extinction also represents the irreplaceable loss of unique biological material.

It should be pointed out, of course, that some vertebrate animals have increased dramatically within the last century, but these are usually domestic animals such as dogs, cats, cattle, goats, and so forth, or wild animals that have learned to become commensal with man; that is, animals which have successfully adapted and even exploited the human environment. This includes the domestic rats and mice such as the Norway rat, the house mouse, the Asian bandicoot rat (*Bandicota bengalensis*), the Asian house shrew (*Suncus murina*), the starling, the house sparrow, and the pigeon (*Columba livea*). Even the herring gull (*Larus argentatus*) is adapting to man's urban environment successfully by feeding on garbage dumps, and gull populations on both United States coasts have increased in recent years.

Apart from those species which seem to be headed towards a permanent decline and eventual extinction, and those which thrive on man-made environments, the fact remains that most animal populations fluctuate substantially from year to year.

It is often a matter of common knowledge among farmers and sportsmen that "this is a good year for pheasants," or "there were rabbits all over the place last year, but not so many this year," or "this is the worst year for mosquitos we've ever had," and so forth. These are all popular expressions of population fluctuations. Scientifically, these fluctuations are not often recorded in a systematic and continuous fashion, but some data do exist to document the types of fluctuations which are common in animal populations.

Figure 13-5 shows the annual fluctuations of cod fish (*Gadus morhua*) taken from Oslofjord between 1870 and 1960. Figure 13-6 shows the annual fluctuations in numbers of red grouse (*Lagopus lagopus*) shot on three moors in Scot-

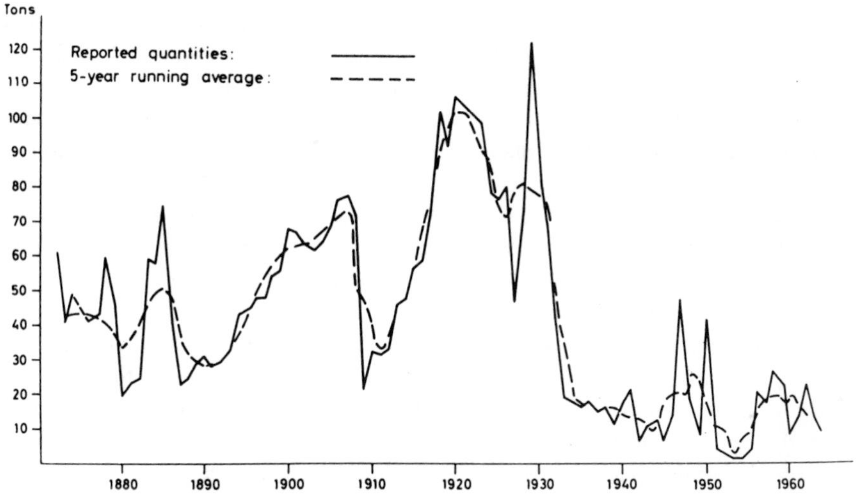

Figure 13–5 Annual landings of cod from the Oslofjord inside Drøbak. (*After Ruud, 1968.*)

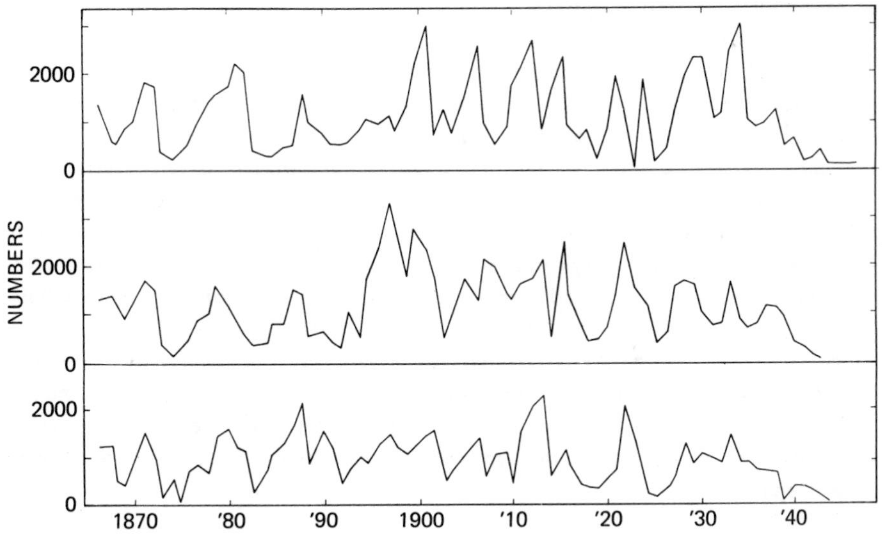

Figure 13–6 Fluctuations in the numbers of red grouse shot on three moors in Scotland. (*After Mackenzie, 1952.*)

land between 1865 and 1953. Numerous other examples of this type of fluctuation could be cited. There is always the problem, of course, of translating harvest data into actual population figures, and the relationship may not always be precise, but we can usually assume that the harvest data do represent major changes in population size and density. There are relatively few long-term population studies on vertebrates, other than harvest data, which show population trends over several generations of animals. Figure 13-7 shows population changes in three species of small mammals in central England over a 12 year period. All three species showed considerable changes in numbers from year to year. Figure 13-8 shows population fluctuations in thrips (*Thrips imaginis*), a small insect, living on rose bushes directly in the flowers.

It is, of course, always in the interest of man to understand what produces these changes, and to manage them for human benefit. Again the problem becomes one of maximizing the populations of beneficial species and minimizing the populations of harmful species. This is almost entirely an empirical or hit-or-miss affair unless we understand the basic biology underlying these population patterns.

In trying to manage animal populations, we tend to do the most obvious thing, which would logically seem to produce results, and often does, but in the long run, may be quite detrimental to the total ecological balance of a community. Thus, in controlling insect populations, we use residual poisons such as DDT, but are now discovering after some 30 years, that this has long term side effects which are quite dangerous, and we are also discovering that many insects have become resistant to it in any case. In controlling predators, such as wolves (*Canis lupus*) or cougars (*Felis concolor*), we often employ modern methods of destruction including poisoning and shooting from aircraft, to later discover that those predators played a vital role in the animal community in controlling the population levels and general health of their prey species.

In enhancing the abundance of desirable species, we have often poured money into artificial breeding and restocking programs of animals such as pheasants, wild turkeys and trout, only to find that in the meantime, the habitat has so deteriorated through pollution that the transplants have little or no chance of success, and the recovery rate per transplant is almost nil. This is not intended as a blanket condemnation of all of our control and management programs, but by and large the balance sheet is not favorable, mainly because we have based our management on trial and error rather than on sound basic knowledge of the ecological factors involved in population growth and regulation.

POPULATION CYCLES

The most dramatic population fluctuations are those which are cyclic: that is, in which the population reaches peak numbers at fairly regular intervals. Some

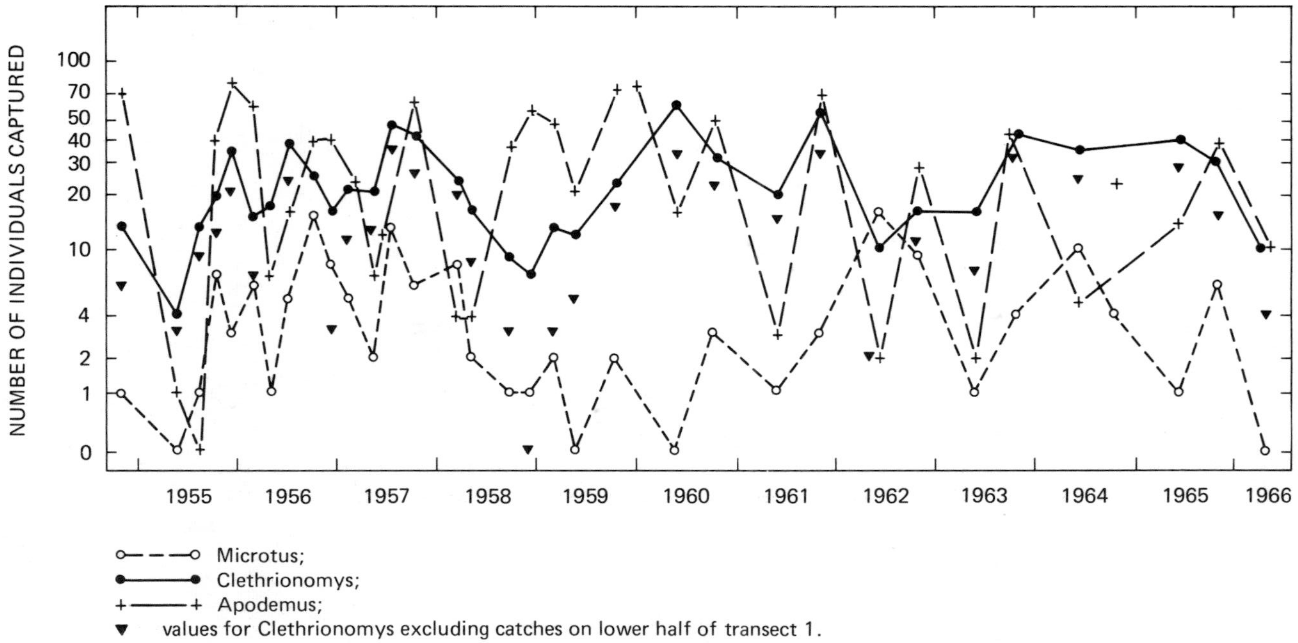

Figure 13–7 Variations in the number of different individuals caught at the periods of trapping on the lower half of transect 1, from 1954 to 1965. The vertical scale is logarithmic based on n + 1. (*After Ashby, 1967. By permission.*)

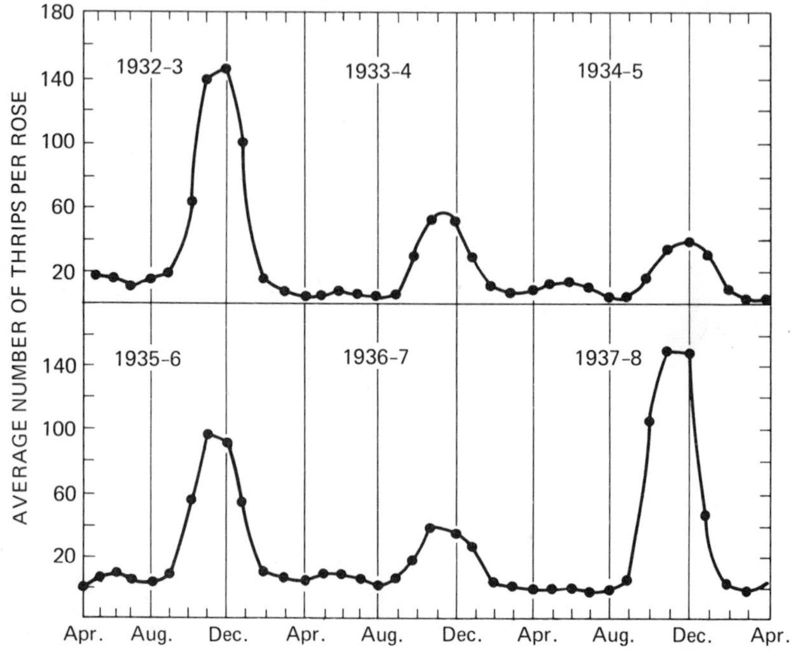

Figure 13–8 Seasonal changes in a population, of adult thrips living on rose flowers. *(From Andrewartha and Birch, 1954.)*

animal populations, such as those of the snowshoe hare (*Lepus americana*) have a periodicity of abundance of 9 or 10 years. Figure 13-9 shows snowshoe hare population estimates from Hudson Bay fur returns and from a variety of other sources for nearly 150 years. Starting about 1865, the beginning of good data, peak populations occurred every 9 to 10 years. Animals which tend to show these 9 to 10 year cycles are the snowshoe hare, the ruffed grouse (*Bonasa umbellus*), and some of the predators of these species, the lynx (*Lynx canadensis*) and the red fox (*Vulpes fulva*). To a lesser extent, marten (*Martes americana*), fisher (*Martes pennanti*), mink (*Mustela vison*), and muskrat (*Ondatra zibethica*) may show 10 year cycles, but not so clearly. Figure 13-10 shows the 10 year cyclical pattern of the snowshoe hare and of the lynx, its most prominent predator. The predator cycle follows the prey cycle, lagging slightly behind it in most cases. The ecology of ten-year cycles of animal populations has been reviewed by Keith (1963).

Other animal populations have a cyclical periodicity of 3 to 4 years. These include the Norway lemming of Scandinavia (*Lemmus lemmus*) and the collared lemming of North America (*Dicrostonyx hudsonicus*), the field mouse (*Microtus agrestis*) of Europe, and some of the predators of these two, notably the colored fox (*Vulpes fulva*) (Figure 13-11) and the snowy owl (*Nyctea scandiaca*). Not

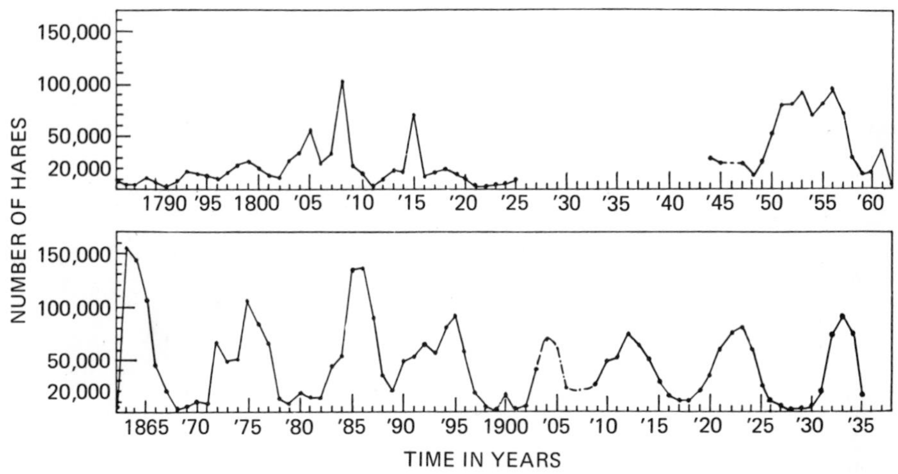

Figure 13–9 Population trends of the varying hare in the Hudson's Bay watershed. (*From MacLulich, 1937.*)

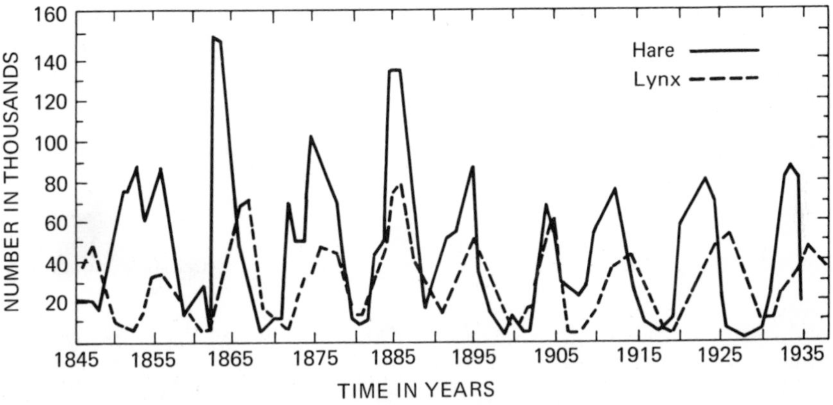

Figure 13–10 Changes in the abundance of the lynx and the snowshoe hare, as indicated by the number of pelts received by the Hudson Bay Company. This is a classic case of cyclic oscillation in population density. (*From MacLulich, 1937.*)

all populations of these animals always fluctuate cyclically. When they appear to do so, sophisticated statistics may be required for proof of real cyclic periodicity.

These animal population cycles have been intriguing and puzzling to biologists for many years. Considerable research has been centered around trying to understand the causes of cycles, and numerous theories have been proposed to account for them. The final answers are still not clear, though many biological

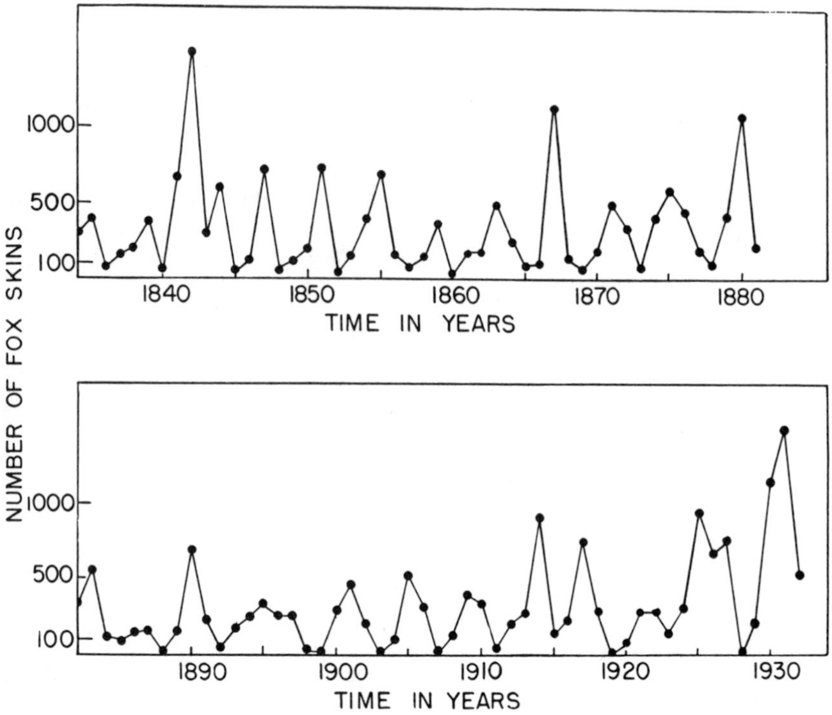

Figure 13–11 Population cycles of colored foxes in Labrador computed from pelt returns reported by Moravian Missions. (*From Elton, 1933.*)

phenomena have been discovered and analyzed in this search for understanding.

In general, two types of factors have been considered as the basic causes of animal population cycles: (1) *extrinsic factors;* that is, those outside of the population itself, such as climate and weather, cosmic events, competition between species, predator-prey relationships, and food supply; or (2) *intrinsic factors;* that is, those within the population, such as disease or parasitism interacting with the population, or behavioral, social and physiological responses within the animals as population increases occur and they become crowded.

Several of these factors can be dismissed rather quickly, since extensive research has failed to show any clear relationships. For example, climatic and meteorological changes have not been satisfactorily correlated with regular long-term cycles of population change in any animal. Sometimes population peaks coincide with years of great rainfall, or some other notable aspect of weather, but these occurrences are by no means consistent.

At one time, sunspot cycles were considered to be correlated with population cycles. Sunspot maxima have occurred every 10 to 11 years, and they were correlated with peak populations of hares and lynx in the late eighteenth and early nineteenth century, but in the last 100 years, they have been entirely out of phase (Figure 13-12).

Changes of populations of predators such as the lynx, fox and snowy owl are clearly correlated with cycles of their prey. But they cannot be shown to cause the prey cycles, for prey cycles often occur in areas devoid of predators. For example, snowshoe hare populations have cycles in areas where no lynx are found.

There has been some attention given to disease and parasitism as the cause of long-term cycles. Twenty years ago there was interest in hypoglycemic shock in snowshoe hares. Animals at peak periods of abundance seemed to have low blood sugar levels and go into states of shock and death. This was interpreted as the alarm reaction of Selye's general adaptation syndrome, which is discussed later. There has also been considerable work on tuberculosis and toxoplasmosis in wild voles (Chitty, 1954; Findlay and Middleton, 1934), and although high disease rates have occasionally been associated with die-offs of peak populations, these associations have not been consistent. Many rodent

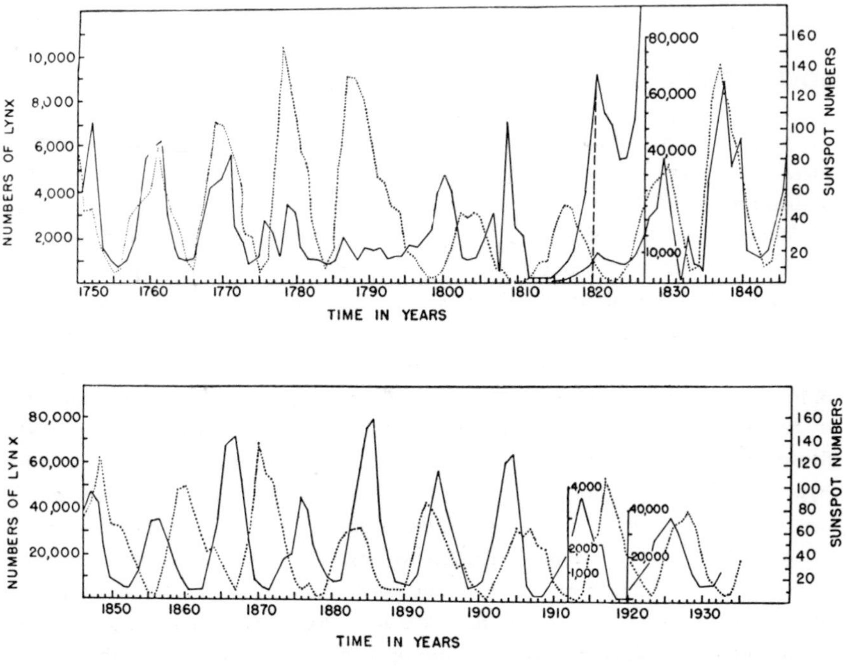

Figure 13–12 Population trends of the lynx (solid line) graphed against sunspot numbers (dotted lines). (*From MacLulich, 1937.*)

declines have occurred without evidence of an infectious disease as the cause. In general, however, there is a need for much more study of infectious diseases in wild animal populations. Most ecologists have not been well trained in pathology, bacteriology and virology, nor have scientists in these disciplines extended their work satisfactorily to population ecology.

Food supply has often been considered important in the decline of animal populations. Lack (1954) proposed that food is the main limiting factor in bird populations as it is for deer populations in North America. Most field studies of rodents have demonstrated, however, that even at peak periods of abundance, ample food is present. Chitty (1960) studied grass during and after high vole plagues and found ample food. There are exceptions; for instance, during lemming peaks food becomes severely depleted and this is one of the factors that is thought to start lemming migrations (Clough, 1965). Pitelka (1958) felt that insufficient attention has been paid to qualitative study of food. The estrogens and steroids in growing plants, particularly grasses and legumes, are now being studied. It has been observed in the Orient, for example, that upsurges of rodent populations are often timed with the flowering and sprouting of bamboo (Tanaka, 1956). Then there is not only increased supply, but the young growing plants are more nutritive and estrogens are increased. The estrogens have direct influence on fertility and fecundity of the animals feeding on them.

SOCIAL STRESS AND PHYSIOLOGICAL RESPONSE

In the last 20 years there has been increasing interest in the role of behavioral, social and physiological factors in control of population size. The observation of hypoglycemic shock in dense populations of snowshoe hares led Christian (1950; 1963) to develop the social stress theory of population limitation. Christian's theory states that population increase results in crowding and increased behavioral interactions, particularly social strife and fighting. Eventually these begin to operate as stressors in individuals. This starts the manifestation of Selye's general adaptation syndrome, and as a result two major changes occur in the population: fecundity is reduced and mortality is increased. The operation of Selye's general adaptation syndrome (G.A.S.) is outlined in Figure 13-13. Stress operates through the sensory pathways to affect the hypothalamus, neurohypophysis, and adenohypophysis so that the endocrine production of the anterior pituitary gland is changed. There is decreased production of the gonadotropic hormones FSH (follicle-stimulating hormone), LH (luteinizing hormone), and LTH (luteotropic hormone), and this decrease is responsible for the decline in fecundity. There is increased production of ACTH (adrenocorticotropic hormone), which stimulates the adrenal glands to increase production of the adrenocorticosteroids such as cortisone. This enables the animal to maintain homeostasis under stress. Blood glucose levels and sodium and potassium balances are kept

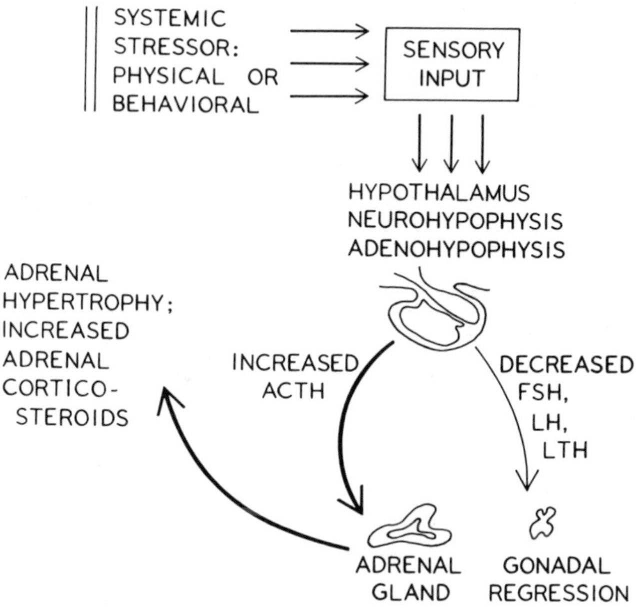

Figure 13–13 General outline of the Stress Syndrome or General Adaptation Syndrome (GAS).

within normal limits and other physiological demands are met. Stress continued for a long time leads to Selye's diseases of adaptation, then adrenal exhaustion and death. Some of these features can be demonstrated in laboratory animals. A rat or mouse exposed to a simple measurable stressor, such as cold temperature, trauma, or surgical injury, will show various stages of the general adaptation syndrome (G.A.S.). Figure 13-14 shows development of the G.A.S. under prolonged stress, and some of the typical symptoms which may develop in experimental animals.

Christian thinks that the phenomenon operates in wild mammal populations and starts population declines. This is based on his work with laboratory populations of mice in crowded conditions. Crowded mice in some experiments showed adrenal hypertrophy and gonadal regression. Chitty (1958; 1960) and Louch (1958) have shown these conditions in some dense field and laboratory populations of *Microtus*, and Christian and Davis (1964) in Norway rat populations in Baltimore City. However, most studies on crowded populations in nature have failed to find such changes in the adrenals and gonads. These studies include the work of McKeever (1959) on vole populations in the United States, Chitty (1961) on wild voles in Canada, Southwick (1958) on wild rodents in England, and Negus et al. (1961) on the rice rat on Cape Breton Island. Thus, field observations do not confirm that the Selye-Christian phenomenon operates

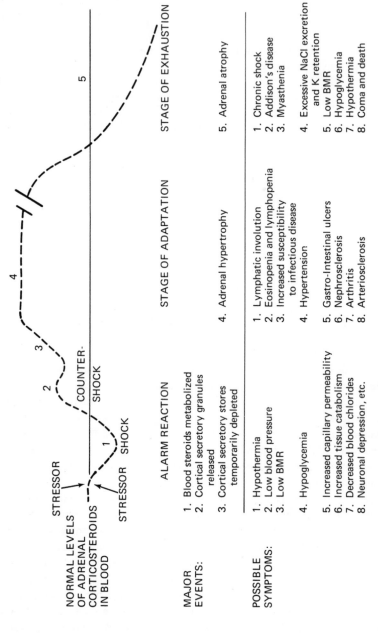

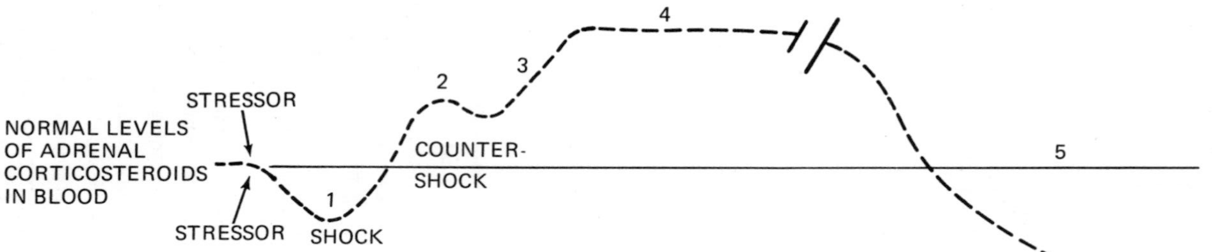

	ALARM REACTION	STAGE OF ADAPTATION	STAGE OF EXHAUSTION
MAJOR EVENTS:	1. Blood steroids metabolized 2. Cortical secretory granules released 3. Cortical secretory stores temporarily depleted	4. Adrenal hypertrophy	5. Adrenal atrophy
POSSIBLE SYMPTOMS:	1. Hypothermia 2. Low blood pressure 3. Low BMR 4. Hypoglycemia 5. Increased capillary permeability 6. Increased tissue catabolism 7. Decreased blood chlorides 8. Neuronal depression, etc.	1. Lymphatic involution 2. Eosinopenia and lymphopenia 3. Increased susceptibility to infectious disease 4. Hypertension 5. Gastro-Intestinal ulcers 6. Nephrosclerosis 7. Arthritis 8. Arteriosclerosis	1. Chronic shock 2. Addison's disease 3. Myasthenia 4. Excessive NaCl excretion and K retention 5. Low BMR 6. Hypoglycemia 7. Hypothermia 8. Coma and death

Figure 13–14 Sequential development of Selye's General Adaptation Syndrome.

commonly in natural populations. Most field studies have indicated that other factors such as food supply, predation, aggression, aberrant parental behavior, and cannibalism operate before symptoms of the G.A.S. are evident. Anderson (1961), writing about rodent populations, concluded: "The gonadal-adrenal-pituitary relationship explored by Christian would be involved only rarely in these phenomena, and its operation in the regulation of rodent population seems to be abnormal or secondary to some other function. It thus appears that under most commonly encountered environmental conditions social and behavioral mechanisms alone may control population density, without immediate involvement of the endocrine system."

Negus et al. (1961) in a field study of population ecology of the rice rat, *Oryzomys palustris*, reported: "The adreno-pituitary hypothesis states that the growth of mammalian populations is controlled by a socio-physiological feedback system that responds to changes in population density. We find this hypothesis difficult to accept. In fact, direct response to climate and food and often other extrinsic factors offers a more plausible explanation of population behavior than does the concept of social stress."

Other ecologists, however, have felt that social stress may play an important role in population regulation. In an eight-year study of population dynamics and social stress in the California ground squirrel (*Citellus beechyi*), Adams et al. (1971) concluded that "an important component of population control is the behavior involved in the dispersal of subordinate animals from population centers. But dispersal is a stressful experience, so stress is also a component of population control." They considered stress and dispersal as two of many possible factors in mammalian population regulation.

Christian's theory cannot be completely rejected, but it apparently occurs only as a secondary mechanism of population regulation in most natural populations.

Parkes and Bruce (1962) elucidated another aspect of behavioral physiology which may limit population growth. They showed that the fertility rate in female house mice is very sensitive to the presence of strange males. When recently impregnated females are exposed to strange males, there is an 80 to 90 percent failure of implantation. This failure does not occur if the sense of smell has been blocked. Physical contact is not necessary for this olfactory block to pregnancy to occur in intact mice; only odor is required.

In natural populations female house mice tend to associate with other females and during pregnancy and nesting they become territorial, excluding males from their nest sites. In crowded populations this territorialism breaks down and there is increased contact with different mice. Hence, it seems likely that the Parkes-Bruce phenomenon may operate in crowded populations and result in a declining fertility rate. It is not uncommon to find in dense rodent populations that there are normal rates of male and female fecundity (that is, the males are producing viable sperm, and the females are ovulating), but pregnancy rates still decline significantly. Thus, there must be some impairment in fertilization and/or embryonic development. No one has yet demonstrated olfactory block phenomena in natural populations, but the theory is compatible with some field data where the Christian social stress theory does not fit.

SUMMARY

The study of animal populations is important for understanding many basic problems of population biology.

Various factors limit animal populations. Among the most important extrinsic factors are food, cover, predation and disease. Interacting with these are intrinsic mechanisms, operating through behavioral and physiological feedback systems to exert control. In feedback mechanisms a change proceeds far enough in one direction to bring into play another group of factors which counteract the change and return the system to its former state. Feedback systems in most populations appear to be so beset by lags that steady control is rarely achieved. As a result, the populations fluctuate within wide ranges. In most populations these fluctuations are irregular and sporadic, but in some populations they fall into fairly regular cycles.

Two prominent feedback mechanisms that have been suggested as factors limiting vertebrate population growth are the Christian-Selye social stress theory, and the Parkes-Bruce olfactory block phenomenon. Both can be demonstrated in the laboratory; neither has as yet been well documented in the field.

It is apparent that most animal populations have a reproductive potential which substantially exceeds the carrying capacity of the environment. This potential may often produce excessive numbers of individuals which are then susceptible to decimation by a variety of mortality factors. At any given time, the triggering factor which increases mortality may be a shortage of food, an infectious disease, a direct behavioral response, or an indirect physiological response mediated through any of the above. At the same time, further reproduction may be impaired in various ways. Thus, it seems illogical to look for any one consistent cause of population regulation, since a great variety of factors are interacting simultaneously and they vary individually in their intensity and significance in different times, places and species.

No completely satisfactory explanation of animal population cycles is available, but it seems likely that the basic motive force and *zeitgeber* is the reproductive potential of the species. This potential regularly tends to produce excessive population levels at periodic intervals. In lemmings, the intervals are usually 3 to 4 years, and in snowshoe hares they are usually 9 to 10 years. Whenever the populations become excessive, they are then vulnerable to decimation through emigration, social strife, abnormal behavior, physiological upsets, disease, or any of these acting in concert.

14

Social Behavior and Population Dynamics

TWO behavioral phenomena which are widespread throughout the animal kingdom, and which have considerable significance in population ecology, are territorialism and dominance hierarchies. Territorialism is essentially a social pattern of spatial utilization by which individuals or groups have control over certain units of space. Dominance hierarchies are rank-ordered systems which determine individual access and priorities to natural resources. Both of these phenomena relate to population ecology intimately because they represent systems of behavioral control which affect the abundance and distribution of animals, their reproductive patterns and their mortality patterns. Many studies of territorial or rank-oriented species have shown, for example, that individuals without territory and low-ranking individuals have less reproductive success and higher mortality rates than those who hold territory and/or have higher rank. Examples of this will be given later.

TERRITORIALISM

A territory has been defined as any area which is defended against other members of the same species (Barnett, 1967; Carpenter, 1958). It may be occupied and defended by an individual, as in the case of a male stickleback fish (*Gasterosteus aculeatus*); by a pair, as in the case of many birds; or by a social group, as in the case of vervet monkeys (*Cercopithecus aethiops*) and gibbons (*Hylobates lar*). It may encompass most or all of the occupant's home range including his nesting and foraging areas, as in primate territories; it may include only a limited area around a nest site or feeding site, as in many birds; or it may involve only a very small plot of ground on which mating occurs, as in the Uganda kob (*Adenota kob*), an African antelope. Hence, territorialism is a highly variable and complex phenomenon, and involves a wide spectrum of behavior patterns. In all cases, however, the essential feature is that individual animals, or groups of animals, have controlling ownership over a given piece of property. Within this property they have the rights of utilization over the space and the resources contained therein. Figure 14-1 shows a typical pattern of territorial spacing as seen in the male song sparrows (*Melospiza melodia*).

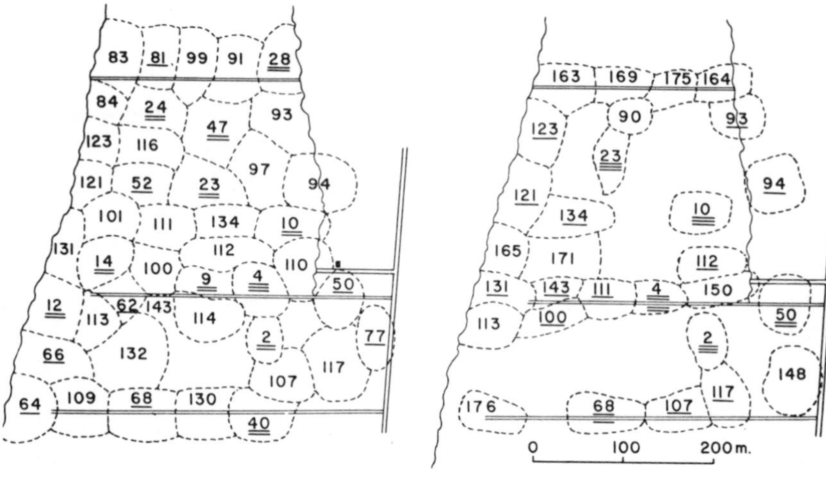

1932 1933

Figure 14–1 Territories of male song sparrows on a brushy floodplain meadow near Columbus, Ohio. The size of individual territories remained approximately constant regardless of population density, in contrast to other animals in which territory sizes expand or contract in relation to density. A bird present in the general area the preceding year is underlined, and a line is added for each subsequent year. (*After Margaret Morse Nice,* Studies in the Life History of the Song Sparrow, *Dover Publications, Inc., New York, 1937. Reprinted through permission of the publisher.*)

Territories are established and maintained by a great variety of behavior patterns and communicative displays. Initially, the establishment of a territory may involve active aggressive behavior and fighting. The territorial combats of Siamese fighting fish (*Betta splendens*), jungle fowl (*Gallus gallus*), elephant seals (*Mirounga angustirostris*) and wild antelope are well known. These combats are often vigorous and are true tests of strength. Injuries are common, and in rare cases, death may occur. Most frequently, however, the vanquished animal merely retreats and vacates the territory.

Once established, territories are most often maintained by communicative signals and displays. These signals may be visual, auditory, or olfactory in nature. Most fish have visual territorial displays which consist of spreading the fins and brightening the colors (Figure 14-2). Some fish have active sound production in territorial calls, as for example, the Atlantic toadfish (*Opsanus tau*). Most birds which are territorial have a combination of vocalizations and visual displays. The male song bird of many species perches in a conspicuous spot and repeatedly gives forth his song, which serves more often as a territorial threat than as a mating attraction (Figure 14-3). Mammals typically use a combination of auditory and olfactory signals as communicative displays in the maintenance of territory. Some of the arboreal primates, such as the gibbon and the howler monkey (*Alouatta palliata*) have an advanced vocal pattern which conveys information about the position of the group and indicates that the territory is occupied. Many carnivores, such as the wolf, lion and tiger (*Felis tigris*) have territorial calls which can be heard for several miles (Figure 14-4), and they also have olfactory signals in scent glands and urine which are used to mark the boundaries of territories (Figure 14-5). The scent signals may persist for days or even weeks, and thus serve as more stable cues than vocalizations which fade rapidly.

Territorialism is common among many higher invertebrates (crustacea, insects, spiders, etc.), and throughout all vertebrate classes. Several authors have suggested that it is an instinctive pattern of behavior in vertebrates (Ardrey, 1966). Ardrey pointed out that clear examples of territorialism can be found in fish, amphibia, reptiles, birds and mammals, which seem to operate on much the same principles and have basically the same functions. Although this is true, it must not obscure the fact that there are many animals which do not show territorialism. Many species do not stake out specific territorial claims and do not actively defend given tracts of habitat. Often, one can find closely related species which are territorial in one case, and nonterritorial in another. In ungulates, for example, the Indian blackbuck (*Antilope cervicapra*) is territorial, whereas the Indian swamp deer (*Cervus duvacelli*) is not. In primates, the gibbon and the howler monkey are clearly territorial, whereas the rhesus monkey (*Macaca mulatta*) and the chimpanzee (*Pan troglodytes*) are not.

Even within the same species, there are circumstances in which the individuals of some populations are territorially distributed and other circumstances in

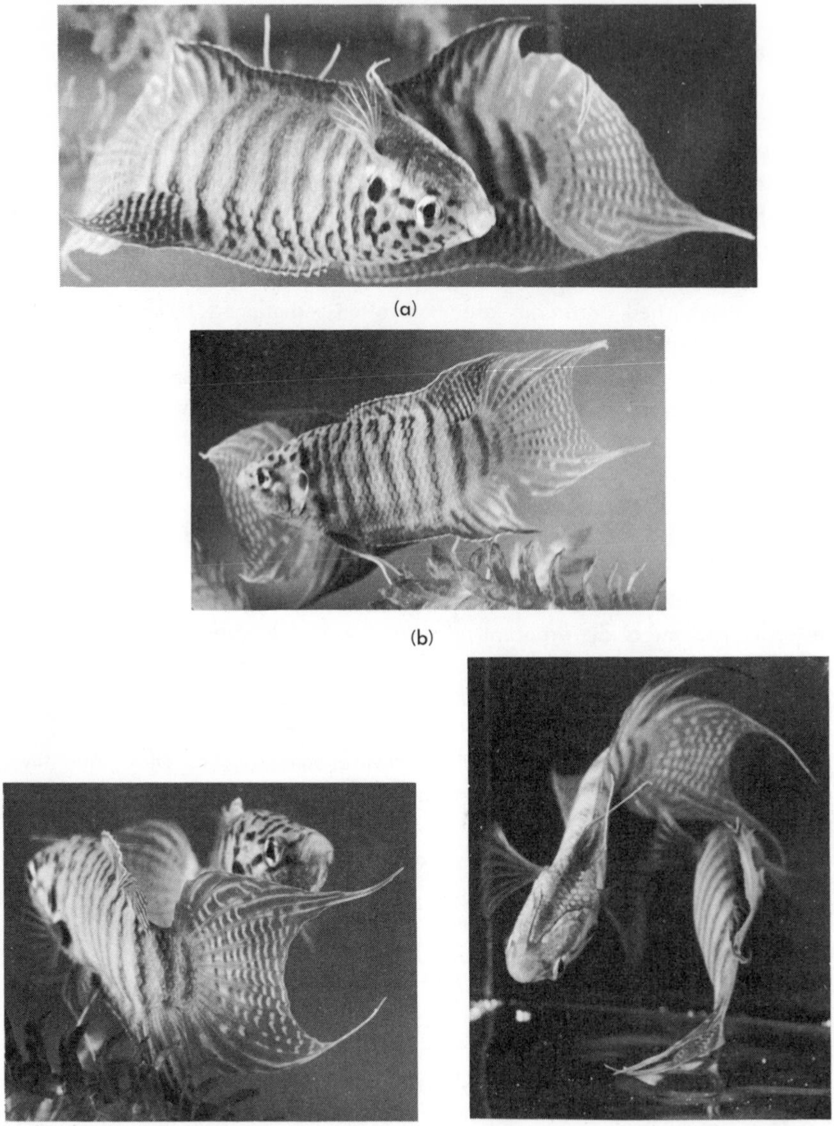

(a)

(b)

(c) (d)

Figure 14–2 (a–d) Aggressive territorial display between male paradise fish. The major components of this display are: (1) a head to tail orientation; (2) full expansion of dorsal and caudal fins; (3) an S-shaped body posture; (4) partial rotation around a vertical axis; and (5) a quiver or vibration of the body accompanied by very rapid beating of the pectoral fins. In these photos, taken sequentially of a single display, the foreground fish is rotating to the left and into the background, whereas the background fish is moving to the right and into the foreground. The beat of the pectoral fins can be seen in "**d.**" (*From Southwick and Ward, 1968.*)

Figure 14–3 Territorial song and posture of male song sparrow. (*From Margaret Morse Nice,* Studies in the Life History of the Song Sparrow, *Dover Publications, Inc., New York, 1937. Reprinted through permission of the publisher.*)

which they are not. This is true in ungulates, for example, many of which are territorial during the rut or mating season, but the same individuals are non-territorial at other times of the year. Apart from this obvious correlation with breeding, however, there is evidence in some species that territorialism is related to the population density and/or environmental circumstances. In the Japanese salmon (*Plecoglossus altivelis*), individual fish are territorial at low population density, but as density increases, individuals give up their territories and form a school (Miyadi, 1956). Wild house mice (*Mus musculus*) show similar behavior with territorialism at low population densities, and mass group behavior at high population densities (Davis, 1958). This variability is related to environmental factors and differences in social behavior of key individuals.

In view of these facts, it does not seem warranted to consider territorialism an instinctive pattern of behavior. Territorialism is certainly common and widespread, but by no means universal within limited animal groups or even within a single

Figure 14–4 Territorial howling of wolf pack. (*From Murie, 1944.*)

Figure 14–5 Territorial marking—a tigress sprays scent against a clump of grass. Kanha, India, March 1, 1964. (*From G. B. Schaller,* The Deer and the Tiger, *© 1967 by the University of Chicago. By permission.*)

species. It seems more appropriate to consider territorial behavior a flexible ecological adaptation—a social and behavioral means of utilizing space.

DOMINANCE HIERARCHIES

Whereas territories function to allot resources on a spatial basis, dominance hierarchies allot resources on an individual priority basis within the same physical space. Hierarchical systems have been observed in all vertebrate classes and in many invertebrates. Table 14-1 shows a typical dominance hierarchy in white-tailed deer (*Odocoileus virginiana*).

TABLE 14-1 Dominance Hierarchy of the White-tailed Deer. Aggressive-submissive Interactions between Dominants and Subordinates[a]

Dominants	*SUBORDINATES*								
	B_1	*WN*	*Bn*	*Sc*	*Dnd*	*Pn*	*Dnl*	*B_2*	*Sp*
B_1	—	4	31	24	15	24	5	27	14
Wn	—	—	30	18	20	21	2	7	9
Bn	—	—	—	23	7	16	11	13	18
Sc	—	—	—	—	11	11	9	19	7
Dnd	—	—	—	—	—	15	8	4	7
Pn	—	—	—	—	1	—	1	12	11
Dnl	—	—	—	—	—	—	—	6	6
B_2	—	—	—	—	1	—	—	—	4
Sp	—	—	—	—	—	—	—	—	—

[a] The scores indicate the number of times the individual deer in the vertical left-hand column was dominant to the individual in the horizontal row. B_1 = dominant 8-point buck; B_2 = young 8-point buck; Sp = spike buck; all others were females; from Collias, 1950.

Dominant animals have priority access to food, mates, nest sites and resting locations. Dominance is often manifested between animals by simple physical displacement—by the displacement of one individual by another at a feeding or resting site. Occasionally it involves displays or threats, and rarely it may involve direct fighting. The latter is more common in initial encounters in which the dominance rank or social status of each individual is not yet determined, but once established, the hierarchy is then usually maintained by display or by social memory. Once a rhesus monkey has established high rank, it does not have to constantly display this rank in a stable social group. The high rank of some individuals and low rank of others become an accepted social norm within the group. Only when the situation becomes unstable through death, emigration or invasion, does there have to be an overt reestablishment of rank within a dominance hierarchy.

SOCIAL BEHAVIOR IN POPULATION REGULATION

The detailed study of territorialism and dominance hierarchies is really in the realm of ethology and animal behavior, but the functions and consequences of these behavioral systems have a great deal of ecological significance. It has been abundantly demonstrated in many territorial animals that the holders of territory have the greatest breeding success. In fact, in some animals, non-territorial animals usually fail to breed. In the Uganda kob, for example, successful breeding by the male requires that he obtain and maintain a territory (Buechner, 1964). In the red grouse of Scotland, males which fail to obtain territories in the fall do not breed in the following spring (Watson, 1964).

Similarly, in those social groups characterized by dominance hierarchies, the high ranking animals have the greatest breeding success, and the low ranking members have less success or fail entirely in breeding efforts. In chicken flocks, low ranking roosters show frequent courting behavior, sometimes more than dominant roosters, but they have relatively little success in completing copulations (Guhl, 1960). Table 14-2 demonstrates relative breeding behavior and

TABLE 14-2 Sexual Behavior and Mating Success of Plymouth Rock Cockerels in Relation to Social Dominance (flock = 36 pullets)[a]

Rank and Name of Cockerel	*Number of Observed Courtings*	*Completed Matings*	*Eggs Fertilized*	*Viable Chicks*
1. Dominant White	710	112	267	221
2. Recessive White	2,184	54	129	120
3. Rose Comb	71	0	0	0

[a] Data from Guhl, 1962.

mating success of dominant and subordinate roosters in a flock of 36 plymouth rock chickens. The lowest ranking cockerel in a group of three failed to complete any matings. In rhesus monkeys, low ranking males may also show courting and sexual mounting, but they have less success in establishing sexual consort relations and achieving complete copulation (Table 14-3). If they do accomplish mating, it is usually after the peak of estrous in the females—a time in which conception is less likely to occur.

It thus seems clear that the social systems of territorialism and dominance hierarchies have a limiting function in population dynamics; that is, they limit the breeding to those individuals which are ecologically and behaviorally most

TABLE 14-3 Sexual Behavior in Aligarh Farm Rhesus Monkey Group (total group = 12 individuals)[a]

Rank and Name of Male	*Number of Observed Sexual Mounts*	*Number of Successful Sexual Consorts*	*Percentage of Sexual Consorts*
1. Dominant male	25	8	73
2. Subdominant male	5	2	18
3. Peripheral male	8	1	9

[a] Data from Southwick and Siddiqi, 1967.

successful in establishing territory or high rank. Theoretically, as population densities increase, and space becomes more limited, a smaller proportion of individuals within the population is able to achieve territories or high rank, and thus total reproduction is reduced in relation to population size. Ideally, these behaviors are density-dependent population limiting factors; that is, the higher the density, the greater the limiting effect on the population.

The role of social behavior in population regulation has been discussed at great length in a book by V. C. Wynne-Edwards, titled, *Animal Dispersion in Relation to Social Behavior* (1962). Wynne-Edwards proposed that the basic long-term population level of many animals is established by food and resource availability, but that social behavior establishes a more immediate regulating influence on population size and density. This influence of social behavior, he felt, comes into play before critical population levels are achieved; that is, before starvation and/or damaging violent aggression are conspicuous.

Thus, according to Wynne-Edwards, most animals have intrinsic, self-regulatory population mechanisms which operate via social and behavioral feedback systems. These social and behavioral feedbacks are usually ritualized displays which relate to territorialism and dominance hierarchies, and thus convey information about population density. Therefore, regulatory feedback is provided according to the intensity of social competition. In other words, ritualized displays, or conventionalized rivalries, regulate breeding success and thus population numbers.

Wynne-Edwards provides numerous examples of this hypothesis, including the complex display patterns of gallinaceous birds (prairie chickens, sharp-tailed grouse, peacocks, bower birds, et al.), waterfowl, and many song birds. He also considers the vocal choruses of amphibians, the schooling of fish, and the flocking of birds and mammals as examples of complex social behavior which serve the function of conveying information about population density and social competition.

Many aspects of Wynne-Edwards' theory are controversial, especially when he theorizes that most social behavior has actually evolved as a means of controlling population levels. Wynne-Edwards proposed, for example, that territorialism has arisen as a social means of preventing overpopulation. This theory has met substantial criticism from other ecologists and students of behavior (e.g., Brown, 1969; Crook, 1965; Lack, 1966; Wiens, 1966; Williams, 1966). These authors felt that there is little reliable evidence to support this hypothesis even though it is logically very attractive. Brown pointed out that, "Population regulation is never completely under the control of the species by itself, but depends in a complex way on interactions between members of the ecological community." Thus, Brown felt, one must consider the interaction of all factors, including predation, disease, food supply, interspecific competition, etc., in understanding the population dynamics of any given species. Brown found satisfactory evidence that territorialism tends to space individuals over available habitat, that it increases emigration and mortality rates in individuals unsuccessful in finding territories, and that it also influences reproductive success. Thus it clearly has a role in population ecology, but Brown did not find satisfactory evidence that it has actually evolved for the prevention of overpopulation. He believed instead that it has arisen in response to individual selection resulting from aggressive competition.

The more detailed aspects of this controversy are outside the scope of introductory ecology. On the more basic parts of the hypothesis—namely, that social behavior provides some regulatory feedback on population ecology—there is certainly no question that this is true in most animal populations.

Even this concession must not obscure the fact, however, that some animals have such poorly developed and/or variable responses to increased density and crowding that they can readily breed themselves into starvation or conditions of violent aggression and extreme social strife. White-tailed deer populations, for example, have frequently increased to the point of extensive starvation. Lemming and other rodent populations have also increased to the point of mass behavioral strife where social mortality becomes extensive.

Quite probably, human populations are also in this category, providing examples of populations in which social feedback limiting population growth often does not come into operation until the point of starvation arrives (as in China and India), or until the point of violent and destructive aggression breaks forth (as in many wars in which population pressure has had a causative influence), or until social mortality becomes limiting (as in countries where abortion is a significant form of population regulation).

It is scientifically important, therefore, to consider animal populations in which environmental crises and/or behavioral breakdowns occur. In terms of human value judgments, these seem to be cases of population regulation by wasteful and sometimes tragic means. In evolutionary terms, however, many of the populations showing these phenomena are highly successful. They seem to capitalize on their adaptive, highly variable, and highly exploitative natures to meet new environmental situations and expand into available opportunities. Perhaps their lack of rigid, stereotyped behavior and ecology, and their lack of precisely tuned population feedback systems, enables them to seize each new ecologic opportunity to the fullest extent.

Some of the best examples of such species are the domestic rodents, especially the wild house mouse (*Mus musculus*), the wild Norway rat (*Rattus norvegieus*), and the Asian bandicoot rat (*Bandicota bengalensis*). Numerous ecological studies over the last 25 years have shown these species to have the following population characteristics:

1. They have tremendous reproductive potential, and under certain environmental circumstances they often show invasive and eruptive population behavior.
2. Populations can readily exceed the carrying capacity of the environment, and this can result in sudden population decline.
3. Populations fluctuate widely; natural control is often abrupt and "ecologically wasteful," in terms of individual health and welfare.
4. Population regulation often involves social disruption and destructive behavior.

These characteristics have been observed frequently in both natural and artificial populations of rodents over the past 40 years. The natural social behavior of the Norway rat and wild house mouse includes both territorial and hierarchical behavior. Females tend to be territorial around the nest site so that they normally exclude adult males and strange females. Often two or three females nest together in a communal arrangement. Males are also territorial if population density is not excessively high, and high ranking males defend a system of runways and the females living therein (Barnett, 1963; Crowcroft, 1966). The size and patterning of the territories depend upon the rank of the male, the density and social structure of the population, and the nature of the habitat.

In a stable habitat, both population levels and social structures of domestic rats might remain quite stable over a period of time. But the habitats of domestic rodents are characteristically unstable with changing amounts of food (around farms, for example, where harvests and food stores come and go), changing mortality patterns (as control compaigns or disease outbreaks wax and wane), or changing cover conditions (as trash availability rises and falls), so that the populations are often experiencing drastically changing resource conditions. Under these circumstances, the populations respond rapidly with surges or declines of growth.

Under periods of rapid population growth leading to high densities, a variety of phenomena have been observed. There is often rapidly increasing social contact resulting in an increase in aggressive interactions (Brown, 1963; Clarke, 1955; Southwick, 1955). This frequently leads to a breakdown of territorial boundaries, increased crowding at the nest sites, and sharply increased mortality of young. Nests become trampled and destroyed. Females often desert the nests, or actually kill the young. Wounds become frequent, especially on adult males, forming sites for mites to invade the skin, so that acarine dermatitis

TABLE 14-4 Population Attributes of House Mice (*Mus musculus* L.) at Different Stages of Population Growth and Crowding[a]

Attribute	*Stage 1: Early Growth, Uncrowded*	*Stage 2: Mid-Growth, Moderate Crowding*	*Stage 3: Terminal Growth, Definite Crowding*
Population size[b]	8–40	40–100	100–125
Pregnancy rate[c]	33 percent	15 percent	8 percent
Number of young born	51	157	118
Percent infant survival	72 percent	15 percent	12 percent
Aggression score[d]	0.05	0.43	1.17
Number of adults per nest box with litter	1.60	3.2	0.29[g]
Condition of nest[e]	good	fair, but polluted	very poor, heavily polluted
Number of wounded adults[f]	1–3	4–18	26–45
Acarine dermatitis	none	slight	severe

[a] Data from populations shown in Figure 12-10, after Southwick, 1955, 1955a.
[b] Total number of postweaned mice (over 3 weeks of age) in population in area of 150 square feet.
[c] Average percent of adult females pregnant on monthly census checks.
[d] Average number of aggressive interactions per hour per mouse.
[e] Good nests—covered or bowl-shaped; fair nests—shallow bowl; poor nests—flat platforms or no nests.
[f] Number of adult mice with moderate to severe wounds from fighting.
[g] A reflection of widespread nest desertion.

may achieve epidemic proportions. Table 14-4 shows typical data on rodent population growth and changes with crowding. These data demonstrate that crowded populations were characterized by a high prevalence of fighting, reduced pregnancy rates, reduced infant survival, nest desertion, environmental pollution, wounding, dermatitis, and other signs of ill health.

With a breakdown of territorial behavior, bizarre aggregations developed, so that clusters of individuals, all in poor physical states, appeared. Calhoun (1962) referred to such aggregations as "behavioral sinks." They are self-stimulating aggregations of individuals, often collectively showing abnormal behavior patterns. Some individuals become inactive and socially withdrawn. As these conditions progress, only limited portions of the environment are utilized, but these become excessively polluted. Thus, the stage is set for both high social mortality and high disease mortality, and the population often collapses as rapidly as it built up. After the crash, if the environment recovers, the surviving individuals recuperate, and population growth may begin again.

The details of this theme differ considerably in different studies and different species. In some circumstances, there may be a cessation of reproduction, so that no young are born into the crowded population, as occurred in house mouse populations studied by Crowcroft and Rowe (1957). In other circumstances, reproduction may continue, but there may be massive infant mortality of all young born (Brown, 1953; Southwick, 1955). In still other cases, crowding may stimulate excessive emigration and dispersal, as in lemmings (Clough, 1965), and these mass migrations are often accompanied or terminated by high mortality.

In all cases, a similar syndrome prevails. There is usually a drastic change in social organization and behavior. It is these behavioral changes which most often have a major influence on reproduction and mortality. In domestic rodents, these changes characteristically involve a breakdown in social stability, and a disruption of normal territorial, reproductive and parental behavior. Frequently, perinatal processes are most dramatically affected, so that the pre- and postnatal survival of the young is reduced. These ecologic phenomena have been extensively documented and discussed in reviews and monographs by Barnett (1964), Calhoun (1962; 1965, et al.), Christian and Davis (1964), Crowcroft (1966), Sadleir (1969), Southwick (1969), and others.

SUMMARY

This chapter has shown the complex and intimate relationship between social behavior and population ecology. Territorialism and dominance hierarchies have been seen as two types of social systems which play a regulatory role in population dynamics, the first by regulating the use of space, and the second by controlling individual priorities within a common space.

In some animal species and populations, territorialism and dominance hierarchies apparently serve as sensitive feedback mechanisms which influence dispersal, mortality, reproductive success and population size. In other species and circumstances, territorialism and hierarchial structures are often subject to breakdown and collapse. Ecological consequences are then profound and extensive.

In many rodent populations, excessive population growth and crowding is often associated with the following phenomena: a breakdown of stable territorialism and dominance structure, reduced fertility as evidenced by lowered pregnancy rates, high prenatal loss and low infant survival, poor nest construction and poor maintenance, nest fouling and pollution, high aggression and wounding, deteriorating health in the majority of the population, abnormal aggregations of individuals in the "behavioral sink" pattern, and finally population decline.

It is apparent that species and populations differ substantially in the extent and sensitivity of their intrinsic regulatory mechanisms. Some populations may show a conservative and economical cessation of reproduction while still retaining social and ecologic stability within the population. Other populations reveal an ecologically wasteful and tragic mortality while experiencing a collapse of social and ecologic stability. Species characteristic of the latter pattern, however, are often highly adaptable and successful in exploiting varying ecologic conditions and new environmental situations. They are thus characterized by invasive, exploitive, and eruptive population growth—a pattern which might be considered advantageous to the species, but with a high cost to the individual.

15

Interspecific Populations

ALL populations of living organisms exist in a network of interactions with other populations. Many of these interactions are more subtle and complex than direct food relationships. Some of them are cooperative and beneficial to one or more of the interacting populations; others are competitive or limiting to the interacting populations. Cooperative interactions are represented by commensalism and mutualism, which are special types of symbiosis. Competitive or limiting interactions are represented by predation, parasitism (including infectious diseases of all types), interspecific competition, and amensalism or antibiosis. Some knowledge of the ecologic principles involved in these special relationships is necessary to understand the broader aspects of population dynamics. In other words, we cannot fully appreciate population changes in any one species without knowledge of its interactions with other species.

The term symbiosis means "living together,"

and in its broadest sense it refers to a relationship of any type between two or more living organisms. A symbiotic relationship may be beneficial or detrimental to the interacting organisms. In common usage, however, symbiosis has come to mean primarily those relationships which are beneficial or stimulating to one or more of the interacting populations. Two populations interacting in such a way as to be beneficial to one and neutral to the other is referred to as commensalism; two populations interacting in such a way as to be beneficial to both is termed mutualism.

COMMENSALISM

An example of commensalism is the remora-shark relationship where the remora fish (*Echeneis naucrates*) attaches to the skin of the shark by means of a strong sucker disc and is transported widely and rapidly by the shark's motive power. The remora also consumes food remnants cast off from the jaws of the shark. Thus the remora benefits in two ways from this attachment, and the shark is relatively unaffected, although its swimming speed may be slightly impeded.

Many large organisms provide commensal harborage for other organisms externally and internally. Large tropical trees provide habitats for numerous commensal plants and animals. Trunks and branches provide attachment sites for epiphytes such as orchids, and the recesses between buttress roots provide harborage for bats, tree frogs, lizards, insects and many other organisms. Whales provide attachment sites for barnacles, algae, and other sessile marine forms. These plants and animals are not parasitic, that is, they do not extract nutrients from the commensal organism with which they are associated, but they do utilize the habitat provided by the host.

Most animals, including man, also contain internal commensals. The human digestive tract is inhabited by several types of bacteria and protozoa, such as *Endamoeba coli,* which are neither parasitic nor pathogenic, but merely residual in the alimentary canal.

Another common type of commensalism occurs when various species of plants and animals use burrows or nests constructed by another species. Termite nests provide ecologic niches for more than 100 species of other animals including ants, aphids, beetles, millipedes, and isopod crustaceans. Tube-dwelling annelids of the genus *Chaetopterus* which inhabit the tidal zone of the seashore provide habitats for small crabs of the genus *Polyonyx*. The crab benefits by the protection of the tube, and it apparently does no harm to the annelid worm which originally built the tube.

Commensalism may be obligatory, in which one organism depends entirely on another for its habitat, or it may be facultative, in which each organism may live independently, but one is enhanced in the presence of the other. An example of obligatory commensalism is the relationship between the algae, *Basicladia,* and certain fresh-water turtles. The algae grows only on the backs of the turtles. Similarly, the bivalve mollusc *Ostrea* grows almost exclusively on the roots of the red mangrove off the coast of Florida. Attached to the shell of the horseshoe crab, *Limulus,* are several species of molluscs, barnacles, and annelid worms, which again are not parasitic, but only residential, and some of them live only on the *Limulus shell.* The oyster crab is a small crustacean which lives only in the mantle cavity of oysters. These all represent cases in which one species of plant or animal has evolved a dependent relationship upon another species, but the relationship is one of habitat or attachment and does not involve feeding upon the tissues of the host. The possibility exists that this is one evolutionary route toward parasitism, and the distinction between obligatory commensalism and parasitism is often difficult to draw.

Facultative commensalism is represented by the relationship between the burrowing owl (*Speotyto cuniculario*) and the prairie dog (*Cynomys* sp.). The burrowing owl often nests in prairie dog burrows, but is not confined to such burrows. In the north polar regions, the arctic fox (*Alopex Lagopus*) feeds in the wintertime on the remains of seals killed by polar bears (*Thalarctos maritimus*), though it may have other sources of food as well. Probably the majority of commensal relationships are facultative. In the tropical forests of Central and South America, when howler monkeys (*Alouatta palliata*) and white-faced monkeys (*Cebus capucinus*) feed on ripe figs high in the canopy of the forest, they drop many uneaten and partially eaten fruits. Wild deer (*Odocoileus rothschildi*), peccaris (*Tagassu tajacu*), agoutis (*Dasyprocta agouti*), pacas (*Cuniculus paca*) and other terrestrial animals come to such monkey-feeding trees, and harvest the fallen fruit. They do not depend entirely upon monkeys for their food, but they benefit from the feeding behavior of the monkeys. Further details on various types of commensalism are given by Clarke (1954) and Allee, et al. (1949).

It is especially interesting to consider animal populations which form facultative commensal relationships with men. Among vertebrates perhaps the best examples are domestic rodents, such as house mice and Norway rats, which have increased in abundance and distribution by their association with human habitation and agriculture. The house sparrow, pigeon, and Asian house crow, and Asian house shrew or musk shrew have also greatly increased their population sizes through human associations. Cities, towns and farms provide increased food supplies and nesting sites for these animals. The rhesus monkey of India is one of the few primates which has become a facultative commensal of man. The rhesus can readily survive in forests without man, but it became especially abundant around villages and towns of the Gangetic basin where it could live on agricultural produce.

MUTUALISM

Mutualism, that relationship where both interacting populations benefit or are positively stimulated by the association is classically represented by the association of algae and fungi to form lichens. Fungi provide the framework, moisture and attachment sites in which algal cells grow, and the algae provide food production for both itself and the fungi. The association of hermit crabs (*Eupacurus prideauxi*) and sea anemones (*Adamsia palliata*) is also a typical example of mutualism. The hermit crab provides an attachment site and transportation for the sea anemone (on a discarded mollusc shell which the hermit crab has appropriated), and the sea anemone provides camouflage and defense for the hermit crab.

Many types of mutualistic relationships exist throughout the living world. Other well known examples include the small tick birds which accompany rhinos and other ungulates, picking off ticks and ectoparasites from their skin; the wrasses, or cleaning fishes, which feed on surface debris and ectoparasites of larger fishes; and the baboon-impala relationship in which the superior vision of the baboon (*Papio anubis*) provides visual sentry and alarm for the impala (*Aepyceros melampus*), while the superior sense of smell of the impala provides olfactory warning for the baboons. These mutualistic relationships may involve physiological, structural, and/or behavioral parameters, and they may function in the realms of nutrition, protection, transportation, bodily care and maintenance, and a variety of other life support activities.

Like commensalism, mutualism may be obligatory or facultative. An example of obligatory mutualism is provided by the relationship between the flagellate protozoan *Trichonympha,* and wood-eating termites. These species cannot live without each other. The protozoan lives only in the digestive tract of the termite and digests cellulose. The termite provides essential habitat, a constant environment, and the basic food material for the protozoan, and the protozoan provides a vital digestive process for the termite. The stomach and caecum of the horse contains millions of ciliate protozoa and bacteria which digest cellulose for the horse and provide 20 percent of its nitrogen requirement per day. These intestinal microorganisms are essential for the normal growth and health of the horse.

An elaborate form of obligatory mutalism occurs between some ants (e.g. *Cremastogaster lineolata*) and aphids or plant lice (e.g. *Aphis caliginosa*). The ants maintain the aphids in specially constructed nests, and they feed upon a nutrient liquid secreted by the aphids. The aphids have sometimes been called "cows" because they release the secretion when stroked by the antennae of the ants. Thus the ants receive food from the aphids, and the aphids receive protection, harborage and care from the ants. Many other arthropods, including mites, springtails, flies, wasps and beetles, have developed similar complex relationships with ants. These relationships are based on interspecific communication systems through which the different species jointly utilize food, shelter or other resources (Hölldobler, 1971).

Many mutualistic relationships are facultative; that is, both populations can exist without the other, but both are favored when living together. Baboons and impalas, for example, can survive well without each other, but each provides a unique protective warning system for the other when they live in association. Squirrels facilitate the extension and propagation of hickory trees by burying the nuts, but the hickory trees can propagate without squirrels, and the squirrels can likewise survive without hickory nuts, if an adequate supply of other food is available.

Man has capitalized on mutualism as a biological principle in the development of agriculture. The domestication of plants and animals originated in facultative commensalism or mutualism. At the dawn of agriculture, 10,000 years ago, man could live without domesticated plants and animals, but as he developed these commensal relationships his productivity and survival were enhanced. Some of these relationships then evolved into obligatory commensalism and mutualism. Modern man obviously cannot live without the domesticated plants and animals of agriculture, and many of these domesticated forms cannot live without the special husbandry provided by man.

PREDATION

Predation is the capture of live animals for food. It is typically pictured as the relationship between the hawk and the mouse, the bass and the minnow, the lion and the antelope, and so forth. Predators are carnivores in the trophic levels above the herbivores. Thus, a hawk feeding upon a seed-eating bird is in the third trophic level; a hawk feeding upon an insectivorous bird is in the fourth trophic level, and so on.

Predation is such a dramatic relationship between different animal populations, and natural selection has developed such elaborate adaptations around predation, that it has received considerable attention from naturalists and ecologists. Only within this century, however, has modern man come to an adequate appreciation of the ecologic importance of predation. In fact, many people, especially those who promote extensive bounty programs directed toward eliminating hawks, foxes, coyotes, cougar and wolves, have still not realized the real significance of predation.

One of the first clear demonstrations of the ecologic importance of predators in maintaining the balance of animal populations was in the Kaibab story of the 1920's. From 1906 to 1930 an intensive predator control campaign for several years resulted in the mass killing of wolves, cougars, coyotes and lynx. Subsequently the white-tailed deer population erupted, growing from an original herd of 4,000 deer to over 100,000 by 1924. The deer decimated all of the available food in the forest, and in the winter of 1924, an estimated 60,000 died of starvation. It took several decades before the health of the forest and the deer herd were restored (Allen, 1954).

In Yellowstone Park, excessive predator control resulted in an increase of the bighorn sheep and elk population. The habitat became overcrowded, range conditions deteriorated, and the animals became infested with lungworm, pneumonia and hemorrhagic septicemia. These diseases are not entirely understood, but they are thought to be related to poor range conditions and possibly the lack of predator removal of unhealthy individuals.

Several studies on predation have shown that predators selectively remove young, old and diseased or injured individuals from the prey population. Durward Allen and his students have shown that the wolves of Isle Royale National Park prey almost exclusively on young and old moose, particularly those which are weak (Mech, 1966). An adult animal in prime condition is rarely taken by the wolves. Table 15-1 shows the age distribution of 50 moose killed

TABLE 15-1 Age Distribution of Moose Killed by Wolves on Isle Royale, Michigan[a]

Age Class of Moose	*Estimated Age in Years*	*Number of Wolf Kills*	*Percentage of Kills in Age Class*
Calf	1	18	36
I	1	—	—
II	2–3	—	—
III	3–4	—	—
IV	4–7	—	—
V	6–10	3	6
VI	8–15	15	30
VII	10–17	3	6
VIII	—	5	10
IX	—	—	—
IXA	20?	6	12
Totals		50	100

[a] From Mech, 1966.

by wolves on Isle Royale. Ninety-four percent of the moose killed by wolves were calves under one year of age or were old adults, 8 to 20 years of age. Murie (1944) found in an analysis of bighorn sheep mortality, some of which was attributable to wolf predation, that 69 percent of the deaths were of old animals, 16 percent were of lambs and yearlings, and 9 percent were of severely diseased adults. Only 6 percent of the known deaths occurred in healthy adults in the 2 to 8 year age range. Schaller's study on the deer and tiger in India (1967) and the lion and wildebeest in Africa (1969) have produced similar findings, though he did find in India that tigers take some chital deer in prime adult condition. Their diet is not exclusively the young, old and diseased, though it is primarily so.

Thus, predation in animal populations seems to function as a natural method of quality control. By removing those individuals in a population which can be caught, predation tends to eliminate the slow, the weak, and the incapable. The alert, healthy, and well-adapted animals are less likely to fall victim to the predator.

This point has been emphasized by Errington in his lifelong studies of predation (1967). Errington observed in muskrats, quail, grouse and other animal populations which he had studied, that certain segments of the population were highly vulnerable to predation and other segments quite secure. The vulnerable individuals were those which, by some attribute of health or behavior, were most frequently exposed. Behaviorally, it is often the nonterritorial animals or subordinate individuals which roam most widely and traverse marginal habitats most frequently. Thus, they fall prey to predator action. We see again, from a different point of view, the regulatory role of predation in animal populations.

There are, of course, many kinds of predation which are not necessarily so selective. The swallow feeding on moths and midges, the top minnow or trout feeding on mosquitoes, and the anteater feeding on an ant nest is less discriminate in taking individuals. Certainly, many able and healthy individuals are preyed upon. In cases such as these, which involve large numbers, the prey population often shows great productivity which can accommodate this mass mortality. Many fish, insects, and other invertebrates have tremendous production of eggs and larvae, but only a small percentage of these survive the onslaughts of predation and other forms of loss. For example, the horseshoe crab or *Limulus* of estuarine and coastal waters deposits its eggs on sandy beaches where they are consumed in great numbers by flocks of laughing gulls and other shorebirds. Large numbers of eggs also wash out to sea and are eaten by fish, whereas others are pushed up on the beach and dry out. Despite this fantastic loss of eggs and young larvae, *Limulus* continues to survive in large populations and has done so for the last 180 million years. It is one of the oldest animals on earth in terms of its evolutionary history, having outlived trilobites and dinosaurs by millions of years, and it is still incredibly abundant.

There has been considerable interest in predation as a factor in quantitative population control as well as its role in quality control. On one hand, sportsmen have wanted to limit predation, being concerned that predators may take excessive numbers of game birds and wildlife. Some sportsmen's organizations have therefore supported predator control campaigns against hawks, owls, foxes, coyotes, wolves, cougar and other predators in the name of conservation. On the other hand, ecologists have often been interested in promoting predation as a method of biological control of pest animals, particularly insects and rodents, and even as a necessary biological control of desirable animals such as deer, to prevent them from becoming pests.

The scientific literature on predation provides examples in which predation does exert a quantitative control on animal populations and other cases in which it does not. Murie (1944) concluded that wolf predation was one of the primary limiting factors on the numbers of Dall sheep in the Mount McKinley region of Alaska. He also felt that wolf predation was an important check on the caribou population. Apparently, predation was an important limiting factor on the white-tailed deer of the Kaibab National Forest, since this population skyrocketed upward after the extensive predator control programs begun in 1906. Similarly, Schaller concluded that tiger predation was the main limiting factor on the populations of chital (*Cervus axis*), sambhar (*Cervus unicolor*) and swamp deer (*Cervus duvauceli*) in central India (1967). Schaller calculated that an adult tiger requires from 6,300 to 7,800 pounds of prey animals per year in its diet. This usually involves 30 to 60 prey animals per year with an average weight of 50 to 100 kilograms per animal. This is a significant harvest on the prey population and probably represents a major controlling influence.

There are also examples from the entomological literature where predation serves as a significant population control on an animal population. In a classic study of natural population balance in the knapweed gallfly (*Urophora jaceana*), a small dipteran insect, Varley (1947) concluded that one of the major controlling factors was the predation of mice upon insect pupae. His data showed that 22 to 43 percent of the fly mortality was attributable to this predation. One study of tsetse flies (*Glossina* spp.) in Africa, found predatory spiders to be a controlling factor on the abundance and distribution of the flies (Sladen, 1969). Insect predators have also been used in applied biological control of agricultural pests in a few cases. In California, the cottony cushion scale, an hemipteran insect pest of citrus crops, was effectively controlled by a predatory ladybird beetle imported from Australia. In Hawaii, the sugar cane leafhopper has been controlled by the capsid bug (Kendeigh, 1963).

Examples of predator-prey interactions in which the predator does not have a significant controlling influence on the prey population are also abundant. In the classic population cycles of snowshoe hare and lynx, the lynx feeds primarily on the hare, but it does not seem to exert any major controlling influence on the hare populations. The hare populations seem to rise and fall independently of the lynx; in fact, hare populations in areas devoid of lynx are also cyclic. The great controlling influences on the hare populations seem to be food supply and intrinsic population factors in relation to excessive density, crowding and stress.

Similarly, population studies of muskrats and bobwhite quail have shown that predation is not a major population control in these species. Errington (1946) concluded that, "A great deal of predation is without truly depressive influence." He further observed:

> In the sense that victims of one agency simply miss becoming victims of another, many types of loss—including loss from predation—are at least partly intercompensatory in net population effect. Regardless of the countless individuals or the large percentage of population that may annually be killed by predators, predation looks ineffective as a limiting factor.

Wynne-Edwards expressed a similar concept when he wrote (1962):

> Predation is not in its own right a density-dependent process, independently capable of controlling a prey population from outside: the "cooperation" of the prey population, in insuring that its surplus members are specially vulnerable to predators through the operation of the social machine, is almost sure to be the indispensable condition underlying whatever density-dependent, homeostatic influence predation may be found to have.

In other words, Errington and Wynne-Edwards found evidence from their studies that predation takes primarily surplus individuals: individuals that would succumb to some other mortality factor. They might be physically inferior, they might be in submarginal habitat, or they might be behaviorally vulnerable to predation, disease, starvation, or any one of several mortality factors. Errington summarized his views on predation by writing (1967): "Predation belongs in the equation of life." That is, he felt it is an important part of equilibrium and balance in the interacting web of populations.

Another important aspect of predation as an influence in population ecology and natural selection is the great variety of adaptations which have arisen in response to it. Protective coloration, warning coloration and mimicry fall into this category. Also, the morphological and behavioral responses of flight speed, armor, diversionary behavior, freezing behavior and many other specialized patterns have developed in various species as adaptations to reduce mortality through predation. One particularly interesting system which has been studied in some detail is the intimate relationship between certain moths and the bats which prey upon them. The bats locate the moths by emitting ultrasonic pulses and detecting the reflecting echoes from flying moths, a natural sonar system (Novick, 1969). The moths have evolved the ability to detect these ultrasonic pulses of the bats, and, upon perceiving the approach of a feeding bat, they instantly go into complicated diversionary flight spirals to avoid the bat. The interaction becomes an aerial "dog fight" between bat and moth, with the moth trying to avoid the bat. This ability and behavior has evolved in selective response to predation in the same way that cryptic coloration has evolved. In both cases the advantageous mutations have been selected and propagated.

PARASITISM

A parasite is an organism living in or on the body of another organism and deriving nutrition from it. It may be a temporary parasite, such as a wood tick, or it may be a more permanent resident, such as a tapeworm. It may weaken, debilitate, or eventually kill the host, or it may cause relatively little harm to the host.

Parasitism is virtually universal in all plants and animals. In vertebrate animals, endoparasites are found within many of the major organ systems of the body, most commonly in the digestive, circulatory, respiratory and urogenital systems. Ectoparasites occur on or within the skin and its appendages, such as hair and scales. Throughout most of the world, man is beset with many parasitic organisms including intestinal worms (tapeworms, flukes, roundworms, etc.), intestinal protozoa (amoeba, ciliates and flagellates), parasites of the blood (microfilaria, malaria protozoa, etc.), and a variety of ectoparasites (lice, mites, ticks, mosquitoes, etc.). Many infectious microorganisms including bacteria and viruses are also parasites in the sense that they extract nutrients from the host. Some of these are pathogenic, that is, they impair the health and normal functioning of the host, as for example, the bacterial organisms of the genus *Shigella* which cause bacillary dysentery. Others, such as some intestinal flagellates, are nonpathogenic and do not harm the host. Many parasitic organisms may be pathogenic in one individual and nonpathogenic in another. For example, most individuals infected with *Endamoeba histolytica* in the digestive tract do not have amoebic dysentery, and, in fact, may have no disease symptoms at all, but in other individuals, the same organism may produce severe disease and even death. This is a matter of individual ecology and immunology which is not well-understood.

As mentioned previously in the section on commensalism, there is no sharp line of distinction between obligatory commensalism and parasitism. Theoretically, *Endamoeba histolytica* feeding on tissues of the host is a parasite, whereas the flagellate, *Trichomonas hominis,* feeding on digested foods before incorporation into host tissue is not a parasite, nor is *Endamoeba coli* feeding on undigested particles of food and intestinal bacteria.

Neither is there a sharp line of distinction between a parasite and a predator. Horseflies and vampire bats bite and suck blood from livestock, and they might be considered either parasites or predators. Normally, we think of predators as killing their prey in a short time. Man's relation to the cow is predatory when he kills it for beef, but it is parasitic when he milks it for dairy products. There are some relationships, however, in which the distinction is arbitrary, as, for example, in the lamprey feeding upon a host fish. The lamprey may kill the fish in a few days or a few weeks, depending upon its size in relation to the fish, and may be considered either a predator or a parasite.

Ecologically, one of the most interesting questions about parasitism is the effect of the parasite population upon the host population. The oldest and best adapted parasites have little or no pathogenic effect upon the host. Pinworm infections in children are relatively harmless, though they may cause a minor irritation around the anus. In West Bengal, India, studies have shown that over 75 percent of the local population harbors hookworm (*Necator* sp. and *Ancylostoma* sp.), but there is no demonstrable pathology or effect on the people (Chowdhury and Schiller, 1968). The infectious burden (i.e., the number of parasites per person) seems to be small and insignificant. Most wild animals have parasites and are able to maintain excellent health with these parasites. Zebras in Kenya are intensively infected with many internal and external parasites, but they are usually in remarkably robust health.

On the other hand, there are numerous examples where parasitic infections are harmful and debilitating to the host. Certainly, hookworm (*Necator americana*) in the southern United States as it occurred 50 years ago was a debilitating disease, sapping the strength and health of the infected person. Throughout the world, malaria still remains as a major health problem and it causes a great amount of illness and human misery. One of the most important debilitating diseases of modern times is schistosomiasis, an infection of blood flukes of the genus *Shistosoma*. This is a waterborne infection and has spread throughout the tropical world with the expansion of irrigated agriculture.

In animal populations, parasites may also weaken infected individuals. In the bighorn sheep of Wyoming and Idaho, lungworm infections are still a major cause of illness and mortality. In dogs and wolves, heartworms reduce the animal's vitality and hunting success. In meadow vole populations, botfly parasitism causes larger spleen weights in infected mice and reduces their survival when exposed to cold temperatures (Clough, 1965). Periodic reductions of field voles in Wales have been attributed in some years to heavy infections with tuberculosis among the mice, but in other years similar reductions occurred without the infection (Chitty, 1954). Similarly, reductions in populations of red grouse in Scotland have been attributed in some years to the roundworm parasite, *Trichostrongylosus pergracilis,* but this has not been a consistent finding in all years (Committee of Inquiry on Grouse Disease, 1911; see Jenkins et al., 1963). Occasional heavy mortality of gray squirrels in Baltimore has been associated with coccidiosis, caused by a protozoan parasite of the digestive tract (Flyger, 1969). In 1971, a virus disease of wildfowl decimated 90 percent of the pheasants of England, in much the same way as myxomatosis reduced rabbit populations of England and Europe in the 1950's.

Thus, parasitism as an interspecific population relationship may be balanced or unbalanced in regard to its effect upon the host. A seriously unbalanced relationship, in which the host dies, is, of course, selectively disadvantageous to the parasite as well.

Another outstanding feature of parasitism is the interspecific complexity which has evolved in some species. Many parasites have primary and secondary hosts and, in some cases, tertiary hosts for various stages of the life cycle. The malaria organism, *Plasmodium falciparum,* for example, has parts of its life cycle, the asexual phase of reproduction, within the mosquito, and part, the sexual phase of reproduction within the human host. Schistosomes infect man or other vertebrate animals as the primary host for sexual reproduction, and they inhabit aquatic snails for asexual reproduction. The Chinese liver fluke, *Clonorchis sinensis,* infects man as a primary host, aquatic snails as a secondary host, and fresh-water fish as tertiary hosts. Despite the complexity of this life cycle, it is a very successful and widespread parasite, infecting millions of people throughout the Orient (Chandler and Read, 1961).

A few cases are known where ectoparasites have had a limiting effect upon host populations. For example, Moss and Camin (1970) found that bird mites (*Dermanyssus prognephilus*) in the nests of purple martins reduced both the size of the broods which could be reared by the parent birds and the growth rate of the young birds. Parasitized birds produced an average of 3.6 young, whereas unparasitized birds produced an average of 4.2. This difference in productivity was statistically significant. Nest parasites have also been known to cause nest desertion in starlings and excessive mortality of young in phoebes (Moss and Camin, 1970).

COMPETITION

Competition may be defined as the mutual utilization of limited resources. Competition implies that the resources utilized by one individual or one species would have been of value to other individuals or species. Ecologic studies have shown that two different species cannot occupy precisely the same niche; that is, they cannot coexist with identical requirements for food and habitat. This represents a simplified statement of Gause's Principle. Closely related species with a substantial degree of niche overlap often compete for the same resources and frequently one may displace the other. This has been called "interspecific exclusion" or "competitive exclusion" and it is considered to be definitive proof of interspecific competition.

One of the first clear demonstrations of interspecific exclusion was the work of Gause on laboratory populations of *Paramecium caudatum* and *Paramecium aurelia,* two closely related species of ciliate protozoans. When cultured separately in the laboratory, each species grew well on the same medium, utilizing bacteria as food, but when cultured together, *P. aurelia* always displaced *P. caudatum* which died out in approximately 16 days (Figure 15-1). Similar results have been obtained with crustacean and insect populations.

The waterflea, *Daphnia pulicaria,* in mixed populations, causes the extinction of *Daphnia magna,* when oxygen and food are in limited supply (Frank, 1957).

In mixed populations of mosquitoes in laboratory conditions, *Aedes albopictus* always replaces *Aedes polynesiensis,* a result of both larval and adult competition (Gubler, 1969; Lowrie, 1971). This finding may have some practical applications if the same principle also operates in natural habitats. *Aedes polynesiensis* is a vector for filiariasis in South Pacific islands, whereas *A. albopictus* is not. Theoretically, *A. albopictus* could be introduced to small islands to reduce or even eliminate a serious disease vector. Such field trials are now underway, and they may demonstrate the possibility of biological control based on interspecific exclusion.

Sometimes the results of interspecific competition change dramatically with various environmental conditions. The flour beetle, *Tribolium castaneum* displaces *T. confusum* in a competitive interaction if the culture medium is free of parasites, but if the populations are parasitized by the protozoan *Adelina tribolii,* the reverse is true (Park, 1948).

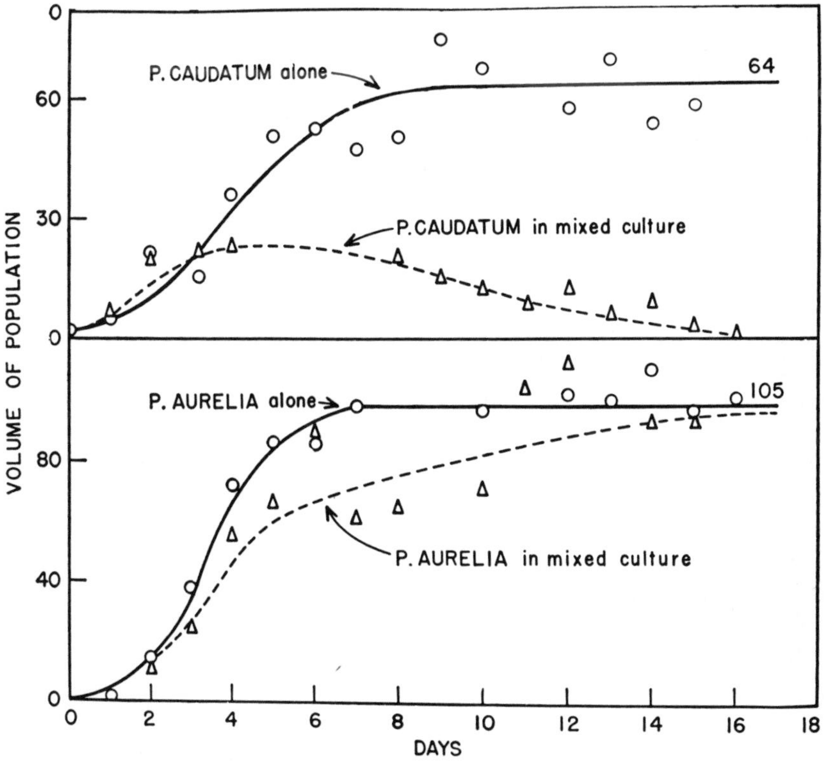

Figure 15–1 Competition between two closely related species of protozoa which have identical niches. When separate, *Paramecium caudatum* and *Paramecium aurelia* exhibit normal sigmoid growth in controlled cultures with constant food supply; when together, *P. caudatum* is eliminated. (*After Allee et al., 1949.*)

In another well-known interaction between competitive species of flour beetles, *Tribolium* displaces a smaller beetle of the genus *Oryzaephalus* if the habitat does not have sufficient complexity and cover in which the smaller beetle can retreat. But if the habitat is provided with small segments of capillary tubes in which *Oryzaephalus* can find refuge, then both species can coexist in the same environment (Crombie, 1946) (Figure 15-2).

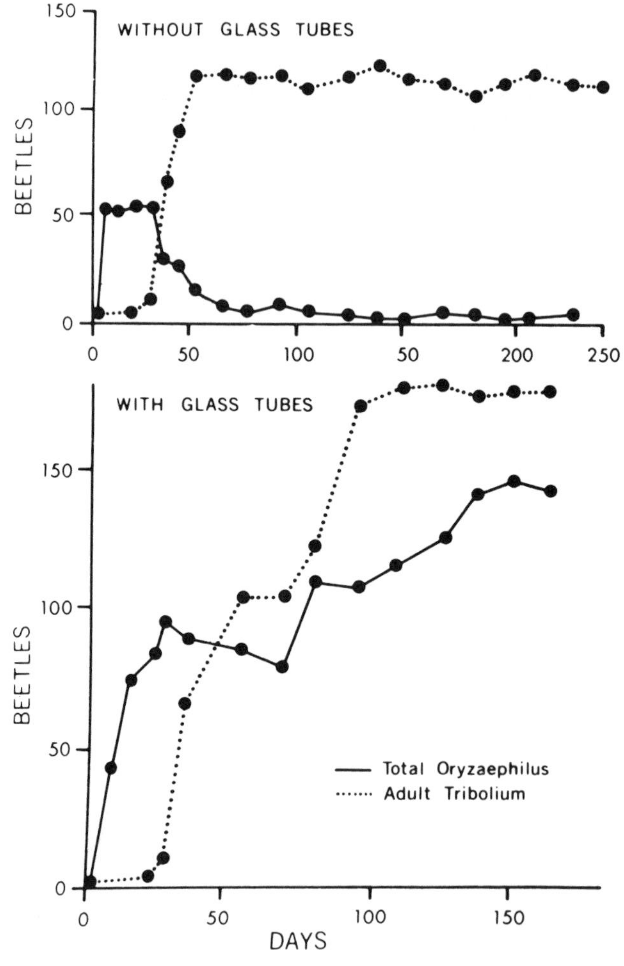

Figure 15–2 Growth of populations of beetles of two closely related genera, *Tribolium* and *Oryzaephilus,* cultured in single vials. In the experiment represented by the upper graph, the medium was plain flour, and *Oryzaephilus* always became extinct. When small pieces of glass tubing were added, both populations survived, as shown in the lower graph. (*After Crombie, 1946. By permission of The Royal Society.*)

The mechanisms of competitive exclusion between closely related species are tremendously varied. The success of one species over another may be as direct as simple aggressive behavior, in which one species drives another away, or it may be due to a subtle difference in biological success. The successful species may have a slightly higher fecundity rate by which it lays more eggs, and thus each generation raises more young to occupy a limited habitat. The successful species may have a slightly enhanced survivorship which could depend upon disease resistance; greater success in finding food, nest sites or mate; or greater success in avoiding competitive confrontation. Conversely, the successful species may be more capable of utilizing and benefiting from cooperative relationships.

Our knowledge of the various routes to ecologic success is just beginning, and it promises to be a very fruitful area for more intensive ecological research. With more satisfactory understanding of why some species displace others, and why some species flourish while others wither and become extinct, we can capitalize on this understanding for the practical management of wild populations. We can also gain insight on critical issues of human survival. Although the human population of the world represents one biological species, it is obvious that there are many variants of this one species which are in competition for the limited resources of the earth. It is biologically and politically unrealistic to believe that all the present forms of man, in economic, social and political terms, will survive with equal success. It is similarly unrealistic to believe that such success will depend upon physical strength or economic resources alone. The study of competitive and cooperative interactions will give us much needed perspective on the ingredients of ecological success and stability.

AMENSALISM AND ANTIBIOSIS

Amensalism is an interspecific relationship in which one population is inhibited while the other is unaffected. A simple example is the shading out of certain plants under tall trees. The trees reduce the available sunshine at the ground level, and numerous species of plants cannot find adequate light in the shade. Hence, only shade-tolerant plants with lower light requirements can survive as ground cover in the forest.

Antibiosis is a specific type of amensalism in which one organism produces a metabolite that is toxic to other organisms. The best known example is the mold, *Penicillium,* which produces an antibiotic substance causing the death of many bacteria. This was observed and identified in the classic work of Florey and Fleming in England in the late 1920's and early 1930's which led to the development of antibiotics in clinical medicine. Penicillin, streptomycin, aureomycin and other antibiotics used against disease organisms all represent a similar ecologic principle, and one which is widespread in nature. Many species of fungi and lichens produce metabolic substances which inhibit bacterial growth. The green algae *Chlorella* produces substances which inhibit the growth of diatoms of the genus *Nitzschia,* and conversely, *Nitzschia* produces a substance which inhibits *Chlorella.*

Many plants also produce substances which are toxic or inhibitory to animals. Old cells of *Chlorella* produce a substance which inhibits the feeding of *Daphnia.* Algal blooms of some blue-green and red algae produce chemicals toxic to fish. The well known Red tide of southern coastal waters, produced by the flagellate *Gymnodinium brevis,* can result in massive fish kills.

Land plants also produce inhibitory substances. The roots of some species of trees produce substances which inhibit the growth of other trees, both of their own species and different species. This may result in a characteristic spacing pattern of trees within a forest, dependent not so much upon competition for sunlight at the canopy level, but upon competition for space, water or nutrients at the subterranean root level. Many plants produce substances toxic to animals which touch them or eat them. In Nevada and Utah the desert bush *Halogeton glomeratus* produces oxalic acid which is poisonous to sheep. Poisonous toadstools and tubers have been well known throughout the history of man. In South America and Africa, many native peoples harvest the tubers of the poisonous manioc plant, but they place it in a series of soaking vats to leach out the toxic chemicals before consuming the starchy tissue of the roots. A small African monkey, the talapoin monkey (*Cercopithecus talapoin*), has learned to steal the manioc from the soaking vats at just the proper stage.

Many of these substances are in a class of compounds known as ectocrines or environmental hormones. Ectocrines are substances produced by one organism which affect other organisms of the same or different species. Not all ectocrines are toxic or inhibitory, of course; many are stimulatory or beneficial. But collectively, they represent chemical messengers by which individuals and species are interrelated.

SUMMARY

Population interactions between species assume a wide variety of forms ranging from mutually beneficial relationships which may be essential for the survival of one or more of the species, to detrimental relationships by which one species kills, inhibits or excludes another. The term symbiosis actually includes any relationship throughout this continuum, but in common practice it usually refers to those relationships which are favorable or beneficial. This continuum of interrelationships is such that it is often difficult or arbitrary to draw precise lines of distinction between some types of relationships, as, for example, between commensalism and parasitism, or parasitism and predation. Borderline examples can be found which emphasize the dynamic nature of interspecific patterns and indicate some of the possible evolutionary relations between these patterns.

Interspecific relations are often major selective forces in evolutionary change. In both cooperative and competitive relationships, specific adaptations to these relationships can be seen. For example, the multitude of adaptations which have arisen to predator-prey relations or to host-parasite relations are clearly evident. These involve specialized patterns of behavior, coloration, adaptive morphology, biochemistry, and, in fact, any biological trait of an organism.

Commensalism is an interspecific relationship in which one population is enhanced or benefited and the other population is unaffected. Mutualism is a relationship in which both populations are enhanced or benefited. Commensalism and mutualism may be either facultative, in which case either population can live without the other, or obligatory, in which one or both populations cannot exist without the other.

Predation is a relationship in which one animal species kills another animal for food. It can exert both a qualitative and quantitative control on the interacting populations. A number of studies on predation have shown that the larger and more dramatic predators, such as wolves and lions, tend to cull out the young, old, and diseased individuals from the prey population. Predation has been considered a regulatory force on some prey populations, as, for example, caribou and moose, but in other prey populations, for example, quail and muskrats, it has been thought to eliminate only surplus individuals which would succumb to some other mortality factor if predation were not present.

Parasitism is an interspecific relationship in which one population derives its nutrition from another, usually without killing the host. It is virtually a universal relationship in plants and animals, and it cannot be always sharply distinguished from some types of obligatory commensalism or predation. Parasites may cause substantial pathology in the host; that is, they may be disease-producing as in the case of the malaria or schistosome parasites, or they may reside in the host with no noxious effect, as in the case of many intestinal amoeba.

Interspecific competition is another powerful ecologic force on populations. Gause's principle states that two species cannot occupy precisely the same niche. Closely related species with very similar niche requirements often interact in such a way that one species displaces another, a phenomenon called interspecific exclusion.

Amensalism or antibiosis is also a widespread phenomena in interspecific population interactions. It is most dramatically seen in the mold *Penicillium,* which produces metabolic products that are toxic to many bacteria.

The analysis of interactions between populations of different species is essential in understanding community structure and dynamics, the subject of the next chapter.

PART 5

Community Ecology

16

Biotic Communities

BIOTIC communities consist of all the plant and animal populations inhabiting a given area. They represent a higher order of biologic organization than populations, yet, since communities refer only to living organisms, they are not as inclusive as ecosystems. In other words, the study of communities involves the entire biology of an area, but strictly speaking, it does not involve the specific study of the interactions with the nonliving components of the environment as does ecosystem analysis. This, of course, is a somewhat arbitrary distinction, and of more value for academic convenience than ecologic reality. As we will soon see, it is impossible to talk about biotic communities without frequent reference to the nonliving environment or total ecosystem.

We have already considered several important aspects of community organization in previous chapters when we discussed food chains and ecologic pyramids. These are convenient ways of analyzing the relationships between producer organisms, herbivores and carnivores in the web of life. The principles of food chains and ecologic pyramids will underlie all of our subsequent considerations on community ecology.

As with populations, it is again convenient to think of communities in terms of structure and dynamics. Structure refers to the spatial organization of communities and the way different populations are related in shape and form. Dynamics refers to the interactional processes, energetic relationships, and patterns of change within communities.

Biotic communities assume a variety of forms as diverse as life itself. If one's imagination alternately pictures a tropical forest, a coral reef, a grassland, a temperate desert, an arctic slope, or the benthic depths of the ocean, one can quickly grasp the tremendous variation which occurs in natural communities. Major terrestrial communities, each characterized by certain types of plants and distinctive life forms, are known as *biomes*. It is obviously impossible to describe in any one book or set of books, the wide range and complexities of all these communities. It is, perhaps, more realistic to consider some principles of community structure and function as they relate to a familiar biome, the temperate forest, and then to briefly describe a few of the common community types found throughout the world.

THE TEMPERATE FOREST COMMUNITY

A good starting point in the study of the forest community is to examine its structure in physical and stratigraphic terms. The forest may be thought of as a layered system with each layer possessing characteristic populations and a typical organization.

A stratigraphic representation of a forest community is shown in Figure 16-1, and some of the typical plant and animal populations of various strata are listed in Table 16-1a.

The structural base of the entire community, of course, is the soil. It supplies the fundamental reservoir in terms of nutrients and water. It is both an aquifer and basic food bank. The soil itself is composed of strata or layers, known as horizons, each with definable properties. The upper layer, or A horizon, commonly called "top soil," consists of the living roots of vascular plants, the mycelia of fungi, a great abundance of bacteria, protozoa, algae, and other microorganisms, and a surprising variety of burrowing animals (nematodes, molluscs, annelids, insects, various other arthropods, amphibians, reptiles, moles, shrews, etc.). Table 16-1b shows general data on the abundance of various soil microorganisms and animals. These figures would differ significantly in various types of soils, but such data are representative of forest and grassland soils. Of nematodes alone, it has been said that if all plant tissues were instantly dissolved and made to vanish, the outline of trees, shrubs,

Figure 16–1 A stratigraphic representation of a forest community. (*Drawing by E. M. Holcomb.*)

grasses and herbs and their roots would still be discernible by the presence of nematodes which live within all plant tissues.

The A horizon of soil also consists of dead and decaying organic material of plant and animal origin. This organic material is undergoing decomposition in the process of humus formation or humification. The A horizon is further subdivided into various layers representing different stages of humus formation, and these layers are commonly called leaf litter, duff, leaf-mold, humus, and leached zone.

The second major stratum of soil, the B horizon or subsoil, consists of a mixture of the underlying parent material and inorganic compounds which have filtered downward from the processes of decomposition in the A horizon. Nitrates, phosphates, acids, bases, salts and other compounds resulting from organic decomposition have moved downward by the leaching action of water into the B horizon. The plant and animal communities of the B horizon are considerably smaller in numbers and species than those of the A horizon.

The third major layer of soil, the C horizon, consists of the basic parent material or substrate. It may be of sedimentary origin resulting from water deposition (alluvial), glacial activity (glacial till), or wind deposition (eolian or loess), or it may also be igneous, metamorphic or volcanic in origin. In general, soils forming on alluvial, glacial or eolian depositions are the richest,

TABLE 16-1a Typical plant and animal populations of a temperate lowland forest community in Maryland

STRATA	TYPICAL PLANTS	TYPICAL ANIMALS
Canopy	maple beech sycamore elm	squirrels vireos warblers flycatchers
Intermediate trees	maple dogwood black gum	squirrels raccoon opossum vireos warblers flycatchers
Shrubs and ground cover	mountain laurel huckleberry honeysuckle poison ivy jack in the pulpit trillium bloodroot	white tailed deer cottontail rabbits opossum raccoon wood mice wood thrush ovenbirds shrews spiders insects
Soil—A horizon	roots of all the above fungi algae	shrews moles larval insects annelids isopods nematodes molluscs
Soil—B horizon	roots of trees and shrubs	annelids nematodes
Soil—C horizon	substrate	

and some of the most productive agricultural areas of the world are found in river deltas or glaciated grasslands.

Soils may be classified in a variety of ways according to their texture, mineral content, organic composition, or profile. Soil profile refers to the depth and vertical patterning of the A, B and C horizons. The A horizon may range in depth from less than an inch in some tropical forests, to several feet in rich delta grasslands. The profile of any region is a function not only of the parent material and geologic history, but also of the climate, topography and biotic community. In temperate forests, for example, the turnover of plants and process of decay is slow, but mineral formation is rapid, so that the

TABLE 16-1b Soil Communities: Typical Abundance of Microorganisms and Invertebrate Animals

Living Organism	*Abundance*	*Reference*
Bacteria	1,000,000,000 per gram	Tepper, 1969
Actinomycetes	5,000,000 per gram	Tepper, 1969
Protozoa	500,000 per gram	Tepper, 1969
Algae	200,000 per gram	Tepper, 1969
Molds	200,000 per gram	Tepper, 1969
Nematodes	175,000 to 20,000,000 per square meter	Clarke, 1954
Molluscs (slugs and snails)	50,000 per acre	Encyclopaedia Britannica
Myriapods (millipedes and centipedes)	1,000,000 per acre	Encyclopaedia Britannica
Annelids (mainly earthworms)	1,000,000 per acre	Encyclopaedia Britannica
Arthropods (insects, woodlice, spiders, etc.)	1,000,000 per acre	Clarke, 1954

organic humus layer of the A horizon remains relatively thin. In grasslands, with rapid turnover of annual plants, rapid decay, but relatively slow mineralization, large amounts of humus are formed each year. Thus, the A horizon becomes very thick. The average humus content of grassland soils has been estimated at 600 tons per acre, compared to only 50 tons per acre in forest soils (Daubenmire, 1947, quoted in Odum, 1959). Tropical forest soils are characterized by both rapid decay and rapid mineralization, so there is relatively little humus formation or organic buildup, rarely more than 5 tons per acre. Thus tropical soils in forest regions are characteristically very thin, and when virgin tropical soils are first exposed they may be deceptively productive for only one or two years. They quickly lose worthwhile productivity unless they are managed very carefully.

The nature of the soil profile obviously has a major influence on the animal community living within it and vice versa. Forest soils tend to have large populations of those organisms adapted for processes of slow organic turnover and decomposition—such organisms as fungi, isopods, millipedes, soft larval insects, and so forth. Grassland soils have higher populations of those organisms which depend upon high rates of organic turnover—bacteria, nematodes, annelids and xerophilic insects.

Animal communities of soil also have major influences on the process of soil formation. Many burrowing animals play a vital role in soil aeration, physical cultivation, hydration and decomposition. Earthworms not only aerate the soil and contribute to its waterholding capacities, but they bring deeper soil materials to the surface and contribute to soil mixing and decomposition. Darwin, in his classic book, *The Formation of Vegetable Mould through the Action of Worms,* published in 1881, estimated that earthworms brought 20 tons of soil to the surface per year. More recent studies have shown that the amount of soil brought to the surface by earthworms may vary from 2.1 tons/acre/year in hot dry soils to 10.7 tons/acre/year in prairie soils, to 107 tons/acre/year in tropical delta soils (Kendeigh, 1961).

Other burrowing animals are also important in cultivating the soil and preventing compaction. In prairie soils, ants may bring over a ton of soil to the surface per acre per year, and prairie dogs may excavate 30 to 40 tons per acre in and around their burrows (Kendeigh, 1961). Not only do all burrowing animals till the soil physically, but they contribute to its chemical maturation by processing organic material through their digestive systems.

Soil formation is a slow process of ecologic maturation. Forest and grassland soils involve many centuries of organic accumulation, and they represent the stored capital of very long-term ecologic processes. The rate at which these processes occur varies greatly according to climate, topography, substrate, biologic community and all those ecologic factors influencing the processes themselves. It is thus impossible to give a precise figure on the rate of soil formation. Some temperate soils are literally millions of years old. In fact, several studies have shown that temperate soils require 10,000 years to develop recognizable soil profiles, and some profiles are not really well-developed in less than 200,000 years. Tropical soils develop much more rapidly, of course, but they are thin and susceptible to destruction. It is thus important for modern man to realize that soil is not a synthetic commodity, or one that can be manufactured in a short time.

The soil-air interface is a particularly rich and active area for living organisms. One can obtain a partial glimpse of the dynamic complexity of this biotic community by carefully examining the leaf litter of the forest floor, turning over a rotten log, or parting the grass and herb cover of a rich meadow. Then one can view a variety of insects, isopods, spiders, and myriapods, but those which are easily seen represent only a small portion of the total community. They are interacting with a great number of smaller forms—collembola or springtails, mites, nematodes, annelids, and so on—and they are also part of the food chain of vertebrates which patrol the area—salamanders, reptiles, shrews, mice and ground dwelling birds. Figure 16-2 is a graphic reminder of the complexity of food chains in a fallen log, and it must be remembered that this figure in itself is a simplification. It includes only arthropods and not vertebrates.

It is also important to realize that the soil-air interface has numerous communities within communities. Fallen logs, woodland ponds or temporary pools, animal nests, and the scats or stools of animals are all examples of complex microcommunities in or near the forest floor.

In proceeding upward from the floor of the forest, the biotic community thins out to a certain extent, animals become more widely spaced in three dimensions, and they become more mobile. The plant community is dominated

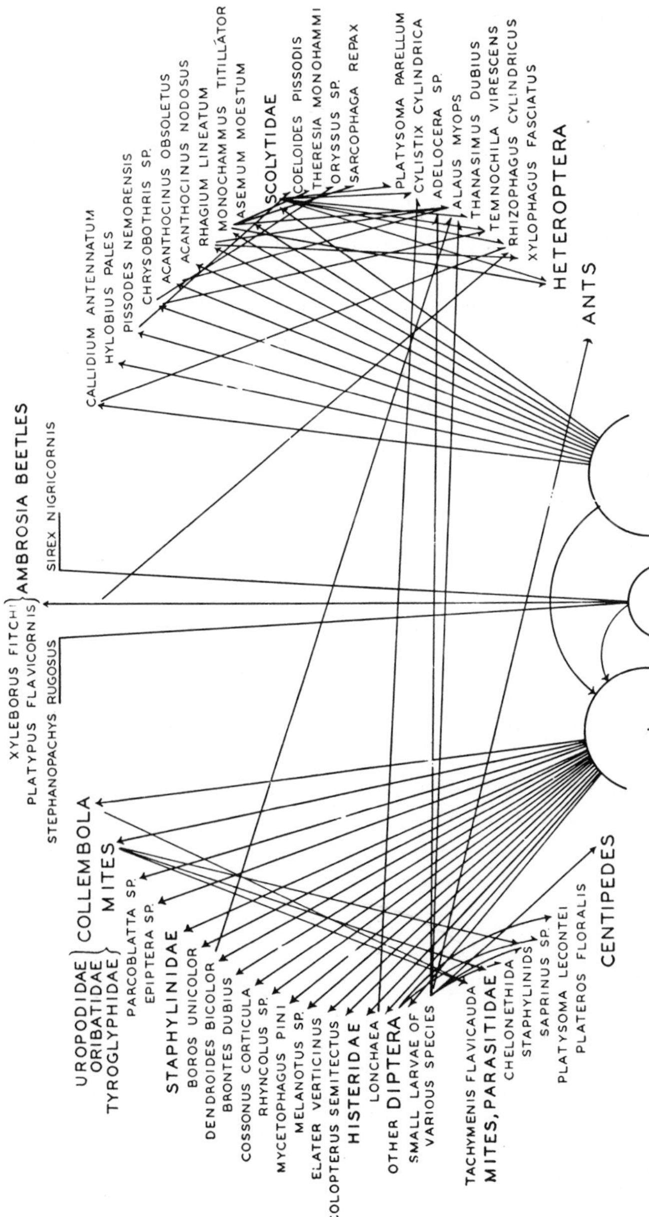

Figure 16–2 Food web of a pine log habitat. *(After Savely, 1939. Copyright Duke University Press. By permission.)*

by herbs and shrubs, and the animal community by insects, birds and mammals. Mammals, including deer, rabbits, mice, squirrels, shrews, raccoon, possum, etc., actively forage through the lower layer of this community.

Many animal taxa including annelids, some molluscs, myriapods, and soil-dwelling arthropods do not enter this realm and are seldom if ever found above the surface of the ground. There are exceptions, of course, such as certain snails (molluscs) which climb trees.

The intermediate and canopy layers of the forest, dominated completely by the foliage of trees and vines, have an even more restricted animal community. This is the realm of birds and insects, and, in temperate forests, only a few mammals penetrate these upper levels, the squirrels (*Sciurus* sp. et al.), opossum (*Didelphis virginiana*) and raccoon (*Procyon lotor*). In tropical forests, a much greater wealth of animal life lives in the canopy including many large mammals —monkeys, sloths (*Choloepus* sp. or *Bradypus* sp.), tayra (*Tayra barbara*), coatis (*Nasua narica*) and ocelot (*Felis pardalis*).

Stratification is evident in bird populations which are obviously capable of ranging throughout the forest from the floor to the canopy, but have definite preferences and tendencies to frequent certain layers. A study of forest bird populations in southern England (Colquhoun and Morley, 1943) showed a definite stratification of bird life: wood pigeons (*Columba palumbrus*) in the upper canopy, tree creepers (*Certhia* sp.), and blue tits (*Parus caeruleus*) in the intermediate zone, marsh tits (*Parus palustris*), and great tits (*Parus major*) in the shrub zone, and robins (*Erithacus rubecula*) and wrens (*Troglodytes troglodytes*) on the ground and in the herbaceous zone (Table 16-2). These patterns of

TABLE 16-2 Relative Stratal Abundance of Birds in Bagley Wood (after Colquhoun and Morley, 1943)[a]

Bird	*Stratum 1*	*Stratum 2*	*Stratum 3*	*Stratum 4*	*Stratum 5*	*Percentage Observed Feeding*
Wood pigeon	333[b]	3	3	—	—	1
Nuthatch	*34*	*34*	1	—	—	27
Blue tit	150	*264*	156	24	6	64
Long-tailed tit	122	*183*	136	18	9	76
Treecreeper	32	*75*	27	17	—	57
Coal tit	45	*108*	78	20	—	79
Marsh tit	15	111	*155*	81	7	74
Great tit	25	74	*197*	103	12	56
Goldcrest	2	10	33	14	—	86
Blackbird	2	7	25	29	47	4
Robin	—	—	29	32	19	21
Wren	—	—	20	*140*	20	10

[a] From p. 489 in Allee, Emerson et al., 1949.
[b] Figures in italics indicate stratum of highest density for each species.

vertical distribution reflect the feeding habits of the birds, and are thus an indication of the distribution of seeds and insects. Morse (1970) has shown that the stratum distribution of many birds within the forest is further limited to specific sites. He found, for example, that the brown creeper (*Certhia familiaris*) and white-breasted nuthatch (*Sitta carolinensis*) forage mainly on the lower part of tree trunks, whereas the downy woodpecker (*Dryobates pubescens*) and Carolina chickadee (*Parus carolinensis*) forage on twig tips high in the canopy. As we learn more about the spatial patterns and movements of animals, we will undoubtedly have many more examples of specific distributions within animal communities.

SPECIES DIVERSITY

In 1958 the eminent ecologist of Yale University, G. E. Hutchinson, delivered an address at the annual meeting of the American Society of Naturalists and used the title, "Homage to Santa Rosalia or Why Are There So Many Kinds of Animals?" He took the title from the Sicilian saint, Santa Rosalia, near whose sanctuary on Monte Pellegrino, Sicily, Hutchinson had found a small pond, rich in animal life, and limestone caves rich with the bones of Pleistocene animals. The setting gave Hutchinson additional inspiration to reflect on the ecology of evolution. Why has natural selection produced such a great variety of living organisms—over a million known species of animals alone, three-quarters of which are insects?

This topic has been a major question in ecologic and evolutionary theory for many years. In discussing the evolution of species diversity, Hutchinson emphasized the mosaic nature of the environment. When the environment is viewed in very small units of space, it may be considered an infinitely variable mosaic of different conditions. Hutchinson pointed out that the small physical size of many animals, such as insects, has permitted them "to become specialized to the conditions offered by small diversified elements of the environmental mosaic, and this clearly makes possible a degree of diversity quite unknown among groups of larger animals." Furthermore, the addition of each new plant or animal to the community increases its diversity, thus creating new conditions on which natural selection can operate. Diversity, then, becomes a self-stimulating phenomenon.

The limiting factors on diversity are the severity of the physical circumstances to which life forms can adjust. All life, having a common biochemical basis, has certain central tendencies in the conditions around which life processes operate with the greatest efficiency. This does not infer that all species of plants and animals have the same environmental requirements, but it does mean that more plant and animal species are going to survive and thrive at environmental conditions approaching the normal requirements for protoplasmic processes. In other words, the biochemical unity of life insures that more plant and animal species are going to do well at 80 to 90° Fahrenheit (27 to 32° Centigrade) than at 32° F (0° C). Plants and animals can, of course, adapt to the latter, but the selection for this will be substantially more rigorous, and the available niches much fewer.

Seen in this light, it becomes reasonable to theorize that those environments whose physical circumstances are harshest and most inimical to life will have the least diversity. Thus we would expect deserts, ocean depths (below the light zone), polar regions and snow-capped mountains to have the least diversity in their biotic communities. This is, indeed, exactly true. These are biotic communities characterized by the lowest numbers of different kinds of plants and animals, though there may be very large populations of these few species.

On the other hand, we would expect those environments with conditions most nearly approaching the ideal for life processes—that is, warm temperatures, ample light and plenty of moisture—to have the greatest diversity of plant and animal life. This also is true. The tropical forests are far richer in number of species of plants and animals than any other biotic community on earth.

Thus, an important ecologic principle is that species diversity of biotic communities generally increases in proceeding from polar regions to the equatorial tropics, with the exception of deserts, mountaintops and ocean depths. Forests of northern Canada often have less than 10 species of trees; temperate forests of the United States often have 20 to 30 species, whereas the tropical forests of Panama usually have over 100 species of trees in relatively small areas. In almost every plant and animal taxa, similar relationships prevail. Canada has 22 species of snakes, the United States has 126 and Mexico has 293. Table 16-3 shows the number of species of breeding birds found in different parts of the western hemisphere. In general, polar regions have less than 80 species of

TABLE 16–3 Number of Species of Breeding Birds in Various Parts of the Western Hemisphere

Region	*Degrees Latitude*	*Breeding Birds: Number of Species*
Antarctica	90–65 (S)	20
Greenland	80–60 (N)	56
Labrador	60–52	81
Newfoundland	47–52	118
New York	41–45	195
Florida	25–30	143
Guatemala	14–18	469
Panama	6–9	1,100
Venezuela	1–10	1,148
Colombia	4–12 (S–N)	1,395

breeding birds, temperate regions have 100 to 200, and tropical regions have 500 to 1300.

Another major factor which affects species diversity is pollution and environmental quality. Pollution usually reduces species diversity. By creating abnormal conditions which require new adaptations, both air and water pollution may reduce the number of species capable of tolerating these new conditions. Figure 16-3 shows the relative numbers of Phyla, Classes, Orders and Species of aquatic invertebrates found living in two streams near Baltimore county. The data were collected in one day by an ecology class and do not represent a thorough survey. One of the streams, Herring Run, passes through a heavily populated area of northern Baltimore, and it is polluted with urban run-off, pesticides from lawns, silt from construction, salt, tar and oil from streets and miscellaneous trash. Only 8 species of invertebrates were found, representing

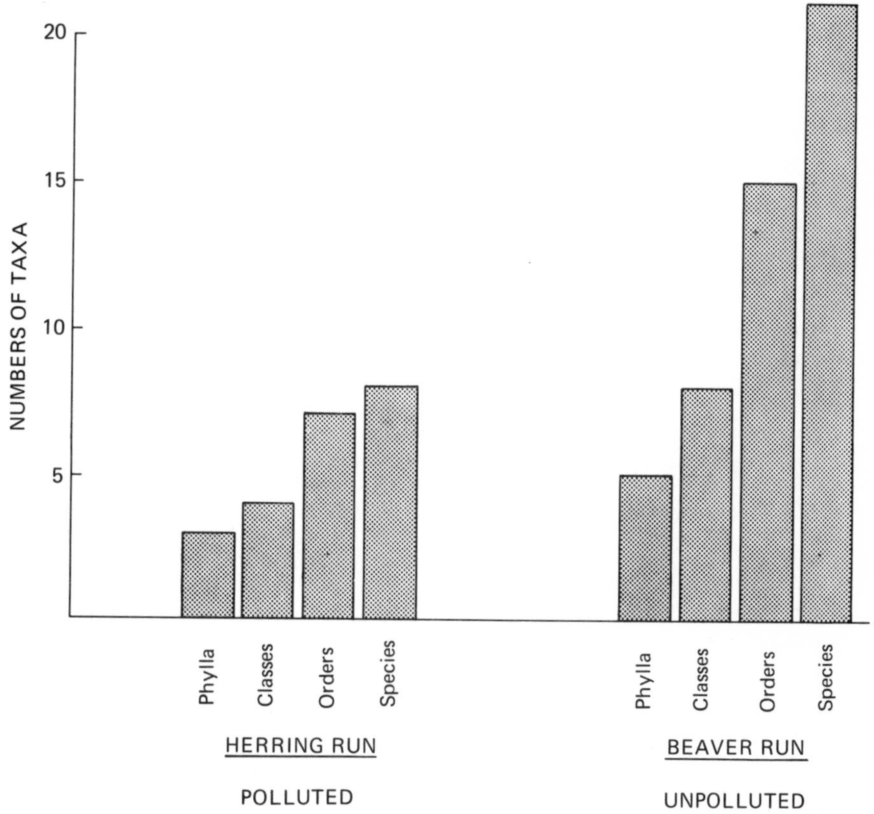

Figure 16–3 Biotic communities of invertebrate animals in two streams in Baltimore County, Maryland.

just 7 orders and 4 classes. Beaver Run, 20 miles north of Baltimore, passes through a well-balanced agricultural region, and does not carry the urban run-off and pesticides of Herring Run. In a short period, the ecology class found 21 species of invertebrates of 15 Orders, and 8 Classes. Several organisms were abundant in Beaver Run that were apparently entirely absent from Herring Run, including dragon fly (*Odonata*) and stone fly larvae (*Plecoptera*). Furthermore, Beaver Run was rich in fish life, whereas Herring Run had no fish. This simple class field trip demonstrated an important aspect of species diversity in relation to domestic pollution.

The relationships between species diversity and community stability have been mentioned previously. In general, the greater the diversity, the greater the stability. Communities with many species have a more complex web of food relationships, and each species has greater diversity in its food and cover resources. Thus each species has alternatives in meeting its life requirements. It has various ways of meeting new environmental shortages and contingencies. There is, in other words, a cushioning network of checks and balances. It must be remembered that we are speaking in broad terms and there are undoubtedly many exceptions to these generalities. Nonetheless, these generalizations are valid as overall principles of community ecology. As a result, any major change in a single plant or animal population in a polar community is likely to have a far greater impact on the total community than a similar change in a tropical community. It thus becomes important to think of communities in terms of their vulnerability to ecologic upset if change occurs in any one of their member populations.

SUCCESSION

Biotic communities change with time as their plant and animal communities change. This process is known as succession, and it involves a sequence of community types from *pioneer* stages to mature or *climax* stages. Each community in the series is known as a *seral stage*. One of the best ways to appreciate succession is to consider the sequential growth and development of biotic communities on a cleared forest.

If a forest is cut, and the land cleared to the soil, a succession of plants will invade and grow on the exposed surface. The first plants which invade are typically those capable of seeding in quickly on disturbed land or capable of germinating from viable roots left within the soil. This depends not only upon the former vegetation, but also upon seeds from surrounding plant communities, as well as the characteristics of the soil and climate. Assuming soil and moisture are present, typical invaders in the United States are herbaceous weeds such as ragweed (*Ambrosia* sp.), lambs quarters (*Chenopodium album*), dock (*Rumex* sp.), horse weed (*Erigeron canadensis*), plantain (*Plantago* sp.), crab grass

(*Digitaria* sp. et al.) and woody vines such as honeysuckle (*Lonicera* sp.) and poison ivy (*Rhus toxicodendron*). As soon as the first plants germinate and become established, the community increases vastly in physical complexity. A surface of bare earth now has shaded areas which differ in light, moisture and humidity. This enables other seeds to grow, and new species become established. Woody plants appear, again those capable of withstanding disturbed environments and harsh conditions, such as black locust (*Robinia* sp.), sumac (*Rhus* sp.), and hawthorn (*Crataegus* sp.). Pioneer animals also arrive, including ants, beetles and flying insects. Birds begin to forage over the new community searching for seeds and insects. Small mammals may venture forth from adjacent forests or grasslands. Each animal traversing the area adds organic nutrients. Within a year or two a surprisingly complex community has arisen, and is growing rapidly. As plants grow, they create more shade, and thus continuously modify the light and moisture conditions on the surface of the ground. Those initial pioneers which thrived on total sun exposure are now less favored and more shade-tolerant seeds can germinate more successfully. Woody shrubs emerge above the herbaceous layer and compete more successfully for the ambient light.

The shade developing beneath the herbaceous and shrubby layers permits the growth of other tree species such as maples, elms or oaks which could not tolerate the initial exposure of sun and wind. This represents the first beginnings of the forest. Further succession continues to involve all the dynamic processes of plant and animal competition. Over a period of years, the intial invaders and pioneers begin to drop out. Trees mature, first the fast growing locust and sumac, but as they reach the limits of their growth and their relatively short life-spans, the slower growing oaks (*Quercus* sp.), maples (*Acer* sp.) and elms (*Ulmus* sp.) begin to take over. The forest may now be 40 to 50 years old, reaching a height of 50 to 60 feet.

The ground cover and intermediate layers have now matured to include those shade-tolerant species capable of living in the forest understory. The insect and bird populations have changed from open field forms to forest species. By 70 to 80 years, the forest may be approaching relative maturity, with tall trees in the 60 to 80 foot range, deep shade and cool, moist conditions on the forest floor capable of supporting the rich array of life described earlier in this chapter. The climax first may be approached in 80 to 100 years, when it becomes a stable and self-perpetuating community. Processes of change still occur. Old forest trees die and crash to the ground, creating a microcosm of renewed successional development, but the main forest continues in equilibrium as a permanent community as long as environmental conditions remain favorable. If it is cut, the process of succession begins again. If it is defoliated, flooded, burned or dehydrated, it will change and attempt to seek a new form of equilibrium with its environment. The tree species that finally emerge as the dominant members of the climax community vary according to the climate, topography and soil type. Beech-maple climax associations are common in the Eastern United States in moist lowland areas, whereas oak-hickory associations are more common on drier slopes and upland soils.

It would be incorrect to assert that the process of succession is always consistent and predictable. It may take a variety of forms, and may become arrested at any of several stages. Early in the shrub stage, the biotic community may become overwhelmed by excessive growth of woody vines such as honeysuckle or poison ivy, so that trees cannot develop adequately, or it may stay in the herb or shrub stage if the land is subjected to continuing disturbance by fire, pollution, or land scarring. Throughout the countryside, especially around cities, one can find many examples of disturbed and unattractive vegetation that never gets out of the weed, shrub or vine stage.

The above discussion has mainly concerned a deciduous forest community, but a similar account could be given of coniferous forests, and, in fact, of all biotic communities. Thus, a prairie grassland, a tropical savannah, a desert arroyo, an estuarine marsh, a cedar bog, a pond, a lake, a delta, a mangrove swamp, or a coral reef all have characteristic patterns of development and succession. In some communities, successional development may involve hundreds or even thousands of years before mature climax conditions are achieved. Primary tropical rain forests, for example, are the culmination of thousands of years of successional development. Once cut, they are not recreated in a few decades or a hundred years.

It should be emphasized that succession is also a naturally occurring process and is not confined to disturbed habitats. Theoretically, each biotic community tends toward maturation and stable climax conditions, but there is sufficient natural change in ecosystems so that succession is a continually recurring process in many situations. The grassland and forest are subjected to natural fire as an ecologic process, the desert arroyo to flash flooding, the marsh, bog, pond, lake, delta and swamp to filling, and the coral reef to the continual processes of construction and destruction. Human activities often increase and accelerate these processes far beyond the natural rates of occurrences. In so doing, man impedes the development of maturity and ecologic stability in biotic communities.

An example of natural succession is the filling of a small pond and its eventual conversion into a terrestrial community. Figure 16-4 shows the primary stages by which this occurs. As the pond ages, it becomes more filled with aquatic vegetation and silt. Emergent vegetation surrounds the edge creating a marsh community. Accumulations of organic material and sedimentary deposits fill in the water volume, soil is formed at the edge, and terrestrial grasses and herbs become established. Eventually, shrubs and trees complete the process. Man accelerates this process by increasing the siltation process and enriching the waters with nutrients so that more extensive plant growth occurs. The process can be retarded by dredging the pond or manipulating the water level to reduce aquatic vegetation. Lakes proceed through some of the same stages,

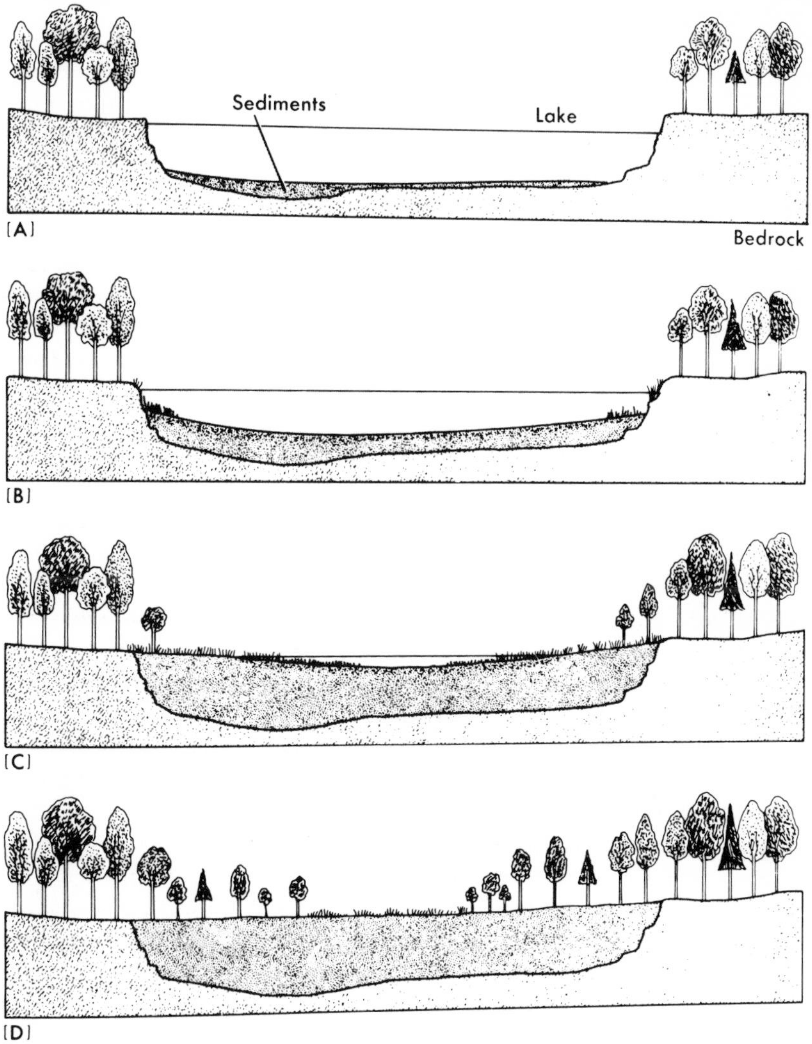

Figure 16–4 Successional stages in the natural aging and filling of a lake. Over many years sediments accumulate, the lake becomes shallow (A and B), and vegetation encroaches from the edges. Finally the lake becomes a marsh (C), a meadow, and a young forest (D.) *(Reprinted by permission of The Macmillan Company from* Communities and Ecosystems, *by Robert H. Whittaker. Copyright © by Robert H. Whittaker, 1970.)*

but if sufficiently deep, a point is reached in lakes where the scouring action of currents slows or even balances the siltation and sedimentation rate.

It is important for man to understand successional processes so that he can manage biotic communities for maximum productivity and esthetic concerns. In agricultural lands and harvested forests, man has learned to maximize productivity for his own economic gain. This is done by artificially maintaining communities in early successional stages, where productivity is greatest, and by channeling all possible production into the species to be harvested. That is, man maintains a cornfield or a wheatfield in an early productive stage by excluding all other pioneer species. He maintains a forest in an early successional stage and directs all productivity into the species of tree he wishes to harvest. This is all well and good provided the cost of such management is not too high in terms of toxic chemicals that must be applied to eliminate other species, or in terms of chemical fertilizers that must be applied to support an artificial single-species community.

Quite another principle prevails in the management of parks and sanctuaries where man wishes to maintain balanced plant and animal communities for esthetic and scientific values. Here it becomes essential to understand and promote natural succession so that mature and stable communities are achieved.

In this day of economic pressures it is increasingly difficult for man to appreciate the value of natural biotic communities. What good is a mature forest if it is not harvested, what good is an estuarine marsh, what good is a relic prairie? These are all questions being asked in greater frequency by citizens focusing on economic progress. They challenge the ecologist to explain the message of ecology in factual terms, and to emphasize the importance of balanced biotic communities to the total life support capabilities of the world. The answers are not easy, and are not always available. Much remains to be learned about the full range of functions of an estuarine marsh, for example, in maintaining the productivity of the oceans. This represents one of the greatest current challenges for ecology; namely, to understand and explain in a meaningful way, without great fanfare and unreasonable emotion, the importance and value of diverse biotic communities. Otherwise economic man will see little reason why he cannot continue to convert the entire world to paved streets, grass lawns, grain fields, and greenhouses.

MAJOR BIOTIC COMMUNITIES

It is worthwhile at this stage to take a brief look at several different communities and highlight some of their outstanding features. This has been done in various degrees of detail by a variety of authors, and no attempt will be made here to be comprehensive or thorough. Descriptions are offered as capsule summaries of six interesting and representative communities: the arctic tundra, northern coniferous forest, desert, grasslands, estuaries and tropical rain forests. This limited list unfortunately omits many other important and interesting communities: savannah, chapparal, temperate rain forests, tropical deciduous forests, fresh-water swamps and marshes, lakes, rivers, coral reefs and open oceans. Volumes could be written about each of these, and many others. Several of these community types are featured in beautifully illustrated books in the Time Inc. series, as for example *The Deserts* by A. S. Leopold (1961) and *The Forest* by Peter Farb (1963), and in the McGraw-Hill series, for example, *The Life of Prairies and Plains* by Durward Allen (1967), and *The Life of the Marsh,* by William Niering (1966).

Arctic Tundra

Arctic tundra is the biotic community occupying the northern latitudes, generally above 60° north latitude. The dominant plants are lichens (the combination of algae and fungi known as "Reindeer moss"), arctic grasses and sedges, and some dwarf shrubs, especially dwarf willow (*Salix* sp.) and cranberry (*Vaccinium* sp.). During the brief summer, the flowering season is abrupt and colorful. The vegetation is underlain by a thick spongy mass of undecayed dead vegetation. Bacterial action is slow due to the low temperatures, and the substrate is often permanently frozen, a condition known as *permafrost.*

The animal community is simplified into a relatively few species of birds, mammals and flying insects. The dominant mammals are the caribou (*Rangifer arcticus*) or reindeer (*Rangifer tarandus,* a domesticated species of caribou), arctic hare (*Lepus arcticus*), lemming (*Lemmus* sp. or *Dicrostonyx* sp.), and arctic fox (*Alopex lagopus*). The dominant birds are ptarmigan (*Lagopus lagopus,* a gallinaceous grouse-like bird), the longspur (*Calcarius* sp., a sparrow-like bird), and the snowy owl (*Nyctea nyctea*). Great numbers of migratory waterfowl (ducks, geese, swans and shorebirds) invade the tundra in the summer for breeding. Insects are dominated by the biting diptera, blackflies and mosquitos.

The arctic tundra is relatively lifeless during long, dark and cold winters, in which plant life is inactive and animals survive by either burrowing beneath the ice and snow or migrating to more favorable climates. In the brief summer, the tundra becomes very productive of both plant and animal life, stimulated by the long day-length and warm temperatures. Perhaps the most important feature of tundra ecology is its short sudden productivity which is dependent upon relatively few species and a very brief season. The values of this productivity extend far beyond the limits of the tundra itself. Waterfowl bred in the tundra migrate southward to temperate zones in fall and early winter. The reindeer and caribou, fattened on tundra lichens and grasses, form the basic food and shelter supplies of polar people around the world. Hence, the tundra is important to man in several ways, and because of its relatively simple biotic community it is especially vulnerable to ecologic imbalance. There is current concern that the oil boom on the Alaska North Slope may destroy certain essential aspects of the tundra community in northern Alaska.

The tundra community can also be found at high altitudes in temperate regions. In the Rocky Mountains and Alps, for example, tundra associations occur around 8,000 to 10,000 feet. At these altitudes, arctic species of plants and invertebrate animals are found.

Northern Coniferous Forest

This is a forest belt at latitudes below tundra, occupying major portions of Alaska, Canada, Scandinavia and Siberia, generally between 50 and 60° north latitude, but with extensions in mountainous regions further south. The plant community is dominated by spruces, firs, pines, and hemlock, with a ground cover of mosses, grasses, sedges and cold-adapted herbs. Decomposition is slow, as in the tundra, and the forest floor tends to accumulate a layer of dead organic material with excellent waterholding capacity.

Animal populations are more varied than in the tundra, but are still characterized by great seasonal change and wide population fluctuations. Typical mammals are the snowshoe hare (*Lepus americanus*), lynx (*Lynx canadensis*), squirrels (*Sciurus* sp., *Tamiasciurus* sp.), marten (*Martes americana*), mink (*Mustela vison*), fisher (*Martes pennanti*), wolverine (*Gulo luscus*), wolf (*Canis lupus*), woodland caribou (*Rangifer caribou*), deer (*Odocoileus* sp.), moose (*Alces americana*) and black bear (*Ursus americana*). Typical birds are ruffed grouse (*Bonasa umbellus*), crossbills (*Loxia* sp.), siskins (*Spinus pinus*), and a multitude of other passerines which migrate in for summer breeding. Winged diptera, such as blackflies and mosquitos, beetles and moths are among the dominant insects.

The northern coniferous forests range from the short, almost scrub forests of spruce and fir in the far north, to the magnificent tall stands of Douglas fir (*Pseudotsuga taxifolia*) and redwood (*Sequoia sempervirens*) in the mountainous forests of the United States. These forests represent the great timber and fur producing regions and some of the finest freshwater resources of the world. Lake Superior in North America, and Lake Baikal in Asia are both located within the northern coniferous forests.

Under improper exploitation, coniferous forests can be reduced to scarred and eroded landscapes. They can then become dangerous fire hazards with excessive water run-off and relatively little value for wildlife or human recreation. Fortunately, man learned a number of important ecologic lessons in regard to forest management in the first half of the twentieth century, and some forests are now being managed on a sustained-yield basis with considerable success. With modern techniques of forest ecology and silviculture, man can achieve tremendous timber yields while still retaining esthetic beauty, wildlife values and watershed capacities.

Grasslands

The grassland community covers vast stretches of land within continental interiors. These are areas characterized by plains and 10 to 30 inches of rainfall per year, generally less than in most forested regions.

The plant community is dominated by grasses, but it contains a great variety of other herbs and forbs as well, particularly legumes, such as lupine (*Lupinus* sp.), trefoil (*Trifolium* sp.), clover (*Petalostemum* sp.), and vetch (*Vicia* sp.); composites, such as aster (*Aster* sp.), goldenrod (*Solidago* sp.), fleabane (*Erigeron* sp.), coneflower (*Rudbeckia* sp.), fireweed (*Epilobium* sp.), and daisy (*Chrysanthemum* sp. or *Bellis* sp.), and members of the Ranunculaceae, buttercups (*Ranunculus* sp.), anemone (*Anemone* sp.), columbine (*Aquilegia* sp.) and larkspur (*Delphinium* sp.). In all, a grassland or prairie flora may contain over 120 species in more than 10 families. The dominant grasses are predominantly of three general sizes: tall grass, often 6 feet in height, which includes tall bluestem (*Andropogon gerardi*), switch grass (*Panicum* sp.) and Indian grass (*Sorghastrum* sp.); medium grass, 2 to 4 feet in height, which includes little bluestem (*Andropogon scoparius*), feather grass (*Stipa* sp.) and brome grass (*Bromus* sp.); and short grass, less than 2 feet in height, which includes buffalo grass (*Buchloe* sp.), bluegrass (*Poa* sp.) and blue grama or mesquite grass (*Bouteloua* sp.).

As previously mentioned, the annual production of organic material in grasslands is great, organic accumulation is rapid, and a thick layer of humus is produced. Grassland soils are among the thickest and richest in the world. Roots of the grasses and forbs penetrate up to 6 feet in these rich soils.

The animal populations of grasslands are also rich and varied. Among the mammals, the ungulates and rodents have flourished with great populations of bison (*Bison bison*), antelope (*Antilocapra americana*), prairie dogs (*Cynomys* sp.), ground squirrels (*Citellus* sp.) and gophers (*Geomys bursarius*), and the mammalian predators are represented by coyotes (*Canis latrans*), foxes (*Vulpes* sp.) and badgers (*Taxidea taxus*). Typical birds are prairie chickens (*Tympanuchus cupido*), sharp-tailed grouse (*Pedioecetes phasianellus*), meadowlarks (*Sturnella* sp.) and a variety of sparrows. Insects are abundant, with grasshoppers and locusts being among the most conspicuous.

Grasslands represent some of the greatest agricultural areas of the world for cattle, corn and wheat farming. Man is now coming into better balance with these areas, but the past records numerous examples of tragic folly in the exploitation of grasslands. In our own North American plains, sudden expansion of agriculture and excessive grazing reduced the productive grasslands to dustbowls and desert-like conditions. When the grass was burned, and the sod broken up with steel plows, with no thought for conservation or wise management, the rich soils soon disintegrated and disappeared under wind and water erosion. A vivid description of this process is given in Paul Sears' book, *Deserts on the March* (1935), and John Steinbeck's classic novel, *The Grapes of Wrath* (1939), deals with the human impact of this tragic era in American history.

What man has now done with the grassland biome is to tremendously reduce its complexity, by replacing 120 species of native prairie flora with just one or two species of domestic plants, wheat and corn. Thus we funnel all of the great productivity of the grassland soils into just one or two species which we can harvest. This is fine for human use, and our economy requires this type of management, but it also requires very intensive care to maintain this artificial state of affairs. By oversimplifying a complex biome to this extent, we greatly exaggerate its ecologic instability. A single species community now becomes vulnerable to massive decimation by a single species pathogen or herbivore. Thus, we must apply intense surveillance and often massive doses of chemicals to control outbreaks of wheat rust or corn smut, both of which are pathogenic fungi, or serious outbreaks of locusts and corn borers, all of which can now explode on the "pure cultures" of their hosts. We obviously cannot return the prairie to its virgin condition, for we now require the great wheat, corn, and beef productivity of our former grasslands, but it is important to be sharply aware of the costs and inherent dangers of modifying any great biotic community to this extent. Fortunately, the United States had the capital wealth to restore the grasslands to agricultural productivity after the dustbowl days of the 1930's, but many countries of the world are decimating their grasslands without the economic potential to restore them. The great grasslands of east Africa, which contain the richest wildlife areas of the world, are under increasing grazing pressure from domestic livestock. Even in some national parks, such as Amboseli Park of Kenya, serious damage from overgrazing is evident. This is a result of both cattle herds of Masai tribesmen entering the park and excessive numbers of wild elephants within the boundaries of the park.

Deserts

Deserts are arid biomes with usually less than 10 inches of rainfall per year. Deserts usually occur in areas of high pressure (Sahara, Australian), rain shadows of mountains (Mohave, Iranian) or high altitudes (Tibetan, Gobi and Bolivian).

Some of the driest deserts may not have any rain for a period of many years. One such area in Chile has not received precipitation for over 20 years. Such deserts may be virtually devoid of living organisms over vast areas. In the Sahara desert of southern Libya, one can travel for hundreds of miles without seeing a living plant or any form of life. Most deserts, however, do have some water resources from occasional rains or from ground water, and they possess a considerable variety of living organisms dependent upon these water resources.

The predominant plants of deserts are succulent species with waxy surfaces, such as cacti, which can conserve water for long periods of time, or deciduous shrubs, also with thick waxy leaves. The growing season is very short, and the flowering period amazingly abrupt, occurring suddenly after flash rains. Many deserts have little or no plant life over extensive areas of sand and rock.

The animal community of deserts is confined, of course, to areas with plant life, and is dominated there by burrowing and nocturnal rodents, reptiles, insects and arachnids (scorpions, spiders, etc.). These animals escape the temperature extremes and desiccation of desert air by living beneath the surface during the day and venturing forth only at night. Most of them have remarkable water conservation adaptations. The desert rodents, such as gerbils (*Gerbillus* sp., *Tatera* sp.), pocket mice (*Perognathus* sp.), and kangaroo rats (*Dipodomys* sp.) utilize metabolic water. They require no free water, but obtain their requirements from the metabolic breakdown of carbohydrates into carbon dioxide and water (the basic respiratory equation). They also produce extremely concentrated urine, and thus expel more wastes per unit volume of water excreted. Many of the insects and arachnids have waxy coats and reduce water loss through the cuticle.

Although water is the primary limiting factor in deserts, excessive soil minerals, salinity and a lack of organic material may also be limiting. It does not follow, therefore, that deserts can be made productive merely by irrigation. In the mid 1950's the United States spent millions of dollars in an irrigation scheme in Afghanistan designed to make the desert bloom. It did bloom for a year or two after the water was applied, but very quickly the scant nutrient supply was used and the high salinity in the soil was concentrated on the surface due to evaporation. Further plant growth was inhibited, and the project became a dismal failure. There was not an adequate evaluation of the total ecology of the area, nor a sufficient appreciation of the limiting factors other than water.

With approximately 18 percent of the world's land surface in deserts, man must make the best possible scientific utilization of these areas. Deserts can be made productive by the proper application of scientific skill, hard work and capital financial support, as has been demonstrated in Israel and several other countries. The Afghanistan debacle showed, however, that desert agriculture requires more than merely "add water, stir and watch the plants grow."

Estuaries

Estuaries are bodies of water where fresh water from the land mixes with and measurably dilutes sea water. They represent dynamic combinations of fresh-water and marine communities, and they have a number of unique characteristics. They are particularly important to man by virtue of these unique biological attributes.

Several different types of estuaries exist, representing various topographic formations and geologic histories. Some develop from coastal submergence and drowned river mouths such as Chesapeake Bay and Delaware Bay. Others develop as coastal lagoons behind barrier sounds, such as Albemarle, Pamlico, and Currituck sounds inside Cape Hatteras. Others arise as steep glaciated fjords such as those along the coasts of Norway, Chile and British Columbia. Finally, many develop within great delta systems such as those of the Nile, Mississippi, Mekong and Ganges rivers.

Some of the outstanding characteristics of estuaries are the following: (1) The water composition is constantly changing with tidal action and water run-off from the land. In different parts of the estuary, the salinity ranges from 1 part per thousand of dissolved salts (almost fresh water) to 34 parts per thousand (sea water is around 35 ppt). (2) Nutrient levels are high due to the flush of organic materials and agricultural chemicals from adjacent land into the water. (3) Temperatures and water currents fluctuate considerably on a seasonal, daily and even hourly basis. (4) Dissolved oxygen and carbon dioxide levels also fluctuate markedly. This list could be extended to point out other characteristics of estuaries, but it will serve to indicate that estuarine systems tend to be highly variable, dynamic and fairly unstable. Because of their high nutrient concentrations and churning action, they are extremely productive. They represent some of the most productive communities on earth for plankton, crabs, clams, oysters and fish. Chesapeake Bay, one of the largest estuaries of its type in the world (150 miles in length, with 4,000 miles of shoreline), produced 11 tons of fish per square mile in 1920 compared to only 3 tons per square mile for Georges Banks off the coast of Nova Scotia (Hildebrand and Schroder, 1927). Furthermore, the productivity of the estuary has value far beyond the limits of the estuary itself. Most estuaries are spawning and nursery grounds for marine fish. Striped bass produced in Chesapeake Bay move as far north as Nova Scotia. Probably 70 percent of the Atlantic coastal fishes depend upon estuaries such as Chesapeake Bay for some stage of their life cycle (Ketchum, 1969).

The plant communities of estuaries are characteristically poorly developed in aquatic vascular plants, and have a predominance of microscopic phytoplankton as the primary producers. There are exceptions to this, of course, and some estuaries have rich plant populations of redheaded grass (*Potomogeton perfoliatus*), eelgrass (*Zostera marino*), wild celery (*Vallisneria americana*), widgeon grass (*Ruppia maritima*), and other flowering plants which serve as fish cover, food and habitat for small invertebrates, and the basic diet of wintering waterfowl populations. The main primary producers of the estuarine community are, however, microscopic algae and diatoms, collectively known as phytoplankton. These are consumed directly by some fish, the so-called pasture fish such as menhaden, but for the most part they form the primary food for small crustacea and rotifers, the zooplankton, which in turn are food for fish.

The estuarine animal community is dominated by a bottom fauna of crabs, shellfish (clams and oysters) and annelids, and a midwater fauna of finfish and jellyfish. Echinoderms (starfish, sea cucumbers and sea urchins) are not well represented in most estuaries, nor are sea anemones or sponges. The fish are predominantly fresh water or anadromous forms. Entirely marine fish enter estuaries for only short periods, and many, such as tuna, marlin, codfish, etc., never come into estuaries. Anadromous fish are those forms which live in salt or brackish water for much of their life, but breed in estuarine or fresh water. Examples are shad (*Alosa*, sp.), alewifes (*Pomolobus* sp.), striped bass or rockfish (*Roccus saxitilis*). They represent important species for both commercial and recreational fishing. Sharks, dolphins, rays and more marine forms also move into estuaries for feeding in certain seasons.

Estuarine animals demonstrate a number of interesting adaptations to the unstable and fluctuating conditions of their environment. They have special capabilities in osmo-regulation; that is, the ability to maintain salt and water balance in the presence of changing environmental salinities. They also often have special adaptations to tidal and wave action. Benthic animals often avoid strong wave action by burrowing into the substrate. Several estuarine fish have delayed development in which the larval fish remain within the egg membranes until their muscles are sufficiently developed to enable them to swim against the tidal action. The eggs of the estuarine fish often have more yolk than comparable marine forms to support this prolonged development.

Estuaries are now under increasing pressure from pollution and commercial exploitation. Many of the world's great cities are on estuaries, e.g., New York, Philadelphia, Baltimore, San Francisco, London, Rotterdam, Lisbon, Cairo, Karachi, Calcutta, Bangkok, Saigon, Tokyo and Seattle, and their adjacent estuarine waters are subject to heavy commercial use in addition to the burdens of dense human populations in their watersheds. Thus, estuarine systems are forced to serve multiple purposes as waterways for ocean transport, dumping grounds for industrial wastes and domestic sewage, cooling waters for power plants, collection basins for surface erosion from scarred watersheds, recreational areas for large populations, and still, remarkably enough, a few of these estuaries remain productive of fish, clams and oysters for human food. Many estuaries have already been lost as productive biological systems—Delaware Bay, for example, once supported a major fish and oyster industry, but this is now a relic of the past. Puget Sound has experienced major damage to its oyster industry by the toxic sulfite wastes of pulp mills. On the other hand, Chesapeake Bay still remains amazingly productive of fin and shellfish, as well as a moderately attractive body of water for recreation. Whether it can remain so is a critical question of the present time, and each year it seems to become a little muddier, its shores a little dirtier, and more prone to algal blooms and oil spills. We would all like to have our cake and eat it too in terms of clean waters and expanding economic development, but there is now a point where choices must be made.

Tropical Rain Forests

The tropical rain forest is a biotic community which has always stirred the human imagination, and rightly so. It is one of the most dramatic and awesome biologic systems on earth. Such forests occur within the Tropic of Cancer and the Tropic of Capricorn (23° 27′ N and S latitude), in areas with more than 80 inches of rainfall per year. Usually such forests have one or more dry seasons per year in which less than 5 inches of rain fall per month. The monsoon forests of southeast Asia are an example, and although it rains throughout the year, concentrations of rainfall occur twice a year in the monsoon periods.

An important characteristic of the tropical rain forest is its relative climatic stability. The temperature variation between seasons is slight, and usually less than the temperature variation between day and night, or between different microhabitats within the forest. For example, in Singapore at the tip of the Malay peninsula, the diurnal variation in temperature is often 20° F (from 75 to 95° F), whereas the variation in temperature from the January mean to the July mean is only 2.5° (from 87° to 89.5°). By contrast, the variation from January to July mean in Dawson City, Alaska is 81.7° F (from −15° F to nearly 70° F).

There is considerable variation in temperature in the tropical rain forest between the top of the canopy and the deep shade of the forest floor, often 10 to 20° F. There are only slight changes in photoperiod in equatorial forests, and humidity remains remarkably constant. Thus, many tropical rain forests have very little climatic seasonality in regard to either temperature or light. This does not mean, however, that there is no seasonality in the lives of plants and animals. Despite the relative year-round uniformity of climate, there is still substantial seasonality and cyclicity in reproduction and other vital functions of many plants and animals. This is not coordinated or synchronized, however, like it is in temperate latitudes. Tropical trees often lose their leaves, but not all trees at once. Individual species, and even individual trees within a species, may be on their own cyclic schedules, so that one of two trees side by side may be barren of leaves and the other fully leaved at the same time. The forest as a community remains green throughout the year, though various individual trees are without leaves at any given time. Some biologic clock other than seasonal climatic influence determines these cycles. Similarly, insects, amphibians, reptiles, birds and even mammals may have their own cyclic schedules of reproduction and growth, though once again, they are not necessarily timed with any striking climatic seasonality.

The outstanding feature of the biotic community in the tropical rain forest is its richness and diversity of species. In almost every major plant and animal group, there are more species than in temperate forests. This was pointed out in the section on species diversity.

A further example of the wealth of the biotic community in tropical rain forests can be seen in the forests of Panama. The island of Barro Colorado in the Panama Canal Zone has been the home of the Smithsonian Tropical Research Institute since the 1920's, and the biotic community of its forest has been well-studied. This community is known to have over 1,200 species of trees, over 30,000 species of insects, 310 species of birds, 32 species of amphibians, 68 species of reptiles, and 70 species of mammals—all on an island of only 6 square miles. A forest of comparable area in the northern United States would have less than one-third this diversity of species.

The tropical rain forest also presents an excellent example of a stratified community. Such forests are often very high, with a canopy 150 to 200 feet above the ground, and some giant trees, known as emergents, extending above the canopy to elevations of 250 feet. There is often a well-defined intermediate layer, 50 to 75 feet in height, and a ground layer less than 20 feet in height. The mature climax tropical rain forest is often open and parklike with relatively little shrubby and herbaceous vegetation. When large trees fall, however, or the forest is cleared, then various forms of secondary growth enter, forming the more typical jungle picture. What is popularly portrayed as a dense tropical jungle is a disturbed plant community in various stages of pioneer and secondary growth.

Extending throughout all layers of the tropical rain forest is a profusion of lianas or woody vines. Many of these grow up with the trees and are of approximately equal age. As the tree achieves heights of 100 feet or more, the lianas often descend directly to the ground as free, flexible shafts. Sometimes they remain entwined around the host tree trunk. Some lianas are entirely free-living except for support, but others show various degrees of parasitism. A common liana in many tropical forests is the strangler fig (*Ficus* sp.) which is basically free living, but often grows so rapidly that it surrounds the support tree which dies leaving a hollow core surrounded by the anastomosing network of vines which continue to live on their own.

The lianas of the tropical rain forests provide important travel avenues for many animals: ants, beetles, and other insects, tree frogs and snakes, monkeys, coatis, tayra, ocelots, etc. They thus add greatly to the three-dimensional complexity and spatial configuration of the tropical forest.

Tropical forests have provided man with economic resources such as Burmese teak (*Tectona grandis*), African mahogany (*Khaya* sp.), Indian rosewood (*Dalbergia* sp.), sandalwood (*Santalum* sp.) and other valuable timbers. They are also important in the total oxygen balance of the world, especially the great forests of the Amazon and Congo basins, Southeast Asia and Indonesia, though their quantitative roles in this are not known at this time.

The tropical forests of the world are being cut with increasing rapidity and efficiency, and set back to pioneer and secondary stages of growth. Under moderate and enlightened management a certain amount of forestry can certainly be sustained, but if ecosystem tolerances are exceeded, there are certainly undesirable repercussions. When tropical vegetation is removed, the water-holding capacity of the landscape is destroyed, and the thin tropical soils erode rapidly. Deforested tropical regions are particularly prone to severe flooding. Many tropical countries are proud of the rate at which they are cutting their jungle, but they fail to realize that this jungle provides the very fabric which holds their land together. They then wonder why they have such severe floods.

We do not know all of the functional relationships of the tropical rain forest in terms of hydrostatic balance, oxygen production and animal community relations, but we do know that these forests are mature and complex communities which have evolved over millions of years to meet the special ecologic conditions

Figure 16–5 The grasslands of central Africa support a rich fauna of ungulates, rodents, carnivores, and other animals. These photos show zebra and wildebeest at a grassland waterhole in Kenya and a nearby group of Grant's gazelles, typical of more than twenty species of wild ungulates in this community.

Figure 16–6 The tropical rain forest in Panama, showing characteristic giant trees with buttress roots, an understory of palm and woody vines or lianas.

of the wet tropical areas of the world. They cannot be destroyed without serious ecologic consequences.

SUMMARY

This chapter has described some of the structural and functional characteristics of biotic communities. Defined as all the plant and animal populations inhabiting a given area, biotic communities may be analyzed in terms of food webs, ecologic pyramids, energy relationships and stratigraphic patterns.

In structural terms, it is convenient to examine communities in reference to strata or layers. In a forest, these can be broadly identified as soil, herbaceous cover, shrubby layer, intermediate layer and canopy. Animal species are often localized in their typical behavior within certain layers, and even to microhabitats within those layers.

It is also valuable to study communities in relation to species diversity; that is, the number of different species coexisting in a community. Tropical communities are characteristically rich and varied, with large numbers of different species of plants and animals. Polar, desert and mountaintop communities are typically low in the number of species in the community. Agricultural man has generally had the effect, either intentionally or inadvertently, of reducing complex communities with a high species diversity to simple communities with very low species diversity. Thus he reduces a complex grassland community with 120 species to one with only corn or wheat. He reduces a complex ungulate community in Africa, consisting of more than 20 species of ungulates, to a cattle community. In so doing, he funnels as much productivity as possible into a readily usable form, but he also increases the instability and vulnerability of the community to drastic perturbations.

Biotic communities undergo change in a process known as succession. Without disturbance, succeeding communities finally develop into a mature and stable community form, the climax, which is self-perpetuating. The forms of climax communities are determined by climatic and geologic factors. Patterns of succession are a function of the pre-existing and surrounding communities, topography and soil type, and climate.

The world provides a vast array of different biotic communities varying from arctic tundra to tropical rain forests. Each terrestrial and aquatic community has more or less definable characteristics: structural configurations, energy relationships and dynamic interactions between its plant and animal populations. Various communities serve unique functions, as, for example, the nursery role of estuaries for marine fish, the oxygen production and water-reservoir capacities of forests, the shelter and land building functions of coral reefs, and so forth. It is increasingly important for man to understand and appreciate these vital processes before he prematurely decides that he can agriculturalize and urbanize the entire world.

17

Ecology and the Future of Man

MUCH of the current controversy about ecology concerns the future. Will man continue to progress in knowledge, health and standards of living, or is he headed toward oblivion in the near future? Is the good life for everyone gradually spreading around the world, and is man coming into better balance with his environment, or is he already on a path of self-destruction?

It is now possible to find numerous writers and scholars expressing opposite views on these questions. Some scientists believe that man's remaining time on earth may be less than 100 years, unless major improvements occur in his basic ecologic knowledge and relationships to his environment (Ehrlich, 1968, 1969; Dorst, 1970; Commoner, 1966; Novick and Cottrell, 1971). Many people believe that the planet earth is already dying from misuse, pollution and overpopulation—trends that must be reversed very quickly if we are to survive. They cite projections on current levels of air and water pollution which could make the earth uninhabitable within the next 50 years unless these processes are sharply controlled.

Other scientists consider these attitudes as overly emotional, fear-producing, and scientifically unjustified. They feel that ecologists cannot support many of their prophesies of doom and gloom, and that they are, in fact, raising unreasonable bogeymen. They point out the great strides in health, welfare, education and social advance which have already occurred, and they feel that there is a good deal of inaccurate sentimentalism about "the good old days." They note that the good old days had no electricity, no easy transportation, long working hours, very little educational and recreational opportunities, and plenty of diphtheria, smallpox, cholera, malaria and tuberculosis. They feel that the alarmist-type ecologists cannot recognize progress when they see it. To this the ecologist replies, "Does progress require polluted rivers and lakes, choking air, birds dying from pesticides, intolerable crowding, ulcers in children, violence on the streets, and heart attacks in young men?" Thus the controversy and arguments can continue almost indefinitely. The dichotomy of these heated arguments has been clearly pointed out by Garrett DeBell (1970).

In general, ecologists have tended toward pessimistic views of the future for more than 30 years. The first scientific books on the ecologic imbalance of man in the world appeared in the 1930's and 1940's (Sears, 1935; Vogt, 1948; Osborn, 1948). Since these early warnings about the ecologic problems of man, these difficulties have generally increased. Ecologists have found ample evidence for their fears. They feel that current events show that world population growth will not be amenable to significant slowing before the world is severely overcrowded. They have found that man has not learned the lessons of history and has continued to destroy his environment. They believe that the future survival of man will be, at best, a close and hectic race between these destructive forces and an enlightened society which can find the knowledge and maturity necessary for solutions. They now see little evidence that man will really achieve these solutions before it is too late.

This ecological view leapt into world attention in 1969 and 1970 in a highly emotional fashion, further antagonizing the optimists. Ecology suddenly became the "in-thing"; the cause of students, politicians, housewives and scientists. It had many of the characteristics of a popular fashion, however, even a cult, and, as such, it clearly attracted resistance and opposition. The editor of *Science,* Dr. Philip Abelson, correctly pointed out that the emotional outburst of 1970 on environmental issues could not be sustained (Abelson, 1971). Although it had substantial elements of fact, it also contained too many half-truths and unwarranted fears. Before 1970 was over, a definite backlash developed in the minds of many people. Speeches and articles began to appear suggesting that ecologists were much too pessimistic and the world wasn't in such bad shape after all. A United States National Commission decreed that the population problem was greatly exaggerated and, in any case, it didn't really apply to the United States. President Nixon asked the nation to declare what was good about America, and not dwell entirely on her problems. A middle class working man in New York was reported to have said, "If those damned ecologists don't shut up, they'll take away our American way of life."

In the terminology of 1970, it was obvious that the credibility of ecology was eroding fast. Much of the progress made in alerting people to environmental and population problems appeared to be fading, and a strong reaction to believe in technologic progress once again asserted itself. The optimists again pointed to the vast potential of the earth when viewed in modern terms. "We've barely begun to capture the wealth of the ocean," they insisted, "to harness nuclear energy, to apply scientific progress to world agriculture, to probe the wonders of the computer age, to capitalize on modern medicine, and to develop the skills and potential of human beings through better education."

Throughout it all, the future has remained fair game for everyone. The prognosticators of gloom and doom on the one hand, and of a technologic utopia for all men on the other, have continued to express their views with equal fervor. The future has attracted not only popular writing (Fuller, 1969; Lancaster, 1967; Toffler, 1970), but scientific writing as well (Medawar, 1959; Handler, 1970). Every conceivable arena of human activity has been considered: from the intentional genetic modeling of man to the exploration of space, from agriculture to the future of warfare, from social and educational problems to computer-driven automated industries.

Some of the difficulties in reconciling diverse points of view arise from the time and area scale of different writers. Some have concerned themselves with the next million years, considering the world as a whole (Darwin, 1953), whereas others have thought only in terms of 25 to 50 years, considering just the United States (Sarnoff et al., 1955). Obviously, one's opinions might differ greatly depending upon the span and extent of one's outlook.

This chapter will not deal directly in the game of forecast and prediction, but it will attempt to highlight some of the major issues and sources of controversy, and the scientific knowledge or lack thereof relating to these issues.

ESSENTIAL QUESTIONS FOR MAN'S FUTURE

The most difficult and essential problems for man's future lie in broad ecologic and social issues. No two questions are more fundamental or more troublesome to the future of man than the ecologists' inquiry,

> "What is the carrying capacity of the earth for man? In other words, what are the limits of the life support capabilities of the earth in terms of human population?"

or the behaviorists' concern,

> "What are the tolerance levels of man to the pressures generated by his own societies? In other words, if man is his own worst enemy, at what point, or by what stimulus, does this lead to his self-destruction?"

The carrying capacity, or life support capability, of the earth for man may be dependent upon a great variety of factors. We can readily identify oxygen, water, food, space and shelter as vital requirements for human life. There might be many others as well, such as mineral resources, energy availability, or the concentration of toxic metabolites which could reach states above human tolerance levels. Most of these, however, would operate through the media of air, water or food. Hence, it is worthwhile to examine each of these briefly.

OXYGEN SUPPLY AND AIR QUALITY

The atmosphere of the earth consists of about 78 percent nitrogen, 21 percent oxygen, 0.03 percent carbon dioxide, and the remainder, other gases. This composition is remarkably stable despite the fact that oxygen is a highly reactive element which readily combines with a great many other elements in chemical processes of oxidation.

The ecologist Lamont C. Cole (1970) has pointed out that oxygen has remained stable in the earth's atmosphere because green plants recycle it in molecular form through photosynthetic processes. Approximately 70 percent of the free oxygen produced each year is thought to come from planktonic plants in the ocean, and the remaining 30 percent from terrestrial vegetation of forests and grasslands. Cole emphasized that if we should seriously pollute the sea, as is already happening in coastal areas, and if we should excessively denude the forests and grasslands of the earth, we shall be in danger of upsetting the oxygen balance in the atmosphere. In other words, we must look upon the oceans, forests and grasslands, as the world's oxygen tank, and we must insure their continued healthy ecologic function.

This point of view is contested by W. S. Broecker (1970), a geochemist at Columbia University. Dr. Broecker points out that each square meter of the earth's surface is covered by 60,000 moles of oxygen gas. Plants in the ocean and land produce only about 8 moles per square meter per year, and this is almost entirely consumed by their own respiration and that of animals and bacteria. Thus other organisms use all but 0.0004 percent of the oxygen produced by plants. Broecker maintains, therefore, that man does not depend upon the oxygen generated by plants, and that there is no current danger at all to his oxygen supply. "In conclusion it can be stated with some confidence that the molecular oxygen supply in the atmosphere and in the broad expanse of open ocean are not threatened by man's activities in the foreseeable future. Molecular oxygen is one resource that is virtually unlimited" (Broecker, 1970). Broecker feels that man's existence will be threatened by other types of environmental pollution long before his oxygen supply becomes endangered, and that we need not worry about total oxygen supply.

This is typical of some of the different opinions which are unfolding between physical and biological scientists. Such differences often indicate the need for more research and better knowledge before we can fully accept one position or another.

Another atmospheric problem of long-term concern is the increase in carbon dioxide in the atmosphere. Between 1860 and 1960, the combustion of fuels increased the CO_2 content of the air by 14 percent of its formerly stable level (Commoner, 1966). Carbon dioxide tends to absorb infrared rays from the sun, and this provides a "greenhouse effect" for the whole earth; that is, it accumulates heat. This will gradually increase the temperature of the earth—an effect which may have long-term implications in climate modification. A rise of just 2 or 3 degrees in annual mean temperature would have a major melting effect on glaciers and ice caps, and this could raise the level of the seas several feet. Such a rise in sea level would inundate many of the world's major cities. Some experts predict it may require 4,000 years, and others feel it will not happen at all. The latter believe that any occurrence of the greenhouse effect due to increasing CO_2 in the atmosphere will be offset by increasing particulate pollution in the air and increasing opacity of the earth's atmosphere. Thus less light energy will enter, and we could actually have a cooling of the earth's surface (Fletcher, 1969; Lansford, 1970). Once again, we see fundamental controversies on major issues—largely because we do not really have the complete answers.

The atmospheric problem may also become acute in terms of various toxic elements released into surface air and waters: carbon monoxide, sulfur dioxide, lead from automobile exhaust, hydrocarbons from industrial operations and numerous other products of modern technological societies. For example, 500 million pounds of lead are discharged into the atmosphere of the United States each year from automobile exhausts alone (Craig and Berlin, 1971), and significant amounts of this become incorporated into living organisms. A recent study in Sweden showed concentrations of lead in mosses to be as high as 300 to 500 ppm along roadsides. Some roadside plants in the United States may have lead levels up to 3,000 ppm. This lead enters the plants via both air and water and can then enter man and animals. A hundred years ago, before the automobile, concentrations in roadside plants were less than 20 ppm. Lead levels over 0.5 ppm have detectable physiological effects in man and animals, and 1.0 ppm levels are often toxic (Chisolm, 1971).

In conclusion, there seems to be little or no immediate threat that we will run out of air for breathing, but several warnings have appeared regarding the quality of our air. We apparently have an adequate supply of total oxygen for many centuries to come, though even this is a debatable point. In some localities the situation may be critical at certain times. If stationary inversions of air trap the polluted atmosphere over some of our largest cities such as Tokyo, London, New York or Los Angeles, toxic air pollutants increase to dangerous levels. In August of 1969, a massive still-air inversion occurred over eastern United States, causing serious air pollution in urban areas from New Orleans and St. Louis, north to Chicago and Milwaukee, and east to Washington and Baltimore (Brodine, 1971). At no time was there an absolute shortage of oxygen, but dangerous levels of carbon monoxide, sulfur dioxide, hydrocarbons and other air pollutants accumulated in several cities. Twenty million people were affected by varying degrees of eye irritation and respiratory distress. Fortunately, this great blanket of smog was dispersed by upper air currents in the first week of September, and no large-scale health tragedies occurred. More serious events of this type are likely to occur in urban areas unless strong measures against air pollution are undertaken.

WATER SUPPLIES

Worldwide water supplies represent a concern for the future of man in terms of both the quality and absolute quantity of fresh water. Many cities have reached or are coming close to their total available fresh water supply and are reaching further and further to obtain new supplies. Los Angeles, for example, is planning tremendous canals and conduits which will bring water from the Pacific northwest over 1,000 miles to the growing megalopolis of southern California.

Two Russian geographers and hydrologists, Kalinin and Bykov, have pointed out (1969) that "The world's water resources form a single entity." The basic hydrologic cycle of the world (Figure 17-1) depends upon solar energy, atmospheric circulation, heat reflectivity and surface retention. The latter is strongly influenced by the biotic community. Kalinin and Bykov point out, on the assumption that the moisture content of the earth remains constant, that the total gains and losses from the land and the sea are in balance. Thus, the total precipitation equals the total evaporative loss from land and sea. However, the relative retention time and thus the total amount of water in terrestrial and fresh-water environments can change substantially in relation to the biota and terrestrial storage capacity. Table 17-1 shows estimates on the world's total water resources. Of 1.46×10^9 cubic kilometers of water in the world, 93 percent is in the oceans, 4.1 percent is in the earth's crust, 2.0 percent is in glaciers and polar ice caps and only 0.052 percent is in freshwater lakes, rivers and atmospheric moisture. The annual production of fresh water by evaporation and precipitation is estimated to be 37,000 cubic kilometers. At

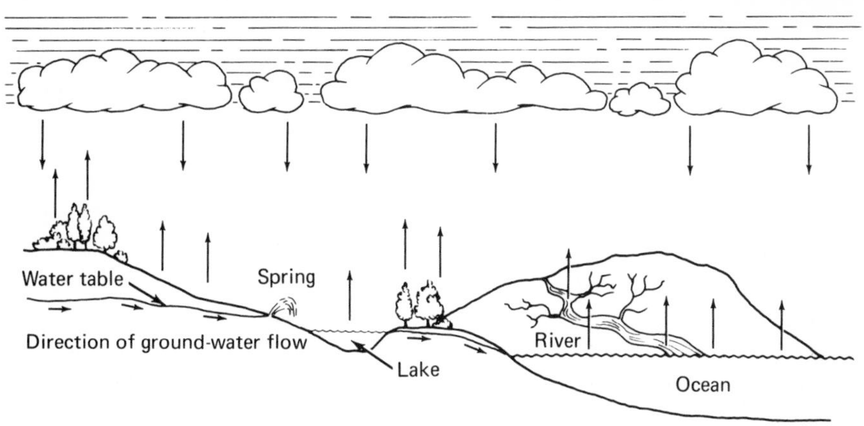

Figure 17–1 Simplified depiction of the hydrologic cycle. Annual evaporation exactly equals annual precipitation. Rain that falls on land which does not evaporate back to the atmosphere descends into the ground water or moves into rivers and streams; the groundwater flow and stream flow eventually return this water into the oceans. *(From Kalinin and Bykov, in* The Impact of Science on Society, *Vol. XIX, No. 2 (1969), p. 135. © UNESCO.)*

the present time, the world's total water use is less than 9,000 cubic kilometers, but this is estimated to increase to approximately 18,700 cubic kilometers by the year 2,000 (Kalinin and Bykov, 1969). In other words, within 30 years, man will be using 50 percent of the total annual fresh water production on the earth. This still remains a relatively small proportion of the total available fresh water in the world (less than 5 percent) but much of this fresh water is locked into glaciers, polar ice caps and earth crust deposits, where it is difficult for man to obtain. Considering the technologic potential of desalination, however, it does not seem likely that man will run short of an absolute supply of fresh water in the next few centuries.

This must be qualified, however, by two practical considerations, that of distribution and quality. A large proportion of the world's fresh water falls and resides in relatively unpopulated portions of the globe, particularly Siberia and northern Canada. Here, man's activities are limited by other factors, and he cannot readily obtain these vast stores of water. Secondly, the matter of water quality is of immediate concern. The principle of water reuse is now becoming accepted, so that people recognize much of their water has been used previously by other municipal systems. Each reuse cycle, however, is increasingly costly in terms of treatment, and each reuse cycle requires disinfectants and other industrial chemicals. Combining these problems of distribution and quality, it seems that the water problem of the future is not going to be one of a

TABLE 17-1 World's Water Resources[a]

Resource	*Volume (W) (in thousands of cu km)*	*Annual rate of Removal (Q) (in thousands of cu km and process)*	*Renewal Period* $\left(T = \frac{W}{Q}\right)$
Total water on earth	1,460,000	520—evaporation	2,800 years
Total water in the oceans	1,370,000	449—evaporation	3,100 years
		37—difference between precipitation and evaporation	37,000 years
Free gravitational waters in the earth's crust (to a depth of 5 km)	60,000	13—underground run-off	4,600 years
(Of which, in the zone of active water exchange	4,000	13—underground run-off	300 years)
Lakes	750	—	
Glaciers and permanent snow	29,000	1.8—run-off	16,000 years
Soil and subsoil moisture	65	85—evaporation and underground run-off	280 days
Atmospheric moisture	14	520—precipitation	9 days
River waters	1.2	36.3[b]—run-off	12(20) days

[a] From Kalinin and Bykov, 1969.
[b] Not counting the melting of Antarctic and Arctic glaciers.

global water shortage on a planetary basis, but one of local crises due to the concentrations of people, industry and water use. Certainly, the immediate concern is finding local water resources of adequate quality to meet the burgeoning needs of urban populations. To the city dweller in New York or Hong Kong whose water tap runs dry, or to the refugee in the Middle East who is dying of thirst, it will be little comfort to realize that 29,000,000 cubic kilometers of fresh water are locked in glaciers and polar ice caps.

WORLD FOOD SUPPLIES

In the United States it is difficult to conceive of a world food shortage. One of the most striking things about returning to the United States after living in Asia or Africa is to visit an American supermarket. It is an incredible experience to see the wealth and variety of food stacked upon the shelves and counters. To most of the world's peoples, this is an unattainable experience; to an American it is a commonplace event.

Nowhere is scientific controversy more evident than on the issue of world food supplies. Some scientists predict famines and mass starvation within the next 10 years (Paddock and Paddock, 1967; Ehrlich, 1970). They feel the stage is irrevocably set for millions of people to die of starvation in the 1970's. In fact, Ehrlich pointed out (1970) that we already have famine in many parts of the world, if one defines famine as people dying of lack of food. Even at the height of the Green Revolution in Asia in 1969 and 1970 in which new strains of rice and wheat were producing phenomenally larger yields, famine conditions existed in parts of Rajasthan in India, in sections of Burma and in southern Thailand. Starvation is now a grim reality in much of the world and it may well get worse despite the Green Revolution.

Other scientists refute this pessimism and find evidence for a hopeful outlook in the world food situation (Boerma, 1970; Hardin, 1969; Clark, 1963; McCormack, 1970). They feel that the Green Revolution of 1968, 1969, and 1970 which was based on new strains of rice and wheat in Asia and Africa, and which produced a doubling and tripling of yields in one or two years, shows what can be done with modern agriculture. For the first time in many decades, agricultural output increased more rapidly than population growth. India began to approach self-sufficiency in food production, and the Philippines began to export rice again in 1968 for the first time since 1900 (Bell, Hardin and Hill, 1969). Some economists have been sufficiently impressed with agricultural potential that they feel the world can feed 45 billion people if modern technology is utilized (Clark, 1963). Boerma (1970) has pointed out that less than one-half of the world's total cultivatable land is now under agricultural production (only 512 million hectares in use in 1962 out of a total of 1.1 billion available).

Many others take a more cautious attitude and, while recognizing technologic potential, they realize the practical difficulties (Brown, 1967; Harrar and Wortman, 1969). These are difficulties of education, ingrained social customs, distribution and warehousing, and numerous other mundane problems which spell the difference between success and failure.

A word of caution about the Green Revolution is also necessary. It is quite true that new strains of agricultural crops can sometimes dramatically increase yields, but often these new strains require more careful management, more intensive care, more water, more fertilizer and more insecticides. They may be more vulnerable to plant viruses or fungi. Although the plant geneticists try to breed in every possible disease resistance, these new strains do not have the centuries of natural selection and regional adaptation behind them as do the local strains which the villagers have always used. In West Bengal and Orissa, in 1969, there were reports that some of the newly introduced high-

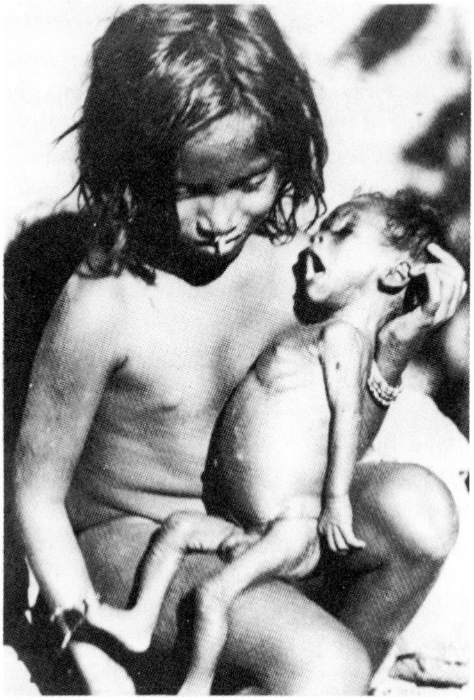

Figure 17–2 Despite the progress of the Green Revolution in 1969 and 1970, starvation conditions still prevailed in parts of India, Pakistan, and other areas of Asia (photo by P. Pittet, courtesy Food and Agricultural Organization, United Nations).

yield strains of rice were succumbing to a disease that was not affecting the old traditional strains. If this happens to a major extent, the Green Revolution could collapse and a severe blow would be dealt to the villager's faith in the agricultural scientist.

Another factor in the world food picture is one of pure climatic good fortune. The Green Revolution of 1968 and 1969 in India was accompanied by the best monsoon seasons in over 10 years. A long series of droughts had broken in southern Asia and adequate rains fell in the right pattern. The possibility exists, therefore, that a good part of the success of the Green Revolution was due to good weather, combined with new strains and more fertilizer. This is a question that will be more adequately answered in future years, to see if increasing agricultural production can be maintained in years of poor rainfall.

In India in 1971, too much rain came and it fell too early. The wheat crops of northern India were magnificent in April and May of 1971 just before harvesting and threshing, but heavy premature rains in many areas caused the wheat to rot before it could be threshed. Bihar had serious crop losses from flooding. Once again, this was a tragic and dramatic demonstration that even the Green Revolution remains dependent on climatic factors outside the control of man.

Throughout all of these arguments on how many people the world can feed, one must ask the realistic questions: "Is it necessary that there always be a race between population and food production? Is it desirable to be teetering on the bring of starvation at all times? Is it the best representation of the human condition to be struggling in this manner to feed more and more people at a subsistence level?"

One also has the impression in reviewing the literature on world food supplies that too much emphasis is placed on future predictions, and not enough on current conditions. We already know that starvation and malnutrition exist throughout the world, even in the most advanced countries. We furthermore know that malnutrition of any kind has profound effects upon growth, development and behavior. Children growing up with malnutrition from early infancy definitely show reduced growth rates and possible retardation of physical and mental development. They furthermore show increased susceptibility to infectious disease. Thus, we return to the issue of the quality of life, and the realization that merely preventing death by starvation does not by any means insure a quality nutritional base for adequate growth and human development. It is apparent that the problem of world food supplies will continue for some time as a problem of both quantity and quality, and especially one of distribution. Two of the scientists who helped to achieve the Green Revolution, Dr. D. S. Athwal of the International Rice Research Institute in the Philippines, and Dr. N. E. Borlaug, of the Rockefeller Wheat Research Institute in Mexico, have both warned that the Green Revolution will fail to meet the world's food needs unless it is accompanied by a drastic slowing of the world's population growth (Athwal, 1971; Borlaug, 1968).

SPACE ON EARTH

In 1958 Paul Sears wrote an important essay entitled, "The Inexorable Problem of Space," in which he concluded that "Our future security may depend less upon priority in exploring outer space than upon our wisdom in managing the space in which we live." With a shortage of space crowding in upon us in the 1970's and 1980's, his words gain validity every year.

The earth's total surface is 196,940,400 square miles, of which about 30 percent or 57,230,000 square miles is land surface. Of this total land surface, approximately 5,000,000 square miles are in Antarctica and are too cold to be easily inhabited except by scientists and explorers; 9,000,000 square miles are in deserts, and 11,000,000 square miles are in mountainous regions. Of deserts and mountains, at least one-fourth of their land area cannot be inhabited, so this eliminates another 5,000,000 square miles of land surface for human habitation. We are left with approximately 47,000,000 square miles of land surface, at the most, for the world's population.

The world now has about 3.5 billion people, and will probably reach 7 billion within the next 35 years. This means we now have about 75 people for every square mile of inhabitable land surface on the face of the earth, and this will approach 150 people per square mile by the year 2000 A.D.

As with water, air and food, it is important to distinguish between the total or potential amount of any given resource on earth, and that amount which is actually utilized or practically available to man. It was pointed out in the sections on water and air that the earth's total amount is apparently adequate for many years to come, but the supply in any given city may not be of sufficient quantity or quality to support the standard of living we would like. So it is with space. Although we theoretically have one square mile of land for every 75 people (about 9 acres per person), the human population is not evenly distributed. In the great urban concentrations of the world, densities often exceed 50,000 people per square mile. The amount of living space per person in some cities such as Calcutta is less than 20 square feet per person. Theoretically, we could redistribute these crowded people to provide more even spacing around the world. Obviously, this cannot be done, and if it were attempted it might wreak havoc with agricultural production.

The economic and behavioral needs of man are such that he must cluster together, but in so doing, he creates superdensities such as Calcutta, London, New York, Paris, Tokyo, Shanghai and Buenos Aires—the slums of which are definitely deleterious to human health and welfare. Why, we might ask, does man insist on this degree of crowding and then complain about it so bitterly? Why do people continue to pour into the cities of the world when they are already known as centers of pollution, crime and illness?

Perhaps we need to know more about the real spatial and social needs of man. What are his requirements for space and privacy, and how do these interact with his needs for social stimulation and group activity? Does man become conditioned to crowding in early childhood and then seek further crowding throughout his life? Does he require the social pressure of a crowd because he feels a compulsion to be in touch with the group, and thus be able to predict what the group is going to do next? Have we produced, in our urban centers, self-destructive systems which so modify the behavior of their inhabitants that these inhabitants exaggerate their own problems and can no longer achieve rational solutions?

Figure 17–3 Crowding is not confined solely to cities. Large areas of rural Asia are seriously crowded as shown in these typical scenes of a village in West Bengal (top) and a market in Uttar Pradesh of northern India.

These are all tough questions that cannot be answered in simple or easy terms. The answers will require the best intelligence we can muster from many disciplines. Some sorts of answers will certainly be necessary before we can even begin to solve the ecologic and behavioral problems associated with our own use of terrestrial space.

DISTRIBUTIONAL INEQUALITY

With all of the last four life-support requirements of man—air, water, food and space—we have seen that the major problems of immediate importance lie in distribution more than in total amount. Technologists therefore look upon these problems as soluble by technological means. They feel they are essentially matters of economics and transportation. Behavioral and environmental scientists see them as more basic, however, relating to man's fundamental ecologic behavior. They feel that certain basic social and ecologic tendencies of man prevent equitable distributions of the earth's resources. Obviously this type of argument can go on forever, and it is perhaps necessary to take a pragmatic approach and look at the current picture for the best glimpse of the future.

Figure 17–4 The pressure of population is so great in most of the world that the landscape is often intricately carved and remodeled to produce more food, as shown in this view of wheat terraces in the Himalayan foothills.

Many of the industrial nations of the world, especially the United States, the nations of northern Europe and Japan, have enjoyed rising prosperity for the last 25 years. In ecologic time, of course, this is but a flash in the pan, but let us assume that it can be sustained for another 25 or 50 years. It is worth examining where the material and energetic support for this prosperity arises.

Rienow and Rienow (1967) pointed out that the United States, comprising less than 7 percent of the world's population, uses 35 percent of the world's energy resources, and 50 percent of its total products. One-fourth of the world's people possess 75 percent of the world's wealth (McNamara, 1968). The incredibly high standard of living enjoyed by many Americans and Europeans draws raw materials and manufactured goods from the entire world. Every American baby born in 1970, if he lives for 70 years, will require 56 million gallons of water, 21,000 gallons of gasoline, 10,000 pounds of meat, 28,000 pounds of milk and cream, and other thousands of pounds of grain, sugar and specialty foods (Rienow and Rienow, 1967). He will require over $6,000 of clothing, $7,000 of furniture, over 10 automobiles, 28 tons of iron and steel, 4,500 cubic feet of wood, and 1½ tons of fiber (Mitchell, 1970). In amazing contrast, the Asian baby will utilize only one-hundredth to one thousandth of these items. The average American can expect to participate in an annual family income of $7,000; the average Asian in an annual family income of less than $100.

This distributional inequality is another major symptom of ecologic instability in the world. One would hope for some narrowing in the gap between the haves and the have-nots, but unfortunately the opposite trend is still occurring. The wealthy nations of the world are far outstripping the poor nations in their rates of economic gain and human betterment. There are, however, self-corrective mechanisms within this system, as, for example, when some segment of the labor market in a wealthy country can no longer compete with its foreign counterpart which does the same work for much less. The world economic enterprise continually strives for the cheapest labor it can find which will do a satisfactory job. Thus, there has been a great expansion of manufacturing in Japan, Hong Kong, Taiwan and Singapore in industries as diverse as steel, electronics, textiles, toys and optics. The Western factory worker can thereby suffer a decline in his economic security and standard of living while the Asian factory worker experiences for the first time an improving situation. The Western industrial worker continues to reduce his own working prospects by striking for higher and higher wages. This will certainly be one mechanism by which wealth is more adequately balanced between nations.

Still the mass of poverty throughout much of Asia, Africa and Latin America is so great, it is difficult to imagine a painless equilibrium. The differential distribution of wealth remains sharp and represents a type of ecologic situation which cannot last indefinitely without severe consequences and reverberations. To the ecologist it means either a successional stage which is not yet mature, or a system headed for a major breakdown. Economic and ecologic maturity should ideally be in the direction of broader and more equitable distribution of the earth's resources—a narrowing of the gap between the haves and the have-nots, both within and between nations, provided the total system does not pass beyond the state of corrective change.

VIOLENCE AND WAR

Any ecological system or biotic community can be overloaded to the point of collapse by a variety of factors. Certainly one of the factors for man of most immediate and realistic danger is that of social violence. Either by war or by civil riot, man threatens his own existence.

To maintain a sense of physical security, man has built an incredible stockpile of destructive power (York, 1970). The nations of the world spend more than 200 billion dollars per year on armaments—more than on health, education, welfare, agriculture, child care and care for the aged combined. Twenty tons of explosives are stored in the world for every man, woman and child. Russia and the United States have stockpiled 6,000 nuclear warheads, possessing thousands of megatons of explosive force. Each megaton has more destructive power than that which existed in any of the previous wars of history. The world spends almost 8,000 dollars per year training each soldier, and less than 100 dollars per year educating each child.

The question is, of course, how long can man continue to preside over armament like this without using it? And the answer seems to be, not very long. The fact is, we do use part of this power in an attempt to achieve social goals. It is being used daily in various parts of the world to maintain territory and dominance ranks. It is justified as a spare and deliberate use in hopes of avoiding the all-out effort. How long this can continue is anyone's guess, and the ecologist can claim no more omniscience than the political scientist. The ecologist points out, however, that the ecologic and social pressures which generate international military tension are currently increasing at an awesome rate. It is not likely that man can solve the problems of war and violence until he approaches and understands their basic origins in terms of the behavioral and ecologic relationships between social groups and their environment.

ADAPTIVE CAPABILITIES OF MAN

Many scholars feel that one of the great problems of the present and the future is the rapid rate of change which is occurring in society and technology. There is not, however, agreement on the consequences of this rapid rate of change. Toffler (1970) feels that changing life patterns are too rapid for man to adapt to successfully. Mankind is so overstimulated by the increase in new sights and sounds, so bombarded by worldwide mass media, and so constantly beset with novel situations and new problems that he will break down psychologically and physically. His responses will be maladaptive and pathologic. He will thus collapse under sensory overload.

On the other hand, Dubos (1965; 1970) feels that man can and will adapt to the most rapid change. He sees this adaptational ability as a major part of the problem. Man will manage to survive with deplorable conditions, abysmal pollution and poverty, fantastic crowding and computerized automation. Man will somehow learn to live with all of these problems he is creating, but in so doing he will lose much of his distinct humanity.

It is quite possible that each of these points of view has large elements of truth in it. In the present world one can find many examples of both. Our mental institutions are already filled with those who could not adapt to sensory overload and who experience a personal breakdown of physical and mental health. Yet one can find slums in any country of the world where people are living a subhuman existence—where the quality of life is so low that the amount of human pain and misery seems unbearable. Yet man has adapted successfully to these inhuman conditions.

In many ways, the future is with us now. We can find throughout the world an abundance of high quality life in high quality environments, but we can also find an abundance of low quality life in impoverished environments. Many of the questions about the future revolve, therefore, around the question of relative balance and proportions. Which way are we heading, and which way will the scales tip?

A major aspect of the future centers around our sense of responsibility to posterity. As Professor Powers of the University of Maryland Law School has pointed out (Hardin, 1970), no representative of posterity is here to speak for that segment of society. How do the present populations of mankind regard their obligations to future generations? Are we willing to be more conservative in our life requirements, and are we, in fact, willing to reduce our rates of personal economic growth to insure a better environment and a greater abundance of natural resources for future generations? This may be one of the central issues of the next twenty years, and one on which the quality of present and future environments may depend.

References

Aaronson, T. Mercury in the environment. *Environment,* **13**(4):16-27 (1971).

Abelson, P. H. Changing attitudes towards environmental problems. *Science,* **172**:517 (1971).

Adams, L., J. A. Finn and R. E. Wetmore. Stress and behavior as components of a population control mechanism. *Bull. Ecol. Soc. Amer.,* **52**(2):48–49 (1971).

Allee, W. C. *The Social Life of Animals.* Rev. Ed., Boston: Beacon Press, 1958.

Allee, W. C., A. E. Emerson, O.Park, T. Park, and K. P. Schmidt. *Principles of Animal Ecology. Philadelphia:* Saunders, 1949.

Allen, D. L. *Our Wildlife Legacy.* New York: Funk & Wagnalls, 1954.

Allen, D. L. *The Life of Prairies and Plains.* New York: McGraw-Hill, 1967.

Altmann, S. A. A field study of the sociobiology of rhesus monkeys, *Macaca mulatta.* Ann. N.Y. Acad. of Sci., **102**:338–435.

Altshuller, A. P. Air Pollution: Photochemical aspects. *Science,* **151**:1104–1106 (1966).

Anderson, P. K. Density, social structure, and nonsocial environment in house mouse populations and implications for the regulations of numbers. Trans. N.Y. Acad. Sci., Ser. 11, **23**:447–451 (1961).

Andrewartha, H. G. and L. C. Birch. *The Distribution and Abundance of Animals.* Chicago: Univ. of Chicago Press, 1954.

Anzou, G. *The Formation of the Bible.* St. Louis and London: Herder, 1963.

Ardrey, R. *The Territorial Imperative.* New York: Atheneum, 1966.

Ashby, K. R. Studies on the ecology of field mice and voles in Houghall Wood, Durham. *J. Zool.* (London) **152**:389–513 (1967).

Athwal, D. S. Semi-dwarf rice and wheat in global food needs. *Quart. Rev. of Biology,* **46**:1–34 (1971).

Barnett, S. A. *The Rat: A Study in Behavior.* Chicago: Aldine, 1963.

Barnett, S. A. Social stress. In *Viewpoints in Biology,* J. D. Carthy and C. L. Duddington (eds.) London: Butterworths, 1964, pp. 170–218.

Barnett, S. A. *Instinct and Intelligence: Behavior of Animals and Man.* Englewood Cliffs, N.J.: Prentice-Hall, 1967.

Bartsch, A. F. and M. O. Allum. Biological factors in the treatment of raw sewage in artificial ponds. *Limnol. and Oceanogr.* **2**:77–84 (1957).

Bartsch, A. F., R. J. Callaway, R. A. Wagner, and C. E. Woelke. Technical approaches toward evaluating estuarine pollution problems. In *Estuaries,* George H. Lauff (ed.), Washington, D.C.: AAAS, 1967, pp. 693–700.

Bates, M. *The Forest and the Sea. A Look at the Economy of Nature and the Ecology of Man.* New York: Menton Books, 1960.

Bates, M. Crowded animals. *Natural History,* **77**(7):20–24 (1968).

Bates, M. Crowded people. *Natural History,* **77**(8):20–25 (1968).

Battan, L. J. *The Unclean Sky: A Meteorologist Looks at Air Pollution.* Garden City, N.Y.: Doubleday, 1966.

Becker, W. H., F. J. Schilling, and M. P. Verma. The effect on health of the 1966 Eastern Seaboard air pollution episode. *Archives of Environ. Health,* **16**(3): 414–419 (1968).

Bell, D. E., C. M. Hardin, and F. F. Hill. Hope for the Hungry: Fulfillment or Frustration. Chap. 4 In *Overcoming World Hunger. C. M. Hardin* (ed.) Englewood Cliffs, N.J.: Prentice-Hall, 1970, pp. 137–170.

Bennett, L. J. *The Blue-winged Teal: Its Ecology and Management.* Ames, Iowa: Collegiate Press, Inc., 1938.

Bitancourt, A. A. Expressao matematica do crescimento de formigueiros de *"Atta sexdens rubropilosa"* representado pelo aumento do numero de olheiros. *Arch. Inst. Biol.* **12**:229–236 (1941).

Blair, W. F. Population structure, social behavior, and environmental relations in a natural population of the beach mouse (*Peromyscus polionotus leucocephalus*). Contributions from the Laboratory of Vertebrate Biology, Univ. of Michigan: No. 48, June, 1951.

Boerma, A. H. A World Agricultural Plan. *Scientific American,* **223**(2):54–69 (1970).

Bolin, B. The carbon cycle. *Scientific American,* **223**:124–135 (1970).

Borlaug, N. E. Wheat breeding and its impact on the world food supply. In Proc. 3rd Int. Wheat Genet. Symp. K. W. Findlay and K. W. Shepherd (eds.), New York: Plenum Press, 1968.

Boughey, A. S. *Ecology of Populations.* New York: MacMillan, 1968.

Boughey, A. S. *Contemporary Readings in Ecology.* Belmont, Calif.: Dickenson Press, 1969.

Boulding, K. E. War as a public health problem: Conflict management as a key to survival. In *Behavioral Science and Human Survival,* M. Schwebel (ed.), Palo Alto, California: Science & Behavior Books, Inc., 1965, pp. 103–110.

Bourliere, F. *The Land and Wildlife of Eurasia.* New York: Time-Life Books, 1964.

Boyle, R. New pneumatical experiments about respiration. Phil. Trans. Roy. Soc. of London, **5**(62):2011–2032; (63):2035–2056 (1670).

Brock, T. D. *Principles of Microbial Ecology.* Englewood Cliffs, N.J.: Prentice-Hall, 1966.

Brodine, V. Episode 104. *Environment,* **13**(1):2–27 (1971).

Broecker, W. S. Man's Oxygen Reserves. *Science,* **168**:1537–1538 (1970).

Brown, H. *The Challenge of Man's Future.* New York: Viking Press, 1954.

Brown, J. Territorial behavior and population regulation in birds. *The Wilson Bulletin,* **81**(3):293–329 (1969).

Brown, L. R. The World Outlook for Conventional Agriculture. *Science,* **158**:604–611 (1967).

Brown, R. B. *Biology.* Boston: D.C. Health, 1956.

Brown, R. Z. Social behavior, reproduction and population changes in the house mouse (*Mus musculus* L.). *Ecol. Monogr.,* **23**:217–240 (1953).

Bryerton, G. *Nuclear Dilemma.* New York: Ballantine Books, 1970.

Buechner, H. Territoriality in the Uganda Kob. *Science,* **133**:698–699 (1961).

Buechner, H. Territoriality as a behavioral adaptation to environment in the Uganda Kob. Proc. XVI Int. Cong. Zool. **3**:59–62 (1963).

Buell, P., J. E. Dunn, Jr., and L. Breslow. Cancer of the lung and Los Angeles-type air pollution. *Cancer,* **20**:2139 (1967).

Bullard, E. The Origin of the Oceans, *Scientific American,* **221**(3):66–75 (1969).

Burt, W. H. Territorial behavior and populations of small mammals in southern Michigan. Museum Zool. Misc. Publ. Univ. Mich. **45**:1–58 (1940).

Burt, W. H. *A Field Guide to the Mammals.* Boston: Houghton Mifflin, 1952.

Calhoun, J. B. Population density and social pathology. *Scientific American,* **206**:139–148 (1962).

Calhoun, J. B. A behavioral sink. In *Roots of Behavior,* E. L. Bliss (ed.), New York: Harper, 1962, p. 295–315.

Candolle, A. P. de. De la germination sous le degrés divers de témperature constante. Biblio. Univ. Rev. Suisse, **14**:243–282 (1865).

Carpenter, C. R. Territoriality: A Review of Concepts and Problems. Chap. 11 In *Behavior and Evolution,* Anne Roe and George Gaylord Simpson (eds.), New Haven: Yale Univ. Press, 1958, pp. 224–250.

Casarett, L. J., G. C. Fryer, W. L. Yanger, Jr., and H. W. Klemmer. Organochloride pesticide residues in human tissue—Hawaii. *Arch. of Environ. Health,* **17**(3):306–311 (1968).

Champion, H. G. A preliminary survey of the forest types of India and Burma. *Indian Forest Records* (Silviculture Series), Vol. 1, No. 1, 1936.

Chandler, A. C. and C. P. Read. *Introduction to Parasitology.* New York: Wiley, 1961.

Chandler, D. C. Plankton and certain physical-chemical data of the Bass Islands Region, from Sept. 1938 to Nov. 1939. *Ohio Jour. Sci.* **40**:291–336 (1940).

Chisholm, J. J., Jr. Lead poisoning. *Scientific American* **224**:15–23 (1971).

Chitty, D. Tuberculosis among wild voles: with a discussion of other pathological conditions among certain mammals and birds. *Ecology,* **35**(2):227–237 (1954).

Chitty, D. Self-regulation of numbers through changes in viability. *Cold Spring Harbor Symp. Quant. Biol.* **22**:277–280 (1958).

Chitty, D. Population processes in the vole and their reference to general theory. *Can. J. Zool.* **38**:99–113 (1960).

Chitty, D. Variations in the weight of the adrenal glands of the field vole, *Microtus agrestis. J. Endocrinol.* **22**:387–393 (1961).

Chow, T. J., and J. L. Earl. Lead aerosols in the atmosphere: increasing concentrations. *Science,* **169**:577–580 (1970).

Chowdhury, A. B., and E. Schiller. A survey of parasitic infections in a rural community near Calcutta. *Amer. J. Epid.* **87**:299–312 (1968).

Christian, J. J. The adreno-pituitary system and population cycles in mammals. *Jour. Mammalogy,* **31**:247–259 (1950).

Christian, J. J. Endocrine adaptive mechanisms and the physiologic regulation of population growth. In *Physiological Mammalogy,* ed. by W. V. Mayer and R. G. Van Gelder. New York: Academic Press (1963).

Christian, J. J., and D. E. Davis. Endocrines, behavior and population. *Science,* **146**:1550–1560 (1964).

Clark, C. Agricultural productivity in relation to population. In *Man and His Future.* London: Churchill, 1963, pp. 23–35.

Clarke, G. L. *Elements of Ecology.* New York: Wiley, Inc., 1954.

Clarke, J. R. Influence of numbers on reproduction and survival in two experimental vole populations. *Proc. Roy. Soc.* (B): **144**:68–85 (1955).

Cloud, P., and A. Gibor. The oxygen cycle. *Scientific American,* **233**:110–123 (1970).

Clough, G. C. Physiological effect of botfly parasitism on meadow voles. *Ecology,* **46**(3):344–346 (1965).

Cole, L. C. A study of Cryptozoa of an Illinois woodland. *Ecol. Monogr.* **16**:49–86 (1946a).

Cole, L. C. A theory for analyzing contagiously distributed populations. *Ecology,* **27**:329–341 (1946b).

Cole, L. C. Playing Russian roulette with biochemical cycles. In *The Environmental Crisis,* Helfrich, H. W. Jr. (ed.), New Haven: Yale Press, 1970, pp. 1–14.

Collias, N. E. Some variations in grouping and dominance patterns among birds and mammals. *Zoologica,* **35**:97–119 (1950).

Colquhoun, M. K. and A. Morley. Vertical zonation in woodland bird communities. *J. Anim. Ecol.* **12**:75–81 (1943).

Commoner, B. *Science and Survival.* New York: Viking Press, 1966.

Cox, G. Readings in Conservation Ecology. New York: Appleton-Century-Crofts, 1969.

Connell, J. H., D. Mertz, and W. Murdoch. *Readings in Ecology and Evolution.* New York: Harper & Row, 1970.

Craig, P., and E. Berlin. The Air of Poverty. *Environment,* **13**(5):2–9 (1971).

Crombie, A. C. Further experiments on insect competition. *Proc. Roy. Soc. London* (B), **133**:76–109 (1946).

Crook, J. H. The adaptive significance of social organization and visual communication in birds. *Symp. Zool. Soc. London,* **14**:181–218 (1965).

Crowcroft, P. *Mice All Over.* Chester, Pa.: Dufour Press, 1966.

Crowcroft, P. and F. P. Rowe. The growth of confined colonies of the wild house mouse (*Mus musculus* L.). *Proc. Zool. Soc. London,* **129**:359–370 (1957).

Dansereau, P. *Biogeography: An Ecological Perspective.* New York: Ronald Press, 1957.

Dansereau, P. *Challenge for Survival.* New York: Columbia Univ. Press., 1970.

Darling, F. F. *Wilderness and Plenty.* New York: Ballantine Books, 1970.

Darling, F. F. and R. F. Dasmann. The Ecosystem view of human society. *Impact of Science on Soc.,* **XIX**(2):109–121 (1969).

Darlington, P. J. Jr. *Zoogeography: the Geographical Distribution of Animals.* New York: Wiley, 1964.

Darnell, R. M. Evolution and the ecosystem. *Amer. Zoologist,* **10**(1):9–15 (1970).

Darwin, C. G. *The Next Million Years.* New York: Doubleday, 1953.

Dasman, R. Methods for estimating deer populations from kill data. *Calif. Fish & Game,* **38**:225–233 (1952).

Dasmann, R. *Environmental Conservation.* New York: Wiley, 1968.

Daubenmire, R. F. *Plants and Environment.* New York: Wiley, 1947.

Davidson, Basil. *The Lost Cities of Africa*. Boston: Little, Brown, 1959.

Davis, D. E. The role of density in aggressive behavior of house mice. *Animal Behaviour,* **VI**:207–210 (1958).

Davis, D. E. 1963. Estimating the numbers of game populations. Chap. 5 In *Wildlife Investigational Techniques,* H. S. Mosby and O. H. Hewitt (eds.), Washington, D.C.: The Wildlife Society, pp. 89–118.

Day, N. The Dimensions, Issues and Determinants of Urban Development. Chap. II in *Environmental Problems,* B. R. Wilson (ed.), Philadelphia: J. B. Lippincott Co., 1968, pp. 30–53.

DeBell, G. *The Environmental Handbook*. New York: Ballantine Books, 1970.

Deevey, E. S. Jr. The Human Population. *Scientific American,* **203**(3):195–204 (1960). (Reprinted in The Subversive Science by P. Shepherd and D. McKinley, 1969).

Deevey, E. S., Jr. Mineral cycles. *Scientific American,* **223**:148–159 (1970).

Delwiche, C. C. The nitrogen cycle, *Scientific American,* **223**:136–147 (1970).

Dice, L. R. *Natural Communities*. Ann Arbor: Univ. of Michigan Press, 1968.

Dorn, H. F. World population growth: an international dilemma. *Science,* **135**: 283–290 (1962).

Dorst, J. *Before Nature Dies*. Boston: Houghton Mifflin, 1970.

Dowdeswell, W. H. *Animal Ecology*. New York: Harper, 1961.

Dubos, R. *Man Adapting*. New Haven: Yale University Press, 1965.

Dubos, R. *So Human an Animal*. New York: Charles Scribner's Sons, 1970.

Dunn, F. L. Health and disease in hunter gatherers. Chap. 23 In *Man the Hunter,* Lee and DeVore (eds.), Chicago: Aldine, 1968, pp. 221–228.

Ehrlich, P. R. *The Population Bomb*. New York: Ballantine Books, 1968.

Ehrlich, P. R. Eco-Catastrophe. *Ramparts Magazine,* Sept. 1969. (Reprinted in, *The Environmental Handbook,* G. DeBell (ed.), New York: Ballantine Books, pp. 161–176).

Ehrlich, P. R. Famine, 1975: Fact or Fallacy. In *The Environmental Crisis,* H. W. Helfrich, Jr. (ed.) New Haven: Yale University Press, 1970, pp. 47–64.

Ehrlich, P. R. and A. H. Ehrlich. *Population, Resources, Environment: Issues in Human Ecology*. San Francisco: W. H. Freeman & Co., 1970.

Ehrlich, P. R. and R. L. Harriman. How to be a Survivor. New York: Ballantine Books, 1971.

Einarsen, A. S. Some factors affecting ring-necked pheasant population density. *Murrelet.* **26**:39–44 (1945).

Elton, C. S. *The Ecology of Animals*. London: Methuen, 1933.

Elton, C. S. *The Ecology of Invasions by Animals and Plants*. London: Methuen, 1958.

Emlen, J. Sex and age ratios in the survival of California quail. *Jour. of Wildlife Mgt.*, **4**:92–99 (1940).

Errington, P. Predation and vertebrate populations. *Quart. Rev. of Biol.* **21**(2): 144–177; (3):221–245 (1946).

Errington, P. *Of Predation and Life*. Ames, Iowa: Iowa State Univ. Press, 1967.

Esser, A. H. (ed.) *Behavior and Environment: The Use of Space by Man and Animals*. New York: Plenum Press, 1971.

Ewald, W. R., Jr. *Environment for Man: The Next Fifty Years*. Bloomington: Indiana University Press, 1968.

Farb, P. The Forest. New York: Time, Inc., 1963.

Faris, R. E. L. and W. H. Dunham. *Mental Disorders in Urban Areas: An Ecological Study of Schizophrenia and Other Psychoses*. Chicago: University of Chicago Press, 1939.

Feller, W. On the logistic law of growth and its empirical verifications in biology. *Acta. Biotheoret.* **5**:51–66 (1940).

Findlay, G. M. and A. D. Middleton. Epidemic disease among voles (*Microtus*) with special reference to *Toxoplasma*. *Jour. Anim. Ecol.* **2**:(3)150–160 (1934).

Fisher, J., N. Simon, and J. Vincent. *Wildlife in Danger*. New York: Viking Press, 1969.

Fletcher, J. O. Controlling the planet's climate. *Impact of Science on Society*, **19**(2):151–168 (1969).

Flyger, V. Squirrel mortality in Baltimore City by coccidiosis. *Personal Communication*, 1969.

Fortune, Editors of. *The Fabulous Future: America in 1980*. New York: American Book-Stratford Press, 1955.

Frank, P. W. Coactions in laboratory populations of two species of *Daphnia*. Ecol. **38**:510–519 (1957).

Fuller, R. B. *Utopia or Oblivion: The Prospects for Humanity*. New York: Bantam Books, 1969.

Garlick, J. P. and R. W. J. Keay. (eds.). *Human Ecology in the Tropics*. New York: Pergamon Press, 1970.

Gibb, J. Bird Populations. Chap. XXIV In *Biology and Comparative Physiology of Birds*. Vol. 2. A. J. Marshall (ed.). New York: Academic Press, 1961, pp. 413–446.

Godfrey, A. (ed.) *The Arthur Godfrey Environmental Reader*. New York: Ballantine Books, 1970.

Gofman, J. W. and A. R. Tamplin. *Population Control Through Nuclear Pollution*. Chicago: Nelson Hall, 1970.

Goldsmith, J. R. and S. A. Landow. Carbon monoxide and human health. *Science*, **162**:1352–1359 (1968).

Gordon, M. *Sick Cities: Psychology and Pathology of American Urban Life*. Baltimore: Penguin Books, 1969.

Gottschalk, L. C. and V. H. Jones. Valleys and hills, erosion and sedimentation. *Yearbook of Agriculture: Water*. Washington, D.C.: U.S. Government Printing Office, 1955, pp. 135–143.

Graham, H. D., and T. R. Gurr. *Violence in America*. New York: New American Library, 1969.

Grant, N. Mercury in man. *Environment*, **13**(4):2–15 (1971).

Gray, J. The kinetics of growth. *Brit. Jour. Exp. Biol.* **6**:248–274 (1929).

Gubler, D. J. "A study of comparative biology and interspecific competition between *Aedes* (*Stegomyia*) *albopictus* and *Aedes* (*Stegomyia*) *polynesiensis*." Johns Hopkins University Thesis for Doctor of Science, Baltimore, Md. Feb., 1969.

Guggisberg, C. A. *Man and Wildlife*. New York: Arco, 1970.

Guhl, A. M. Psycho-physiological factors and social behavior related to sexual behavior in birds. *Trans. Kans. Acad. Sci.*, **63**:85–95 (1960).

Guhl, A. M. The behavior of chickens. Chap. 17 In *The Behaviour of Domestic Animals*, E. S. E. Hafez (ed.) 1962, pp. 491–530.

Guthrie, D. A. Primitive man's relationship to nature. *Bioscience*, **21**:721–723 (1971).

Hall, E. T. *The Hidden Dimension*. New York: Doubleday, 1966.

Halstead, B. W. Poisonous and Venomous Marine Mammals of the World. Vol. 2. *Vertebrates*. Washington, D.C.: U.S. Government Printing Office, 1967.

Handler, P. (ed.) *Biology and the Future of Man*. New York: Oxford University Press, 1970.

Hardin, C. M. (ed.) *Overcoming World Hunger*. Englewood Cliffs, N.J.: Prentice-Hall, 1969.

Hardin, G. *Population, Evolution and Birth Control: A Collage of Controversial Readings*. San Francisco: Freeman, 1964.

Hardin, W. Posterity: Key Issue in Pollution. *Baltimore Evening Sun*, July 30, 1970. Page C-1.

Harrar, J. G. and S. Wortman. Expanding Food Production in Hungry Nations: The Promise and the Problems. Chap. 3 in *Overcoming World Hunger*. C. M. Hardin (ed.) Englewood Cliffs, N.J.: Prentice-Hall, 1969, pp. 89–135. U.S. Bureau of Fisheries. Vol. 43, 1927.

Hawkes, J. *Prehistory*. New York: Mentor Books, 1965.

Hazen, W. E. *Readings in Population and Community Ecology*, 2nd ed. Philadelphia: Saunders, 1970.

Helfrich, H. W., Jr. (ed.). *The Environmental Crisis*. New Haven and London: Yale University Press, 1970.

Hesse, R., W. C. Allee, and K. P. Schmidt. *Ecological Animal Geography*. New York: Wiley, 1951.

Hickey, J. J. (ed.). *Peregrine Falcon Populations: Their Biology and Decline*. Madison: University of Wisconsin Press, 1969.

Hildebrand, S. F., and W. C. Schroeder. Fishes of Chesapeake Bay. Bull. of U.S. Bureau of Fisheries. Vol. 43, 1927.

Hirschhorn, N. and W. B. Greenough, III. Cholera. *Scientific American*, **225**(2): 15–21 (1971).

Hochbaum, H. A. The Canvasback on a Prairie Marsh. Washington, D.C.: The American Wildlife Institute, 1944.

Hoffer, E. *The Ordeal of Change*. New York: Harper & Row, 1963.

Hogben, L. T. Some biological aspects of the population problem. *Biol. Rev.* **6**:163–180 (1931).

Hogetsu, K. and S. Ichimura. Studies on the biological production of Lake Suwa. VI. The ecological studies of the production of phytoplankton. *Japanese J. Bot.*, **14**:280–303 (1954).

Hölldobler, B. Communication between ants and their guests. *Scientific American*, **224**:86–93 (1971).

Horsfall, J. G. The green revolution: agriculture in the face of the population explosion. In *The Environmental Crisis*, H. W. Helfrich, Jr., New Haven: Yale University Press, 1970, pp. 85–98.

Hunt, E. G. Biological Magnification of Pesticides. In *Scientific Aspects of Pest Control*. Publ. 1402, National Academy of Sciences, Washington, D.C., 1966, pp. 251–262.

Hunt, E. G. and A. E. Bischoff. Inimical effects on wildlife of periodic DDT applications to Clear Lake. *Calif. Fish and Game*, **46**(1):91 (1960).

Hutchinson, G. E. Nitrogen in the biogeochemistry of the atmosphere. *Amer. Scientist*, **32**:178–195 (1944).

Hutchinson, G. E. Teleological mechanisms: Circular causal systems in ecology. *Ann. New York Acad. Sci.* **50**:221–246 (1948).

Hutchinson, G. E. The biochemistry of phosphorus. In *The Biology of Phosphorus*, L. F. Wolterink (ed.). East Lansing: Michigan State University Press, 1957.

International Biological Program. Program statement of the Subcommittee on Productivity of Fresh Water Communities and the Subcommittee on Productivity of Terrestrial Communities. Washington, D.C.: National Research Council, Division of Biology and Medicine, 1966.

Jacobson, J. S. and A. C. Hill. Recognition of air pollution injury to vegetation: a pictorial atlas. Air Pollution Control Assoc., Pittsburgh, 1970.

Jaffe, L. S. Photochemical air pollutants and their effects on man and animals: Adverse Effects. *Archives of Environ. Health,* **16**(2):241–255 (1968).

Janssen, W. A. and C. D. Meyers. Fish: serologic evidence of infection with human pathogens. *Science,* **159**:547–548 (1968).

Jenkins, D., A. Watson, and G. R. Miller. Population studies on Red grouse, *Lagopus lagopus scoticus* (Lath) in Northeast Scotland. *J. Anim. Ecol.,* **32**: 317–376 (1963).

Juday, C. The annual energy budget of an inland lake. *Ecol.* **21**:438–450 (1940).

Kalinin, G. P. and V. D. Bykov. The world's water resources, present and future. *Impact of Science on Society,* **19**(2):135–150 (1969).

Kavanagh, A. J. and O. W. Richards. The autocatalytic growth curve. *Amer. Nat.* **68**:54–59 (1934).

Keith, L. B. *Wildlife's Ten-Year Cycle*. Madison, Wisconsin: University of Wisconsin Press, 1963.

Kendeigh, S. C. *Animal Ecology*. Englewood Cliffs, N.J.: Prentice-Hall, (1961).

Kerner Report. U.S. National Advisory Commission on Civil Disorders Report. New York: Bantam Books, 1968.

Ketchum, B. H. Eutrophication of Estuaries. In, Eutrophication: Causes, Consequences, Correctives. National Academy of Sciences, Washington, D.C., 1969, pp. 197–209.

Koford, C. B. Population Dynamics of Rhesus Monkeys on Cayo Santiago. Chap. 5 In *Primate Behavior*. Irven DeVore (ed.). New York: Holt, Rinehart and Winston, 1965, pp. 160–174.

Kohn, A. J. and P. Helfrich. Primary organic productivity of a Hawaiian coral reef. *Limnol. and Oceanogr.* **2**:241–251 (1957).

Kormondy, E. J. *Concepts of Ecology*. Englewood Cliffs, N.J.: Prentice-Hall, 1969.

Krebs, C. J. Demographic changes in fluctuating populations of *Microtus californicus*. *Ecol. Monogr.,* **36**:239–273 (1966).

Krenkel, P. A. and F. L. Parker (eds.). *Biological Aspects of Thermal Pollution.* Nashville: Vanderbilt University Press, 1969.

Lack, D. *Population Studies of Birds.* Oxford: Clarendon Press, 1966.

Lancaster, P. (ed.). *Here Comes Tomorrow: Living and Working in the Year 2000.* Princeton, N.J.: Dow Jones Books, 1967.

Lansford, H. The supercivilized weather and sky show. *Natural History,* **79**(7): 92–113 (1970).

Laycock, G. *The Alien Animals.* Garden City, N.Y.: Natural History Press, 1966.

Lederberg, J. Quoted in News and Comment, based on House Foreign Affairs Subcommittee Hearings on December 2, 1969. *Science,* **166**:1490 (1969).

Lee, R. B. and I. DeVore (ed.). *Man the Hunter.* Chicago: Aldine, 1968.

Leighly, J. (ed.). *Land and Life. A selection from the writings of Carl Ortwin Sauer.* Berkeley: University of California Press, 1967.

Leinwand, G. *Air and Water Pollution.* New York: Washington Square Press, 1969.

Leopold, A. *Game Management.* New York: Scribner, 1933.

Leopold, A. *The Land Ethic.* An essay in *A Sand County Almanac.* New York: Oxford University Press, 1949, pp. 201–226.

Leopold, A. S. *The Desert.* New York: Time, Inc., 1961.

Lindeman, R. L. The trophic-dynamic aspects of ecology. *Ecology,* **23**:399–418 (1942).

Linton, R. M. *Terracide: America's Destruction of Her Living Environment.* Boston: Little, Brown, 1970.

Louch, C. D. Adrenocortical activity in two meadow vole populations. *J. Mammalogy,* **39**:109–116 (1958).

Loucks, O. L. Evolution of diversity, efficiency and community stability. *Amer. Zool.,* **10**(1):17–25 (1970).

Lowenthal, D. (ed.). *Man and Nature,* by G. P. Marsh. Cambridge, Mass.: Belknap Press of Harvard University, 1967.

Lowrie, R. C. "Studies on competition between larvae of *Aedes albopictus* and *A. polynesiensis.*" Johns Hopkins University, Sc.D. Thesis. Baltimore, Md., 1971.

MacArthur, R. and J. Connell. *The Biology of Populations.* New York: Wiley, 1966.

Mackenzie, J. M. D. Fluctuations in the numbers of British tetraonidae. *J. Anim. Ecol.* **21**:128–153 (1952).

MacLulich, D. A. Fluctuations in the numbers of the varying hare (*Lepus americanus*). *Univ. Toronto Stud. Biol. Ser.* **43**:1–136 (1937).

Malthus, T. R. *An Essay on the Principle of Population*. London: Johnson, 1798.

Margalef, R. *Perspectives in Ecological Theory*. Chicago: University of Chicago Press, 1968.

Marsh, G. P. *Man and Nature*. D. Lowenthal (ed.), Cambridge: Harvard University Press. 2nd Printing, 1967. Orig. Publ. 1864.

Martin, P. S. and H. E. Wright. Pleistocene Extinctions. Proc. 7th Cong. Int. Assoc. for Quarternary Res., Vol. 6. New Haven: Yale University Press, 1967.

McCarroll, J. R., E. J. Cassell, W. T. Ingram and D. Wolter. Health and urban environment: health profiles versus environmental pollutants. *Amer. Jour. of Public Health* **56**(2):266–275 (1966).

McCormack, A. *The Population Problem*. New York: Thomas Y. Crowell, 1970.

McCormick, J. *Forest Ecology*. New York: Harper & Bros., 1959.

McCaull, J. Questions for an old friend. *Environment,* **13**:2–9 (1971).

McHarg, I. *Design with Nature*. Garden City, New York: Natural History Press, 1969.

McKeever, S. Effects of reproductive activity on the weight of adrenal glands in *Microtus montanus. Anat. Record.* **135**:1–5 (1969).

McNamara, R. S. *The Essence of Security*. New York: Harper & Row, 1968, pp. 145–158.

Mead, M. The Island Earth. *Natural History,* **79**(1):22, 102–103 (1970).

Mech, L. D. The Wolves of Isle Royale. Fauna of the U.S. National Parks, Series 7. Washington, D.C.: U.S. Government Printing Office, 1966.

Medawar, P. B. *The Future of Man*. New York: Menton Books, 1959.

Mihursky, J. Patuxent Thermal Studies. Nat. Resources Inst., Univ. of Maryland, Spec. Rept. No. 1, Contribution No. 326, 1969.

Milgram, S. The experience of living in cities. *Science,* **167**:1461–1468, 1970.

M.I.T. Report. *Man's Impact on the Global Environment*. Cambridge, Mass.: Massachusetts Inst. of Technology, 1970.

Mitchell, J. G. On the spoor of the slide rule. In *Ecotactics,* J. G. Mitchell and C. L. Stallings (eds.). New York: Pocket Books, 1970.

Mitchell, J. G. and C. L. Stallings. *Ecotactics: The Sierra Club Handbook for Environmental Activists*. New York: Pocket Books, 1970.

Miyadi, K. Perspectives of experimental research on social interference among fishes. In *Perspectives in Marine Biology*. A. A. Buzzati-Traverso (ed.). Scripps Inst. of Oceanography, 1956.

Moller, C. B. *Architectural Environment and Our Mental Health*. New York: Horizon Press, 1968.

Morse, D. H. Ecological aspects of some mixed species foraging flocks of birds. *Ecol. Monogr.,* **40**(1):119–168 (1970).

Moss, F. E. *The Water Crisis.* New York: Frederick A. Praeger, 1967.

Moss, W. W. and Camin, J. H. Nest parasitism, productivity, and clutch size in purple martins. *Science,* **168**:1000–1002 (1970).

Mountain, I. M., E. J. Cassell, D. W. Wolter, J. D. Mountain, J. R. Diamond, and J. R. McCarroll. Health and the urban environment. VII. Air pollution and disease symptoms in a "Normal" population. *Archives of Environ. Health,* **17**(3):343–352 (1968).

Mumford, L. *The Brown Decades: A Study of the Arts in America.* New York: Harcourt, Brace, 1931.

Murie, A. The Wolves of Mount McKinley. Fauna of U.S. National Parks, Series 5. Washington, D.C.: U.S. Government Printing Office, 1944.

Myers, J. K. and L. L. Bean. *A Decade Later: A Follow-Up of Social Class and Mental Illness.* New York: Wiley, 1968.

Myers, K. The effects of density on sociality and health in mammals. *Proc. Ecol. Soc. Australia,* **l**:40–64 (1964).

Neel, J. V. Lessons from a "primitive" people. *Science,* **170**:815–822 (1970).

Negus, N. C., Gould, E., and R. K. Chipman. Ecology of the rice rat, *Orzyomys palustris* (Harlan), on Breton Island, Gulf of Mexico, with critique of the social stress theory. *Tulane Studies in Zool.* **8**:95–123 (1961).

Neville, A. The War Against Suffering. *Johns Hopkins Magazine,* Summer, 1967.

Newell, R. E. The global circulation of atmospheric pollutants. *Scientific American,* **224**:32–42 (1971).

Nice, M. M. Studies in the life history of the song sparrow. Trans. Linaean Soc. of New York, Vols. IV and VI. Dover Publications, Vol. II. 1964.

Nicholson, M. *The Environmental Revolution.* New York: McGraw-Hill, 1970.

Niering, W. A. *The Life of the Marsh.* New York: McGraw-Hill, 1966.

Niering, W. A. The Effect of Pesticides. Chap. 6 In *Environmental Problems: Pesticides, Thermal Pollution and Environmental Synergisms,* Billy Ray Wilson (ed.) Philadelphia: J. B. Lippincott Co., 1968, pp. 101–122.

Nordenskiold, E. *The History of Biology: A Survey.* New York: Knopf, 1932.

Novick, A. *The World of Bats.* New York: Holt, Rinehart and Winston, 1969.

Novick, S., and D. Cottrell. (eds.). *Our World in Peril: An Environmental Review.* St. Louis: Committee for Environmental Information, 1971.

Novick, S. and S. Hopper. Last year at Deauville. *Environment,* **13**:36–37 (1971).

Oberle, M. Forest fires: suppression policy has its ecological drawbacks. *Science,* **165**:568–571 (1969).

Odum, E. P. *Fundamentals of Ecology* (2nd ed). Philadelphia: Saunders, 1959.

Odum, E. P. *Ecology*. New York: Holt, Rinehart and Winston, 1963.

Odum, E. P. The strategy of ecosystem development. *Science,* **164**:262–270 (1969).

Odum, E. P. Fundamentals of Ecology, (3rd ed.). Philadelphia: Saunders, 1971.

Odum, H. T. Trophic structure and productivity of Silver Springs, Florida. *Ecol. Monogr.,* **27**:55–112 (1957).

Odum, H. T. Primary production measurement in eleven Florida springs and a marine turtle-grass community. *Limnol. and Oceanogr.,* **2**:85–97 (1957a).

Ophel, I. L. The fate of radiostrontium in a freshwater community. In *Radioecology*. V. Schulz and A. W. Klement (eds.). New York: Reinhold; Washington, D.C.: A.I.B.S., 1963.

Osborn, F. *Our Plundered Planet*. Boston: Little Brown, 1948.

Owen, D. F. *Animal Ecology in Tropical Africa*. San Francisco: Freeman, 1966.

Paddock, W. and P. Paddock. *Famine 1975! America's Decision: Who Will Survive?* Boston: Little, Brown, 1967.

Park, T. Studies in population physiology. II. Factors regulating initial growth of *Tribolium confusum* populations. *J. Exper. Zool.* **65**:17–42 (1933).

Park, T. Experimental studies of interspecies competition. 1. Competition between populations of flour beetles, *Tribolium confusum* Duval and *Tribolium castaneum* Herbst. *Ecol. Monogr.* **18**:265–308 (1948).

Parkes, A. S. and H. M. Bruce. Pregnancy-block in female mice placed in boxes soiled by males. *J. Reprod. Fertility,* **4**:303–308 (1962).

Patten, B. C. (ed.). *Systems Analysis and Simulation in Ecology*. New York: Academic Press, 1971.

Pearl, R. *The Biology of Population Growth*. New York: Knopf, 1925. (2nd ed. published in 1930).

Pearl, R. *Introduction to Medical Biometry and Statistics*. Philadelphia: W. B. Saunders, 1930.

Pearl, R. On biological principles affecting populations, human and other. *Amer. Naturalist,* **71**:50–68 (1937).

Pearl, R., L. J. Reed, and J. F. Kish. The logistic curve and the census count of 1940. *Science,* **92**:486–488 (1940).

Pearson, O. P. History of two local outbreaks of feral house mice. *Ecology,* **44**:540–549 (1963).

Perkins, D. Jr. Fauna of Catal Huyuk: Evidence for early cattle domestication in Anatolia. *Science,* **164**:177–179 (1969).

Peterson, R. Social Behavior in Pinnepeds. Chap. 1 in *The Behavior and Physiology of Pinnepeds*. Ed. by R. J. Harrison. New York: Appleton-Century-Crofts, 1968.

Peterson, W. *Population.* (2nd ed.) London: The Macmillan Co., 1969.

Petrusewicz, K. Investigation of experimentally induced population growth. Ekologia Polska (Seria A), **5**(9):281–309 (1957).

Pitelka, F. Some aspects of population structure in the short-term cycle of the brown lemming in northern Alaska. Cold Spring Harbor Symp. Quant. Biol. **22**:237–251 (1958).

Powers, G. *Baltimore Evening Sun,* July 29, 1970.

Quetelet, A. Sur le Climat de la Belgique. Des phénomenes périodiques des plantes. Ann. de l'Observatoire Royal de Bruxelles, **5**:1–183 (1846).

Rabinowitch, E. I. *Photosynthesis and Related Processes.* **II**(1):603–1208. New York: Interscience, 1951.

Red Data Book. International Union for the Conservation of Nature and Natural Resources. Survival Service Commission. Vol. 1. Compiled by Noel Simon, 1970.

Reed, L. J. Population growth and forecasts. Ann. Amer. Acad. Polit. and Soc. Sci. **188**:159–166.

Reich, C. *The Greening of America.* New York: Bantam Books, 1971.

Rienow, R., and L. T. Rienow. *Moment in the Sun.* New York: Ballantine Books, 1967.

Riley, G. A. The carbon mechanism and photosynthetic efficiency of the earth. *Amer. Sci.,* **32**:132–134 (1944).

Riley, G. A. Oceanography of Long Island Sound, 1952–54. IX. Production and utilization of organic matter. *Bull. Bingham* Oceanogr. *Coll.,* **15**:324–344 (1956).

Riley, G. A. Phytoplankton of the north central Sargasso Sea. *Limnol. and Oceanogr.,* **2**:252–270 (1957).

Ruud, J. T. Introduction to the studies of pollution in the Oslofjord. Helgolander Wiss. Meeresunters, **17**:455–461 (1968).

Ruud, J. T. Changes since the turn of the century in the fish fauna and fisheries of the Oslofjord. Helgolander Wiss. Meeresunters, **17**:510–517 (1968).

Sadleir, R. M. S. F. *The Ecology of Reproduction in Wild and Domestic Animals.* London: Methuen, 1969.

Saggs, H. W. F. *The Greatness that was Babylon.* New York: Hawthorn Books, 1962.

Sang, J. H. Population grown in *Drosophila* cultures. *Biol. Rev.* **25**:188–219 (1950).

Sarnoff, D. et al., and the editors of Fortune. *The Fabulous Future. America in 1980.* New York: American Book-Stratford Press, 1955.

Sauer, C. O. Theme of plant and animal destruction in economic history. *Jour. of Farm Econom.,* **20**:765–775 (1938).

Sauer, C. O. Land and Life. *A selection from the writings of Carl Ortwin Sauer.* John Leighly (ed.). Berkeley: University of California Press, 1967.

Savely, H. E. Ecological relations of certain animals in dead pine and oak logs. *Ecol. Monogr.* **9**:321–385 (1939).

Schaller, G. B. *The Deer and the Tiger.* Chicago: University of Chicago Press, 1967.

Schaller, G. B. Life with the king of the beasts. *Nat. Geographic,* **135**(4):494–519 (1969).

Schneider, H. A. Ecological ectocrines in experimental epidemiology. *Science,* **158**:597–603 (1967).

Schultz, V., and A. Klement. *Radioecology.* Proc. 1st International Symp. on Radioecol. Colorado: Colorado State University Press, 1963.

Schwebel, M. *Behavioral Science and Human Survival.* Palo Alto, California: Science and Behavior Books, 1965.

Sculthorpe, C. D. *The Biology of Aquatic Vascular Plants.* London: Edward Arnold, 1967.

Sears, P. B. *Deserts on the March.* Norman: University of Oklahoma Press, 1935.

Selye, H. *Stress.* Montreal: Acta, 1950.

Shepherd, P. *Man in the Landscape.* New York: Knopf, 1967.

Shepherd, P. and D. McKinley. *The Subversive Science. Essays Toward an Ecology of Man.* Boston: Houghton Mifflin, 1969.

Siegel, B. J. Defensive cultural adaptation. Chap. 22 In *Violence in America.* Graham and Gurr (ed.). New York: Signet Books, 1969, pp. 743–764.

Simpson, D. The Dimensions of World Poverty. *Scientific American,* **219**:27–35 (1968).

Singer, C. J. A *History of Biology: a general introduction to the study of living things.* New York: Henry Schuman, 1950.

Sladen, B. K. The Ecology of Animal Communities. Chap. 7 In *Biology of Populations,* B. K. Sladen and F. B. Bang (eds.). New York: American Elsevier, 1969, pp. 87–100.

Sladen, B. K. and F. B. Bang. *Biology of Populations.* New York: American Elsevier, 1969.

Sladen, W. J. L. Social organization in some birds and mammals. Chap. 19 in Sladen, B. K. and F. B. Bang, *Biology of Populations.* New York: American Elsevier, 1969, pp. 264–283.

Sladen, W. J. L., Menzie, C. M., and W. L. Reichel. DDT residues in Adelie penguins and a crabeater seal from Antarctica. *Nature,* **210**:670–673 (1966).

Slobodkin, L. B. *Growth and Regulation of Animal Populations.* New York: Holt, Rinehart and Winston, 1961.

Southwick, C. H. The population dynamics of confined house mice supplied with unlimited food. *Ecology,* **36**:212–225 (1955a).

Southwick, C. H. Regulatory mechanisms of house mouse populations; social behavior affecting litter survival. *Ecology,* **36**:627–634 (1955b).

Southwick, C. H. Population characteristics of house mice living in English corn ricks; density relationships. *Proc. Zool. Soc.* (London). **131**:163–175 (1958).

Southwick, C. H. An experimental study of intragroup agonistic behavior in rhesus monkeys (*Macaca mulatta). Behaviour,* **28**:182–209 (1967).

Southwick, C. H. Population Dynamics and Social Behavior of Domestic Rodents. Chap. 20, In *Biology of Populations,* edited by B. K. Sladen and F. B. Bang (eds.). New York: American Elsevier, 1969, pp. 284–298.

Southwick, C. H. The biology and psychology of crowding. *Ohio Jour. of Sci.,* **71**(2):65–72 (1971).

Southwick, C. H., B. Sladen, and A. Reading. Patterns of Survivorship in Laboratory populations of fairy shrimp, *Eubranchipus vernalis. Amer. Midland Nat.,* **72**:133–141 (1964).

Southwick, C. H., and M. R. Siddiqi. The role of social tradition in the maintenance of dominance in a wild rhesus group. *Primates,* **8**:341–353 (1967).

Southwick, C. H., M. R. Siddiqi, and M. F. Siddiqi. Primate populations and biomedical research. *Science,* **170**:1051–1054 (1970).

Southwick, C. H. and R. W. Ward. Aggressive display in the Paradise Fish, (*Macropodus opercularis* L.). *Turtox News,* **46**:(2):57–62 (1968).

Spengler, J. J. Population problem: In search of a solution. *Science,* **166**:1234–1238 (1969).

Spillett, J. J. *The Ecology of the Lesser Bandicoot Rat in Calcutta.* Bombay: Leaders Press Private Ltd., 1968.

Srole, L., T. S. Langner, S. T. Michael, M. K. Opler, and T. A. C. Rennie. *Mental Health in the Metropolis.* New York: McGraw-Hill. Vol. 1, 1962.

Steemann-Nielsen, E. The production of organic matter by phytoplankton in a Danish lake receiving extraordinary amounts of nutrient salts. *Hydrobiologia,* **7**:68–74 (1955).

Stoddard, H. L. *The Bobwhite Quail: Its Habits, Preservation and Increase.* New York: Scribner, 1932.

Storer, J. H. *The Web of Life.* New York: The New American Library of World Literature, 1953.

Tamiya, H. Mass culture of algae. *Ann. Rev. Plant Physiol.,* **8**:309–334 (1957).

Tanaka, R. Fluctuations in vole populations following the widespread synchronous flowering of bamboo-grasses on Mt. Turuji. *Bull. Kochi Women's College,* **4**(2):61–68 (1956).

Tepper, B. S. Population growth of bacteria. Chap. 2 In *Biology of Populations,* B. K. Sladen and F. B. Bang (eds.). New York: American Elsevier, 1969.

Terao, A., and T. Tanaka. Influence of temperature upon the rate of reproduction of the water flea, *Moina macrocopa* Strauss. *Proc. Imper. Acad.* (Japan) **4**:553–555 (1928).

Thomas, W. L. (ed.). *Man's Role in Changing the Face of the Earth.* Chicago: University of Chicago Press, 1955–1956.

Toffler, A. *Future Shock.* New York: Random House, 1970.

Tsai, Chu-fa. Effects of sewage pollution on fishes in upper Patuxent River. *Chesapeake Sci.* **9**(2):83–93 (1968).

Udall, S. L. *The Quiet Crisis.* New York: Holt, Rinehart and Winston, 1963.

Udny-Yule, G. The growth of a population and the factors which control it. *J. R. Statist. Soc.,* **88**:1–58 (1925).

United Nations. *U.N. Statistics of Hunger Report.* New York, 1968.

U.S.P.H.S. *Environmental Health Problems.* Rockville, Md.: Environmental Service, 1970.

Van Dyne, G. M. (ed.). *The Ecosystem Concept in Natural Resource Management.* New York: Academic Press, 1969.

Varley, G. C. The natural control of population balance in the knapweed gall-fly (*Urophora jaceana*). *J. Animal Ecol.* **16**:139–187 (1947).

Verduin, J. Primary production in lakes. *Limnol. and Oceanogr.,* **1**:85–91 (1956).

Vogt, W. *Road to Survival.* New York: William Sloane Associates, 1948.

Waggoner, P. E. Plants and polluted air. *Bioscience,* **21**:455–459 (1971).

Walford, C. The famines of the world: past and present. *Roy. Statist. Soc. Jour.,* **41**:433–526 (1878).

Wallbank, W. T. *A Short History of India and Pakistan.* New York: Scott Foresman, 1958.

Watson, A. Aggression and population regulation in Red Grouse. *Nature,* **202** (4931):506–507 (1964).

Watt, K. E. F. (ed.). *Systems Analysis in Ecology.* New York: Academic Press, 1966.

Watt, K. E. F. *Ecology and Resource Management.* New York: McGraw-Hill, 1968.

Wayne, L. G. and L. A. Chambers. Biological effects of urban air pollution. V. A study of effects of Los Angeles atmosphere on laboratory rodents. *Arch. Environ. Health,* **16**(6):871–885 (1968).

Welty, J. C. *The Life of Birds*. New York: Knopf, 1963.

White, L. The historical roots of our ecologic crisis. *Science,* **155**:1203–1207 (1967).

Whittaker, R. H. *Communities and Ecosystems*. New York: Macmillan, 1970.

Whyte, W. H. *The Last Landscape*. Garden City, New York: Doubleday, 1968.

Wiens, J. A. On group selection and Wynne-Edward's hypothesis. *Amer. Scientist,* **54**:273–287 (1966).

Williams, G. C. *Adaptation and Natural Selection.* Princeton: Princeton University Press, 1966.

Wilson, E. B. and R. R. Puffer. Least squares and laws of population growth. *Proc. Amer. Acad. Arts and Sci.* **68**:285–382 (1933).

Woodwell, G. M. Toxic substances and ecological cycles. *Scientific American,* **216**(3):24–31 (1967).

Woodwell, G. M. Radioactivity and fallout: the model pollution. *Bioscience,* **19**:886(1969).

Wright, Q. 1942. *The Study of War*. Chicago: University of Chicago Press, 1942.

Wrigley, E. A. *Population and History*. New York: McGraw-Hill, 1969.

Wynne-Edwards, V. C. *Animal Dispersion in Relation to Social Behavior*. New York: Hafner, 1962.

York, H. *Race to Oblivion: A Participants View of the Arms Race*. New York: Simon and Schuster, 1970.

Young, G., and J. P. Blair. Our ecological crisis: Pollution, threat to man's only home. *Nat. Geographic,* **138**:737–781 (1970).

Young, L. B. (ed.). *Population in Perspective*. New York: Oxford University Press, 1968.

Zinsser, H. *Rats, Lice and History*. Boston: Little, Brown, 1935.

Glossary

The following is an abbreviated list of important terms used in the text. A much more extensive list of ecological terms appears in Dictionary of Ecology *by H. C. Hanson. New York: Philosophical Library, Inc., 1962.*

ABIOTIC—without life; an absence of living organisms

AEROBIC—requiring oxygen for life processes

ALLERGEN—substance producing a pathogenic sensitivity within an organism

AMBIENT—existing conditions; encompassing on all sides

ANADROMOUS—referring to animals which live in the sea and breed in fresh water, such as salmon

ANAEROBIC—capable of living without oxygen

ANTIBIOSIS—a process fatal or injurious to any form of life

ANTIBODY—a chemical produced in a living organism in response to a foreign substance entering the organism

ANTIGEN—a foreign substance which produces an antibody response in an organism

AUTECOLOGY—the study of individuals, or species, in response to environmental conditions. cf. SYNECOLOGY

AUTOTROPHIC—refers to organisms capable of producing organic materials from inorganic chemicals by means of energy conversion; e.g., green plants

BENTHIC—refers to the bottom layer of any body of water, and the organisms therein

BIOLOGICAL CLOCK—the rhythmic occurrences of processes within organisms at periodic intervals

BIOMASS—the total weight of living organisms per unit area

BIOME—a major terrestrial biotic community characterized by distinctive life forms; e.g., temperate deciduous forest, grassland, coniferous forest, etc.

BIOMETRY—the application of statistics to biology

BIOSPHERE—the portion of the earth and its atmosphere capable of supporting life

BIOTA—all the living organisms in an area

CALORIE—the amount of heat energy capable of raising the temperature of one gram of water 1° centigrade (small calorie or g-cal); or one liter of water 1° (1 kg-cal or large calorie)

CARRYING CAPACITY—the ability of landscape to support any given animal or plant species or groups of species

CLIMAX—the mature biotic community capable of self-perpetuation under prevailing climatic and topographic conditions

COMMENSALISM—the living together of two species, usually with benefit to one

COMMUNITY—the group of organisms living within a definable area

COMPETITION—the mutual utlization of limited resources; intraspecific—between members of the same species; interspecific—between members of different species

CONSUMERS—organisms which ingest organic food (usually particulate) or other organisms

CYBERNETICS—the study of communication and control systems in living organisms and machines

DDT—an insecticide consisting of dichloro-diphenyl-trichloroethane; one of a group of chlorinated hydrocarbons

DEMOGRAPHY—the mathematical study of populations

DENSITY-DEPENDENT FACTOR—an ecologic influence which varies in relation to population density

DENSITY-INDEPENDENT FACTOR—an ecologic influence independent of density

DOMINANCE, ECOLOGIC—the condition in biotic communities in which one or more species exert a major or controlling influence

DOMINANCE, SOCIAL—the condition by which an individual or group of individuals have priority access to environmental resources

ECOLOGY—the study of the interrelationships of living organisms to one another and their environment

ECOSYSTEM—the biotic community and its non-living environment as an interacting system

ECOTONE—a transition zone between two adjacent and differing communities

ECOTYPE—a group of living organisms adapted to a certain set of environmental conditions

ENDEMIC—referring to a group of organisms native to a given region

ENTROPY—the loss or degradation of energy; that portion of heat energy which is unavailable for work

ENVIRONMENT—all the conditions and influences surrounding and affecting an organism or group of organisms

EPIDEMIC—the occurrence in greater numbers than usual of a pathogen, parasite or pest

ETHNOLOGY—the study of the divisions of mankind

ETHOLOGY—the comparative and biological study of behavior

ETIOLOGY—the cause or developmental history of a condition

EUTROPHICATION—the process of enrichment and aging in an ecosystem

FAUNA—a collective term for all the animals in a given region or geological period

FECUNDITY—the production of sex cells

FEEDBACK—information by which a system returns toward a former condition

FERTILITY—the capability to produce living offspring

FOOD CHAIN—a series of organisms dependent upon one another for food

GAUSE'S PRINCIPLE—the concept that two species cannot occupy exactly the same ecologic niche within a community

GENUS—a group of closely related species; the first part of a scientific Latin name for an organism

HABITAT—the natural abode of an organism including its total environment

HIERARCHY—a social rank order of animals; dominance hierarchy

HOMEOSTASIS—the maintenance of equilibria in an organism or biological system

INDICATOR ORGANISM—an organism which indicates the presence or absence of certain environmental conditions

INVASIVE—the tendency to invade or enter a new location or niche

IRRUPTIVE—the tendency to increase suddenly in numbers

ISOTOPES—forms of the same element which differ in atomic weight and organization of the atomic nucleus

LEACHING—the removal or downward movement by water of soluble chemicals from the soil or other materials

LIMITING FACTOR—the ecologic influence which limits or controls the abundance and/or distribution of a species

LOGISTIC CURVE—an S-shaped growth curve describing the growth of an organism or a population

MALTHUSIAN—refers to the doctrine of Malthus that living organisms tend to increase faster than their means of support

METABOLISM—the total of chemical processes occurring within an organism or biological system

MIMICRY—a condition by which an organism has selective advantage against predation by appearing dangerous, inedible, or inconspicuous

MORTALITY—death

MUTUALISM—an interspecific relationship of benefit to two or more interacting species

NATALITY—the production of offspring by organisms

NATURAL SELECTION—the process of evolution by which some organisms give rise to more descendants than others

NICHE—the ecologic role of an organism in its ecosystem; its relationships to its biotic community and total environment

ORGANISM—an individual unit constituted to carry on the activities of life

PARASITE—an organism which derives its nutrition by feeding upon another living organism

PATHOGEN—an organism which causes disease

PELAGIC—refers to the open ocean

PESTICIDE—an agent which kills pests; e.g. insecticide

PHYLUM—the major subdivisions of the plant and animal kingdoms

PHYTOPLANKTON—plant plankton, primarily algae

PIONEER—an organism which first appears in a newly exposed or altered environment

PLANKTON—floating or weakly swimming organisms in aquatic environments, primarily microscopic

POPULATION—a group of interacting individuals in a definable space

ppm—parts per million, a measure of concentration, the same as micrograms per milliliter or milligrams per liter

PREDATION—the process of one organism killing another for food

PRIMARY PRODUCTION—the process by which organic compounds are formed in photosynthesis or chemosynthesis

RADIOISOTOPE—an isotope that is unstable and disintegrates, emitting radiation energy

SERE or SERAL STAGE—the stage in a successional series

SMOG—polluted air and water vapor

SOIL—the aggregate of decaying organic materials, living organisms and weathered substrate

STRESS—a condition which forces a deviation from homeostasis

STRESSOR—a stimulus or factor producing stress

SUCCESSION—the changes in vegetation and animal life by which one population or community is replaced by others

SYMBIOSIS—the living together of two or more species

SYNECOLOGY—the study of the environmental relations of communities

SYSTEMATICS—taxonomy or the classification of living organisms

SYSTEMIC—refers to the entire body of an organism

TERRITORY—an area occupied by an animal or group of animals and defended against others of the same species

THRESHOLD—the level (duration or intensity) of a stimulus required to produce an effect

TOPOGRAPHY—characteristics of the ground surface in terms of physical features

TROPHIC LEVEL—one stage in a nutritive series including producers or various levels of consumers

TROPHIC STRUCTURE—the organization of a community in terms of trophic levels

VIRUS—a microscopic living particle, usually smaller than bacteria, which is parasitic on living cells and incapable of reproducing outside of living cells

WATERSHED—the total land area contributing surface or ground water to a lake, river, or drainage basin

XERIC—refers to a dry environment

YIELD—the part of productivity utilized by man or other animals

ZOOPLANKTON—animal plankton

Bibliography

The following bibliography includes both academic and popular books on ecology and human affairs which were published during the rapid rise of ecological awareness from 1969 through 1971. It is not a comprehensive list in any sense of the word, but it provides a sample of the great volume of literature which appeared in this short period. Other more specific citations are included in the References.

Adams, L. *Population Ecology*. Belmont, Calif.: Dickenson, 1970, 160 pp.

Anderson, P. K. Omega: *Murder of the Ecosystem and Suicide of Man*. Dubuque: Wm. C. Brown, 1971, 447 pp.

Boughey, A. S. *Man and the Environment*. Riverside, N.J.: Macmillan, 1971, 480 pp.

Brookhaven National Laboratory. *Diversity and Stability in Ecological Systems*. Springfield, Virginia: U.S. Dept. of Commerce, 1969, 264 pp.

Chute, R. M. *Environmental Insight*. New York: Harper & Row, 1971, 242 pp.

Connell, J. H., D. B. Mertz, and W. W. Murdoch (eds.). *Readings in Ecology and Ecological Genetics*. New York: Harper & Row, 1970, 397 pp.

Cox, G. W. (ed.). *Readings in Conservation Ecology*. New York: Appleton-Century-Crofts, 1969, 595 pp.

Crowe, P. K. *World Wildlife: The Last Stand*. New York: Scribner, 1970, 308 pp.

Dansereau, P. (ed.). *Challenge for Survival*. New York: Columbia Univ. Press, 1970, 235 pp.

DeBell, G. (ed.). *The Environmental Handbook*. New York: Ballantine Books, 1970, 365 pp.

Dickinson, R. E. *Regional Ecology. The Study of Man's Environment*. New York: Wiley, 1970, 199 pp.

Dorst, J. *Before Nature Dies*. Trans. by C. D. Sherman. Boston: Houghton Mifflin, 1970, 352 pp.

Ehrenfeld, D. W. *Biological Conservation*. New York: Holt, Rinehart and Winston, 1970, 226 pp.

Ehrlich, P. R. and A. Ehrlich. *Population, Resources, Environment*. San Francisco: Freeman, 1970, 383 pp.

Falk, R. A. *This Endangered Planet: Prospects and Proposals for Human Survival*. New York: Random House, 1971, 485 pp.

Graham, F. J. *Since Silent Spring*. New York: Houghton-Mifflin, 1970, 333 pp.

Hazen, W. E. (ed.): *Readings in Population and Community Ecology*. 2nd ed. Philadelphia: Saunders, 1970, 421 pp.

Helfrich, H. W., Jr. (ed.). *Agenda for Survival. The Environmental Crisis—II*. New Haven: Yale Univ. Press, 1970, 234 pp.

Holdren, J. P. and P. R. Ehrlich (eds.). *Global Ecology. Readings toward a Rational Strategy for Man*. New York: Harcourt, Brace, Jovanovich, 1971, 296 pp.

Hynes, H. B. N. *The Ecology of Running Waters*. Toronto: Univ. of Toronto Press, 1970, 555 pp.

Jackson, W. *Man and the Environment*. Dubuque, Iowa: Wm. C. Brown, 1971, 322 pp.

Johnson, C. E. (ed.). *Eco-Crisis*. New York: Wiley, 1970, 182 pp.

Kinne, O. *Marine Ecology Vol. I. Environmental Factors*. New York: Wiley, 1970, 681 pp.

Klopfer, P. H. *Behavioral Ecology*. Belmont, Calif.: Dickenson, 1970, 229 pp.

Kormondy, E. J. *Concepts of Ecology*. Englewood Cliffs, N.J.: Prentice-Hall, 1969, 209 pp.

Linton, R. M. *Terracide*. Boston: Little, Brown, 1970, 376 pp.

McHale, J. *The Ecological Context*. New York: George Braziller, 1970, 188 pp.

McLaren, I. A. (ed.). *Natural Regulation of Animal Populations*. New York: Aldine, 1971, 195 pp.

Mayr, E. *Population, Species and Evolution*. Cambridge: Harvard Univ. Press, 1970, 453 pp.

Milne, L. and M. *The Nature of Life. Earth, Plants, Animals, Man and Their Effect on Each Other*. New York: Crown, 1970, 320 pp.

Mitchell, J. G. and C. L. Stallings (eds.). *Ecotactics: The Sierra Club Handbook for Environmental Activists*. New York: Pocket Books, 1970, 288 pp.

Nicholson, M. The Environmental Revolution. New York: McGraw-Hill, 1970, 366 pp.

Odum, E. P. *Fundamentals of Ecology*. 3rd ed. Philadelphia: Saunders, 1971, 574 pp.

Odum, H. T. *Environment, Power and Society*. New York: Wiley Interscience, 1970, 336 pp.

Owen, O. S. *Natural Resource Conservation*. Riverside, N.J.: Macmillan, 1971, 576 pp.

Patten, B. C. (ed.). *Systems Analysis and Simulation in Ecology*. New York: Academic Press, 1971, 607 pp.

Petrusewicz, K. and A. MacFayden. *Productivity of Terrestrial Animals*. IBP Handbook No. 13. Philadelphia: F. A. Davis Co., 1970, 190 pp.

Pielou, E. C. *An Introduction to Mathematical Ecology*. New York: Wiley Interscience, 1969, 286 pp.

Revelle, R., A. Khosla, and M. Vinovskis. *The Survival Equation. Man, Resources, and His Development*. Boston: Houghton Mifflin, 1971, 512 pp.

Rienow, R. and L. T. Rienow. *Man Against His Environment*. New York: Ballantine Books, 1970, 307 pp.

Sax, J. L. *Defending the Environment*. New York: Knopf, 1971, 252 pp.

Singer, S. F. (ed.). *Global effects of environmental pollution*. AAAS Symposium. New York: Springer-Verlag, 1970, 218 pp.

Sladen, B. and F. B. Bang (eds.). *Biology of Populations*. New York: American Elsevier, 1969, 449 pp.

Vernberg, F. J. and W. B. Vernberg. *The Animal and the Environment*. New York: Holt, Rinehart and Winston, 1970, 398 pp.

Wagner, R. H. *Environment and Man*. New York: Norton, 1971, 491 pp.

Watson, A. (ed.). *Animal Populations in Relation to Their Food Resources*. Oxford: Blackwell Sci. Pub., 1970, 477 pp.

Whittaker, R. H. *Communities and Ecosystems*. New York: Macmillan, 1970, 158 pp.

Index